Springer Proceedings in Earth and Environmental Sciences

The series Springer Proceedings in Earth and Environmental Sciences publishes proceedings from scholarly meetings and workshops on all topics related to Environmental and Earth Sciences and related sciences. This series constitutes a comprehensive up-to-date source of reference on a field or subfield of relevance in Earth and Environmental Sciences. In addition to an overall evaluation of the interest, scientific quality, and timeliness of each proposal at the hands of the publisher, individual contributions are all refereed to the high quality standards of leading journals in the field. Thus, this series provides the research community with well-edited, authoritative reports on developments in the most exciting areas of environmental sciences, earth sciences and related fields.

Joan Cecilia Casila · Anupam Khajuria ·
Saim Memon
Editors

Proceedings of The 3rd International Conference on Climate Change and Ocean Renewable Energy

Climate Change and Ocean Renewable Energy

Springer

Editors
Joan Cecilia Casila
Land and Water Resources Engineering
Division, IABE, CEAT
University of the Philippines
Los Baños, Philippines

Anupam Khajuria
Institute for the Advanced Study
of Sustainability
United Nations University
Tokyo, Japan

Saim Memon
Department of Industrial R&D in Vacuum
Insulation Energy Technologies
Sanyou London
London, UK

Department for Engineering, School
of Engineering and the Built Environment
Birmingham City University
London, UK

ISSN 2524-342X ISSN 2524-3438 (electronic)
Springer Proceedings in Earth and Environmental Sciences
ISBN 978-3-031-93886-3 ISBN 978-3-031-93887-0 (eBook)
https://doi.org/10.1007/978-3-031-93887-0

This Springer imprint is published by the registered company Springer Nature Switzerland AG
The registered company address is: Gewerbestrasse 11, 6330 Cham, Switzerland

Preface

The 3rd International Conference on Climate Change and Ocean Renewable Energy (CCORE 2024) was convened virtually on November 2–3, 2024, in response to the increasing environmental challenges brought about by climate change and the pressing need for sustainable and circular solutions. With presentations from distinguished experts and researchers throughout the globe, the online format of the conference ensured compliance with prevailing guidelines while fostering the importance of collaboration and knowledge-sharing across borders.

CCORE 2024 exemplified a commitment to advancing discourse in ocean renewable energy, climate resilience and circular and sustainable practices. The conference provided a unique opportunity to enrich understanding and explore emerging developments in these critical fields. The conference was structured into three sessions, allowing for an engaging exploration of various themes. Distinguished keynote speakers were allocated 30–35 minutes each, delivering presentations that sparked insightful dialogues. Subsequently, faculty, scientists, and research scholars were allotted 15–20 minutes each to share their cutting-edge research, sparking engagement and exchange among participants.

The conference received widespread praise for its role in sharing important knowledge and insights of climate change and its impact. Attendees commended CCORE 2024 for its impact on enhancing comprehension of climate change and ocean energy, recognizing its significance and timeliness through multidisciplinary approaches.

The proceedings of CCORE 2024 reflect the hard work and dedication of every participant and contributor. The diversity of topics covered—ranging from the impacts of climate change on ocean renewable energy to advancements in renewable energy and climate resilient technologies, global environmental change, hydrology, and the principles of a circular economy—demonstrates the multidisciplinary approach essential for understanding and addressing the complexities we face. All papers underwent rigorous review to ensure adherence to international publication standards.

We extend our sincere gratitude to all of the participants and institutions whose support and contributions were instrumental in the success of CCORE 2024. Special acknowledgment is due to the organizing committee, whose leadership and vision were instrumental in preparing a robust program that engaged and inspired researchers and practitioners alike.

We hope that the knowledge shared throughout CCORE 2024 will ignite further discussions and collaborations, fostering dynamic research endeavors and innovative, circular solutions that address our most critical environmental challenges. We trust that the proceedings of CCORE 2024 will serve as an invaluable reference, contributing to the insights and findings inspiring action in the quest for sustainable solutions and innovations and providing a foundation for future endeavors.

The Committee of CCORE 2024

Committee Members

On behalf of the CCORE 2024 Organizing Committee, we would like to take this opportunity to express our sincere gratitude to our participants. We would also like to express our acknowledgements to the committee members who have given their professional guidance. To those who contributed to the success of the conference organization without listing the name below, we would love to say thanks as well.

Honor Committee

M. Dolores Esteban	Universidad Politécnica de Madrid, Spain; European University of Madrid, Spain
Saim Memon	Department of Industrial R&D in Vacuum Insulation Energy Technologies, Sanyou London Pvt Ltd.
Sergio Peña-Neira	Universidad Mayor, Chile
Mike Elliott	University of Hull, UK
José-Santos López-Gutiérrez	Universidad Politécnica de Madrid (UPM), Spain
Hossam A. Gabbar	Ontario Tech University, Canada
Sharifah Rafidah Wan Alwi	Universiti Teknologi Malaysia, Malaysia

Technical Program Committee Chairs

A. M. Hoguane	Eduardo Mondlane University, Mozambique
Jian Zhang	Jiangsu University of Science and Technology, China
M. Dolores Esteban	Environment, Coast and Ocean Research Laboratory (ECOREL-UPM), Universidad Politécnica de Madrid, Spain
José-Santos López-Gutiérrez	Universidad Politécnica de Madrid (UPM), Spain
I. Ljubenkov	Water Development, Croatia
K. L. Priya	TKM College of Engineering, India
Joan Cecilia C. Casila	University of the Philippines Los Baños, Philippines
S. Haddout	Ibn Tofail University, Morocco
K. Komali	University of Central Lancashire, UK
D. Elemam	Damietta University, Egypt

Publication Chair

Joan Cecilia C. Casila	University of the Philippines Los Baños, Philippines
Anupam Khajuria	United Nations University, Institute for the Advanced Study of Sustainability, Japan
Saim Memon	Department of Industrial R&D in Vacuum Insulation Energy Technologies, Sanyou London Pvt Ltd.
Assiya Haddout	Ibn Tofail University, Morocco
S. Haddout	Ibn Tofail University, Morocco

International Technical Program Committee

M. A. Quirapas-Franco	Aquatera Asia Pte. Ltd., Singapore
Huynh Cong Khanh	Can Tho University, Vietnam
Bing Bai	Beijing Jiaotong University, China
Philipp Thies	University of Exeter, United Kingdom
Eckart Schumann	Nelson Mandela University, Gqeberha, South Africa
Tri Retnaningsih Soeprobowati	Universitas Diponegoro, Indonesia
Omar Farrok	Ahsanullah University of Science and Technology, Bangladesh
K. R. Renjith	KSCSTE-Centre for Water Resources Development and Management, India
Bilal Acar	University of Selcuk, Turkey
Gökçen Uysal	Eskişehir Technical University, Turkey
Yossi Mart	Recanati Institute of Maritime Studies, University of Haifa, Israel
Ahmed Mohamed Hassan El-kasaby	Matrouh University, Egypt
Y. Haddout	Ibn Zohr University, Ouarzazate, Morocco
Fuad Mahfud Assidiq	Universitas Hasanuddin, Indonesia
Yousry El Sayed Mattar	Water & Agriculture, Riyadh, Kingdom of Saudi Arabia
Enrico Zacchei	Portugal University of Coimbra, CERIS, Coimbra, Portugal
Z. Bolatova	Almaty University of Power Engineering and Communication, Kazakhstan
A. Haddout	Ibn Tofail University, Kenitra, Morocco
Tuba Hatice Doğan	Ataturk University, Turkey

About This Book

This book is dedicated to presenting a thoughtfully curated compilation of research papers from the esteemed 3rd International Conference on Climate Change and Ocean Renewable Energy (CCORE 2024). This conference acts as a vital forum for the discussion and advancement of cutting-edge research in the fields of climate change and ocean renewable energy. Within this volume, readers will find a wealth of scholarly contributions that address key issues such as climate resilience, innovative adaptation strategies, circular economy, and the latest developments in ocean-based renewable technologies.

Every paper in this volume has undergone a rigorous review process, ensuring that it meets the highest international standards for academic excellence. This evaluation guarantees that readers can rely on the credibility and quality of the findings and insights presented. From thorough examinations of current challenges to pioneering research that charts new territories, these works collectively represent the evolving landscape of climate science and environmental stewardship.

Additionally, this book serves as a crucial resource for scholars, decision-makers, and industry professionals, encouraging cross-disciplinary dialogue and collaboration. It aspires to drive significant progress toward sustainable solutions for our global community by consolidating the latest research and technological advancements. As we face the urgent realities posed by climate change and energy transitions, the insights in this book offer essential perspectives and guidance, informing and inspiring our shared journey toward a more equitable and sustainable future for both humanity and the planet.

The Committee of CCORE 2024

UNIVERSIDADE EDUARDO MONDLANE

POLITÉCNICA UNIVERSIDAD POLITÉCNICA DE MADRID

wd

Contents

Opening Paper

Achieving Net-Zero, Protecting the Oceans, and Embracing Circular Economy: The Urgency of Addressing Climate Change

Joan Cecilia C. Casila[1](✉), Luisito C. Abueg[2], Anupam Khajuria[3], Saim Memon[4,5], Soufiane Haddout[6], and K. L. Priya[7]

[1] Land and Water Resources Engineering Division, IABE, CEAT, University of the Philippines Los Baños, Los Baños, Laguna, Philippines
jmcatubigcasila@up.edu.ph

[2] Department of Economics, CEM, University of the Philippines Los Banos, Los Baños, Laguna, Philippines

[3] Institute for the Advanced Study of Sustainability (UNU-IAS), United Nations University, Tokyo, Japan

[4] Department of Industrial R&D in Vacuum Insulation Energy Technologies, Sanyou London Pvt Ltd, Level 18, 40 Bank Street, Canary Wharf, E14 5NR, London, UK

[5] Department for Engineering, School of Engineering and the Built Environment, Birmingham City University, Millennium Point, Curzon Street, Birmingham B4 7XG, England

[6] Department of Physics, Ibn Tofail University, Kenitra, Morocco

[7] Department of Civil Engineering, TKM College of Engineering, Kollam, Kerala, India

Abstract. Over the past decades, global warming and climate change have been urgent issues that require serious attention. The dual challenges of achieving net zero emissions and protecting our oceans are intricately linked. Nowadays, we are witnessing strong impacts of climate change. It is crucial to understand the interconnected nature of climate change, the implications of inaction, and the steps necessary for a sustainable future. This chapter outlines different strategies for harnessing ocean energy, reaching net-zero emissions, and safeguarding coastal areas from the adverse effects of climate change. It highlights the role of the circular economy as a transformative approach to mitigating climate impacts by promoting resource efficiency, reducing waste, and fostering sustainable consumption and production patterns. The critical impacts of global warming are also highlighted, specifically the alarming trend of glacier melting, sinking islands, forest fire, wildfire, and flash floods. The glacier melting not only contributes to rising sea levels but also disrupts hydropower and freshwater supply for glacier-fed rivers. Further, it can lead to landslides and sedimentation problems, negatively affecting lake and river water quality, disrupting aquatic habitats, and increasing the vulnerability of many communities and ecosystems. These climate change driven disasters pose significant risks to both human settlements and natural ecosystems and also threatening biodiversity, leading to devastating impacts on infrastructure and livelihoods. Additionally, the chapter discusses the challenges related to the construction and adoption of renewable energy, emphasizing how circular economy practices can overcome barriers such as resource scarcity and waste generation in renewable energy technologies. Collectively, these issues underscore

J. C. Casila et al. (Eds.): CCORE 2024, SPEES, pp. 3–13, 2025.
https://doi.org/10.1007/978-3-031-93887-0_1

the urgent need for effective climate adaptation measures including nature-based solutions and comprehensive coastal management practices.

Keywords: Ocean Warming · Ocean Energy · Net Zero · Climate Change · Glacier Melting · Wildfires · Flooding · Circular Economy

1 Introduction

In an ever-evolving planet, where the interplay of ecological forces shapes the very fabric of life, the call for action against climate change has never been more pressing. The dual challenges of achieving net zero emissions and protecting our oceans are intricately linked. Addressing climate change impacts to protect marine ecosystems is urgent because these ecosystems are vital for the planet's health, human well-being, and global biodiversity. Climate change poses immediate and long-term threats to the oceans, and delaying action could result in irreversible damage.

The three primary greenhouse gases (GHGs)—carbon dioxide (CO_2), methane (CH_4), and nitrous oxide (N_2O)—are thought to be the main contributors to global warming and climate change. Fossil fuels, like coal, are limited resources and will run out soon, hence the term non-renewable. When fossil fuels are burned, they release CO_2 into the atmosphere, contributing to the greenhouse effect by increasing CO_2 concentrations and raising air temperatures. The rise in CO_2 levels is closely linked to the extensive use of fossil fuels [1]. While various initiatives have been implemented to reduce emissions, significant challenges still need to be addressed before substantial progress can be made.

One of the ways to reduce GHG emissions is through the utilization of renewable energy. Renewable energy is from natural sources of energy that are replenished faster than they are consumed. The traditional non-renewable energy sources like coal, oil, and natural gas contribute to global warming, air pollution, and depletion of natural resources. Researchers continuously study ways to harness renewable energy sources, specifically ocean energy, and search for cleaner technological alternatives. These initiatives aim to significantly reduce GHG emissions and mitigate the greenhouse effect, which is critical to preserving our planet's health.

Human activities are primarily responsible for the accelerated rate of climate change. The burning of fossil fuels, deforestation, and industrial processes have led to increased concentrations of GHGs in the atmosphere. This, in turn, has resulted in rising global temperatures, melting glaciers, and erratic weather patterns [2]. The urgency of the situation cannot be understated since we are witnessing unprecedented challenges, from damages due to strong typhoons, wildfires, and flashfloods, to the loss of biodiversity and habitat destruction.

The last decade, spanning from 2015 to 2024, has experienced the highest temperatures on record [3]. In 2024, both land and sea surface temperatures reached remarkable levels, along with notable increases in ocean heat. Since 90% of the excess heat from global warming is stored in the ocean, it is a critical indicator of climate change. The intricate ecosystems that thrive within our oceans are not merely beautiful; they are essential for maintaining global ecological balance and supporting livelihoods for millions around the world.

The scientific evidence supporting climate change is overwhelming. More intense and frequent rainfall events occur due to climate change, leading to catastrophic flooding conditions [4]. Additionally, glaciers in polar regions and mountain ranges continue to melt at alarming rates, contributing to rising sea levels, damage to communities and rivers that are glacier fed, and altering ecosystems [5]. These phenomena are not isolated; rather, they are interconnected, creating a multifaceted crisis that affects not only the environment but also social, economic, and political structures globally. Thus, the scientific consensus is clear: human activities are primarily responsible for the accelerated rate of climate change we are currently witnessing.

A challenge is being given to design and develop ways to harness ocean tidal energy considering protection of marine ecosystems to minimize their impact, preserve functionality and performance over time. Nature-based solutions are also put into consideration to contribute to clearer and more functional clean energy production over time [6].

Various strategies for enhancing climate resilience and coastal protection have been proposed in recent literature. These strategies encompass a range of approaches, from restoring natural ecosystems like mangroves and wetlands, which serve as buffers against storm surges, to implementing sustainable urban planning practices that minimize flood risks. Equally important is the reduction of carbon emissions to achieve net-zero targets. These are crucial steps in mitigating further climate effects.

The chapter further elaborates on the consequences of rising global temperatures, linking them to the increased frequency of natural disasters, as well as issues that arise from these catastrophic events. It highlights the urgent need for a collective effort to address climate change, emphasizing that without decisive action, the repercussions for both natural and human systems will be dire and far-reaching. The urgency of shifting towards sustainable net zero emission practices and protecting our oceans cannot be overstated, as they play a vital role in regulating the climate and supporting global biodiversity.

2 Role of Oceans in Climate Regulation

The oceans, which cover over 70% of the Earth's surface, support life, the hydrologic cycle, and many vital functions like transport, trade, tourism, provision of food and minerals. They play a pivotal role in regulating our climate. Oceans absorb about 25% to 30% of the CO_2 emitted into the atmosphere, helping to mitigate the effects of climate change. However, excessive CO_2 absorption leads to ocean acidification, which harms marine ecosystems [7, 8]. Apart from carbon sequestration, oceans absorb over 90% of the excess heat trapped by GHGs. This results in ocean warming, which can disrupt weather patterns and intensify extreme events like hurricanes and typhoons.

Increased CO_2 levels in oceans will lower seawater pH, impairing calcifying organisms like corals, shellfish, and plankton, which are foundational to marine food webs. Thus, heightened temperature and CO_2 levels will force species to migrate, disrupting ecosystems and threatening biodiversity. Some species may not adapt quickly enough to survive. These unprecedented migration serves as a threat to global food security. This is because over three billion people depend on marine ecosystems for their primary source of protein. Climate change jeopardizes fish stocks, aquaculture, and coastal fisheries due to habitat loss, warming waters, and disrupted ecosystems.

Consequences of climate change have socioeconomic implications. For example, marine tourism, which depends on healthy ecosystems like coral reefs, generates billions of dollars annually. Degraded oceans harm industries, economies, and jobs. The destruction of marine ecosystems exacerbates inequality, particularly affecting small island nations and developing countries that depend heavily on ocean resources. Degraded marine ecosystems, such as dying coral reefs and disappearing mangroves, release stored carbon back into the atmosphere, intensifying climate change. - Ocean warming and acidification reduce the oceans' ability to act as a carbon sink, creating a vicious cycle of climate degradation.

While these continued feedback loops on ecosystems exacerbate climate change, there is also limited time to reverse the damage of climate change. For one, many marine species and ecosystems are at a tipping point. Without immediate action, ecosystems like coral reefs and Arctic habitats could collapse within decades. Thus, restoration becomes more difficult and costly the longer action is delayed, and some losses may become permanent.

3 Achieving Carbon Neutrality Through Ocean Energy

When a process for example, transport, manufacture, or use does not add a net CO_2 to the atmosphere, it is called carbon neutral [9]. The goal of carbon neutrality is to maintain stable GHG level and not essentially meaning *not* adding more CO_2 into the atmosphere beyond what it can normally handle.

The ocean has high potential to generate a significant amount of wave energy. Numerous renewable energy advantages have been presented, such as decreasing dependence on fossil fuels and cutting GHG emissions, which contributes to combating climate change [10]. These devices have a minimal carbon footprint since they do not release GHGs during the production of electricity. The ocean energy, harnessing the tides, current, and waves are vast, clean, significant energy sources. They are renewable energy sources that are practically inexhaustible.

One example of ocean energy harnessing marine currents and waves is using wave energy converters (WECs). Aside from reducing reliance to fossil fuels and lowering GHG emissions, WECs diversify the energy mix reducing reliance to imported fuels. When designed and positioned appropriately, WECs also have a limited effect on marine ecosystems. Furthermore, a key benefit of WECs is their ability to reduce coastal erosion by dissipating wave energy. It could also be a beneficial habitat aiding coastal protection and contributing to climate change resilience [6]. The lack of mature and cost-competitive WEC technologies, however, has become a major roadblock for WEC products entering the marketplace [11].

The cost of energy production and per capita consumption are expected to increase. Supply of electricity in remote areas is still considered expensive. Electricity in isolated islands are still irregular and unreliable. The key item therefore in development of renewable energy is production of cheap energy. Ocean energy is available in remote areas but cheap technologies to convert them to electricity remains a significant area for research. Storing energy in batteries is the most preferred but alternative storage and fuel sources are being studied.

When constructing ocean renewable energy structures, it is essential to recognize their significant impact on the environment, natural habitats, local communities, and wildlife. Problems arise when the planning and management are inadequate. Despite the potential negative effects, it is important to emphasize the environmental benefits of using renewable energy. By effectively utilizing ocean resources, we can create a cleaner energy future while protecting marine ecosystems and benefiting local communities.

4 Role of Circular Economy to Mitigate Climate Impact

The circular economy presents a transformative model for addressing the environmental challenges posed by climate change, particularly in reducing GHG emissions and enhancing resource efficiency. Unlike the traditional linear economy—which follows a 'take, make, dispose' approach—the circular economy emphasizes keeping resources in use for as long as possible, extracting maximum value, and then recovering and regenerating products and materials at the end of their service life [12]. This approach not only minimizes waste but also reduces the demand for raw material extraction, a process often associated with significant carbon emissions. For instance, applying circular principles in sectors such as construction and manufacturing can lead to substantial reductions in emissions through practices like material recycling, product life extension, and design innovations that support easy disassembly and repurposing [13]. Moreover, integrating circular economy strategies in waste management, such as promoting composting, recycling, and energy recovery from waste, directly contributes to lowering methane emissions from landfills, thus supporting broader climate mitigation goals.

In the context of ocean protection and coastal management, circular economy practices can significantly reduce the flow of plastic and other pollutants into marine ecosystems, thereby protecting biodiversity and maintaining the integrity of natural carbon sinks like mangroves and seagrasses [14]. The promotion of sustainable product design, extended producer responsibility, and improved waste collection systems reduces marine litter, which not only affects marine life but also undermines the resilience of coastal communities to climate-related hazards. Additionally, the adoption of circular principles in the renewable energy sector—such as the reuse and recycling of materials in solar panels and wind turbines—can alleviate resource constraints and environmental impacts associated with the production and disposal of renewable energy technologies [15]. By embedding circular economy strategies within climate policies and adaptation plans, nations can foster sustainable development pathways that align with both environmental protection and socio-economic resilience, thus supporting the achievement of the United Nations Sustainable Development Goals (SDGs), particularly SDG 12 (Responsible Consumption and Production) and SDG 13 (Climate Action).

5 Transitioning to Renewable Energy While Safeguarding Our Ocean

Transitioning to renewable energy means phasing out fossil fuels, thus making it a good alternative to non-renewable energy resource use. Apart from renewable energy utilization, there are developing ocean-based methods that enhance carbon capture [16, 17]. These methods may be complemented by blue carbon solutions include restoring mangroves, seagrasses, and salt marshes, marine protected areas (MPAs) and sustainable fishing practices [18, 19]. Marine ecosystems, such as mangroves, seagrasses, and phytoplankton, absorb significant amounts of carbon dioxide. Climate change disrupts these ecosystems, reducing their capacity to mitigate GHG emissions.

Hydrogen fuel is being considered to mitigate climate change effect and help industries reduce their carbon footprint. Ships in maritime transport use fuel that can harm the environment through emissions and oil spills. Hydrogen is being considered as an alternative energy source to power ships. However, discussion among researchers mentioned that increase in hydrogen use could affect methane degradation. Hydrogen when burned produces water vapor but when fossil fuels are used to produce it, it is called grey hydrogen with CO_2 emissions [1]. Therefore, more research has to be done in this area.

The pathway to achieving net zero emissions is a collective responsibility that necessitates collaboration across all sectors of society. Governments, industries, communities, and individuals must acknowledge their roles in this global effort. Transitioning to renewable energy sources, enhancing energy efficiency, and promoting sustainable practices are essential components of this transformation.

At the heart of achieving net zero lies the need for innovative technologies and strategies. Carbon capture and storage (CCS), reforestation initiatives, and advancements in battery storage are just a few of the tools at our disposal. Moreover, systemic change in how we produce and consume energy is vital. Sustainable transportation, circular economies, and responsible agriculture can significantly reduce our carbon footprint while promoting resilience against climate impacts.

As we strive for net zero, we must also implement policies and practices that protect our oceans from degradation. In addition, MPAs are a crucial strategy for conserving marine biodiversity and habitats. By designating regions of the ocean as protected, we can promote the recovery of depleted fish stocks and mitigate the impacts of human activities.

Additionally, tackling plastic pollution and overfishing are necessary steps toward ensuring the health of our oceans [20]. Implementing sustainable fishing practices and reducing plastic production and consumption will foster healthier marine ecosystems. Education and awareness are also key—engaging communities and stakeholders in ocean conservation efforts can empower individuals to take meaningful action.

Aside from mitigating climate change, the coastal and marine ecosystems are essential for maintaining the planet's ecological balance, supporting human livelihoods, and mitigating climate change. Protecting them requires urgent action to reduce GHG emissions, preserve biodiversity, and implement sustainable practices. Delaying action on climate change risks permanently losing the benefits of these ecosystems, compounding the climate crisis and jeopardizing life on Earth.

6 Climate-Driven Disasters: Global Warming Impact on Natural Hazards

The relationship between climate change, the oceans, and natural hazards is a complex interplay of factors. Rising global temperatures contribute to the melting of polar ice caps and the thermal expansion of seawater, which leads to rising sea levels. This threatens coastal ecosystems such as mangroves and salt marshes, and results in the sinking of islands.

Mangroves and salt marshes serve as natural buffers against storms and flooding, so their loss increases the vulnerability of coastal communities to extreme weather events. Several islands have already disappeared due to rising sea levels and climate change, including those in India, Papua New Guinea, and the Solomon Islands [21]. Significant land losses from rising sea levels and coastal erosion have forced residents to relocate.

Additionally, places like the Torres Strait in Australia are at risk from increasing coastal erosion, storm surges, and saltwater intrusion. All of these could be linked to glacier melting.

6.1 Global Warming and the Consequences of Glacier Melting

Glaciers are sensitive indicators of climate change, and their disappearance are clear proof of global warming. In October 2024, an extraordinary phenomenon was observed concerning Mount Fuji, Japan's iconic peak. The snowcaps were missing which was unusual and broke a 130 years record. The lack of snow raised concerns among scientists and local residents about climate change and its effects on the environment.

Venice Italy's increasing vulnerability to flooding could be linked to rising sea levels, driven by global warming that melts polar ice and expands seawater [22]. More intense and frequent storms, intensified by warmer ocean temperatures, lead to increased water levels in the Venetian lagoon, worsening instances of high water. Although natural geological processes cause Venice to sink, the impact of climate-related sea level rise speeds up this issue, resulting in unprecedented acqua alta (high water) events [23].

In mountainous regions, the reduction of snow cover and glacial ice has serious repercussions, including a possible rise in the frequency of extreme weather events. As glaciers recede, less sunlight is reflected back into space (because glaciers are high albedo surfaces), the destabilization of mountain slopes occurs more frequently, increasing the risk of landslides and debris flows, which are often triggered by quick snowmelt during periods of intense heat or rain-on-snow interactions. Furthermore, the swift melting of snow can result in flooding during spring, with rapid snowmelt causing flood waves, the transportation of significant quantities of sediments [5]. Sediment deposition in rivers and lakes, alters morphology and impacts water quality [24].

Rivers fed by glaciers provide a steady water flow, especially in the summer, which is essential for consistent hydropower generation. In areas with glaciers, hydropower is a key energy source. The ongoing decline of glaciers presents a serious challenge to this energy source. As glaciers melt, less water flow is available during crucial times, and the likelihood of extreme weather events, like glacial lake outburst floods, increases, endangering hydropower infrastructure. There could be sudden release of large volumes of water, potentially causing severe damage to dams and other facilities, resulting in costly interruptions to energy supply [5].

Glacier melting and rising sea levels threaten inland and coastal communities, as well as various ecosystems. The disruption of marine ecosystems due to warming waters leads to shifts in fish populations and the decline of coral reefs, which are critical to marine biodiversity. To combat these challenges, we must adopt an integrated approach. Policies that address climate change must include provisions for protecting and restoring our ocean ecosystems. This means investing in research, fostering global cooperation, and sharing best practices to ensure that we can address these urgent issues collectively.

6.2 Climate Change Link to Wildfires and Flooding

The ocean being the last sink from terrestrial flows is severely affected by many factors [25]. Natural hazards are heightened nowadays with the overarching influence of climate change. Forests have been subjected to illegal logging, mining, forest fires. Apart from forests, many agricultural areas have been converted to built-up areas and industries to accommodate the ever-growing population, for construction, housing, and for production of materials and needs of the population. The earth's forest cover has the capacity to reduce CO_2 from the atmosphere through photosynthesis. Destruction of forests reduce the removal rate capacity of CO_2.

The reduction in forest covers due to human action have been made worse by the effects of climate change. Higher global temperatures caused by climate change lead to drier conditions, making forests like the Amazon more vulnerable to fires [26]. In some cases, forest fires extend into human settlements, and the high temperatures make it even more challenging to control them. Climate change has disrupted weather patterns in South America, resulting in more frequent and prolonged droughts in the Amazon. These droughts dry out vegetation and increase the risk of fires. Additionally, human-driven deforestation worsens this risk by fragmenting forests, reducing humidity, and exposing them to potential ignition sources.

Deforestation and climate change create a vicious cycle; forest loss leads to higher carbon emissions and reduced rainfall, which worsens the problem. Forest fires like in the Amazon release significant amounts of carbon dioxide, further accelerating global warming. This creates a feedback loop that worsens other climate impacts, such as flooding and rising sea levels.

Meanwhile, the prolonged fires in Los Angeles, California is an example of an extreme event linked to climate change. Elevated temperatures dry out vegetation and soil, creating favorable conditions for wildfires [27]. Extended heatwaves heighten the risk of fires and result in more intense blazes. Climate change worsens droughts, decreasing moisture levels in forests and making them more susceptible to igniting. Los Angeles has faced unprecedented droughts associated with rising temperatures. Consequently, a warmer climate extends fire seasons, as snow melts sooner and rainfall patterns alter.

A warmer atmosphere retains more moisture, causing heavier and more frequent rainstorms. This raises the risk of flash floods and river flooding, as observed in Brazil and Spain. Climate change disrupts regional precipitation patterns, leading to extremes like prolonged droughts followed by heavy rainfall, which can overwhelm soil and drainage capabilities.

The rising frequency and quantity of rainfall leading to floods are exacerbated by increased urbanization and rapid land use shifts. In various regions, urban growth and deforestation amplify flood risks by diminishing natural water absorption and heightening runoff during storms.

Climate change escalates the frequency and intensity of extreme weather events, including droughts, floods, and storm surges. More specifically, each region faces distinct consequences determined by its geography, ecosystems, and socio-economic conditions, but the underlying cause—global warming—remains a shared element. Thus, while each of these occurrences has local contributing factors (such as land use, urban development, and infrastructure), the impact of climate change should not be overlooked. Addressing these challenges necessitates global cooperation to lower GHG emissions, adapt to evolving conditions, and adopt sustainable practices that safeguard ecosystems, communities, and the ocean.

7 Conclusion

As we confront the urgency of climate change and the necessity of protecting our oceans, we must remember that the choices we make today will shape the legacy we leave for future generations. Achieving net zero and safeguarding the oceans is not merely an environmental issue; it is an ethical imperative that encompasses social justice, economic stability, and global cooperation.

The scientific community recognizes that climate change has altered the physical, social, economic, and other spheres of human life, and there is limited time to prevent the earth to reach a point of irreversibility. More specifically, climate change has affected the natural and environmental qualities of our oceans, which have been one integral component of our ecosystem as temperature controller and carbon sink. Our dependence on the oceans for economic value and ecosystem services has been strained and put in significant risks given the effects of increased CO_2 emissions and other GHGs. Hence, society must ramp up its efforts and work collectively to a path of successful mitigation of the negative effects of these GHGs to our climate and our oceans.

The journey ahead is undoubtedly challenging, yet it is one filled with opportunity. By uniting our efforts, embracing innovation, and prioritizing sustainability, we can create a future where both the planet and its oceans flourish. The time to act is now—our collective wellbeing depends on it.

References

1. Tharakee, H., Kadigamuwa, C.C., Perera, U.D.N.T.: Advancement of hydrogen economy through the development of clean energy technologies for blue economy. In: Kurniawan, T.A., Prasad, M.N.V. (eds.) Accelerating the Transition to a Hydrogen Economy, vol. 1, pp. 39–57. Elsevier, Netherlands (2025)
2. Kanna, V.I., Roseline, S., Balamurugan, K., Jeeva, S., Santhiyagu, I.A.: The effects of greenhouse gas emissions on global warming. In: Elsevier eBooks, pp. 143–154 (2024). 10.1016/b978-0-323-93940-9.00216-4

3. World Meteorological Organization: WMO confirms 2024 as warmest year on record at about 1.55°C above pre-industrial level. WMO Press Release. https://wmo.int/news/media-centre/wmo-confirms-2024-warmest-year-record-about-155degc-above-pre-industrial-level last accessed 2025/01/22
4. Casila, J. C.: Philippines Small Hydropower Development Report 2022. In: UNIDO, ICSHP, World Small Hydropower Development Report 2022. United Nations Industrial Development Organization, Vienna, Austria; International Center on Small Hydro Power, Hangzhou, China. https://www.unido.org/sites/default/files/files/2023-08/South-Eastern%20Asia_2022.pdf (2022)
5. Kolerski, T., Pan, J.: Glaciers the dynamic giants of cold regions. In: Findikakis, A., Kolokytha, E. (eds.) Hydrolink Net Zero and Water, vol. 3, pp. 17–19. International Association for Hydro-Environment Engineering and Research (2024)
6. Mossa, M.: Energy supply from sea wave, currents, and tides. In: Findikakis, A., Kolokytha, E. (eds.) Hydrolink Net Zero and Water, vol. 3, pp. 10–12. International Association for Hydro-Environment Engineering and Research (2024)
7. Thangal, S.H., Muralisankar, T., Mohan, K., Santhanam, P., Maran, B.: Biological and physiological responses of marine crabs to ocean acidification: a review. Environ. Res. **248**, 118238 (2024). 10.1016/j.envres.2024.118238
8. Hill, T.S., Hoogenboom, M.O.: The indirect effects of ocean acidification on corals and coral communities. Coral Reefs. **41**(6), 1557–1583 (2022). 10.1007/s00338-022-02286-z
9. Mitra, C.: Hydrogen: a versatile tool for decarbonization. In: Kurniawan, T.A., Prasad, M.N.V. (eds.) Accelerating the Transition to a Hydrogen Economy, vol. 1, pp. 3–22. Elsevier (2025)
10. Samad, A., Suchithra, R.: Marine power technology—Wave energy. In: Sustainable Fuel Technologies Handbook, pp. 241–267 (2020). 10.1016/B978-0-12-822989-7.00009-3
11. Hasan Maheen, M., Yang, Y.: Wave energy converters with rigid hull encapsulation: a review. Sustain Energy Technol Assess. **57**, 103273 (2023). 10.1016/j.seta.2023.103273
12. Ellen MacArthur Foundation: Towards a Circular Economy: Business Rationale for an Accelerated Transition. https://www.ellenmacarthurfoundation.org/towards-a-circular-economy-business-rationale-for-an-accelerated-transition (2015)
13. Geissdoerfer, M., Savaget, P., Bocken, N.M.P., Hultink, E.J.: The circular economy—a new sustainability paradigm? J. Clean. Prod. **143**, 757–768 (2017). 10.1016/j.jclepro.2016.12.048
14. Jambeck, J.R., Geyer, R., Wilcox, C., Siegler, T.R., Perryman, M., Andrady, A., et al.: Plastic waste inputs from land into the ocean. Science. **347**(6223), 768–771 (2015)
15. Stahel, W.R.: The circular economy. Nature. **531**(7595), 435–438. https://www.nature.com/articles/531435a (2016)
16. Hanson, E., Nwakile, C., Hammed, V.O.: Carbon Capture, Utilization, and Storage (CCUS) technologies: evaluating the effectiveness of advanced CCUS solutions for reducing CO2 emissions. Resul. Surf. Interf. **100381** (2024). 10.1016/j.rsurfi.2024.100381
17. Vibbert, H.B., Park, A.A.: Harvesting, storing, and converting carbon from the ocean to create a new carbon economy: challenges and opportunities. Front. Ener. Res. **10** (2022). 10.3389/fenrg.2022.999307
18. Choudhary, B., Dhar, V., Pawase, A.S.: Blue carbon and the role of mangroves in carbon sequestration: its mechanisms, estimation, human impacts and conservation strategies for economic incentives. J. Sea Res. **199**, 102504 (2024). 10.1016/j.seares.2024.102504
19. Hoppit, G., Schmidt, D.N., Brazier, P., Mieszkowska, N., Pieraccini, M.: Are marine protected areas an adaptation measure against climate change impacts on coastal ecosystems? A UK case study. Nat. Bas. Solut. **2**, 100030 (2022). 10.1016/j.nbsj.2022.100030
20. Kurniawan, T.A., Mohyuddin, A., Dzarfan Othman, M.H., Goh, H.H., Zhang, D., Anouzla, A., Aziz, F., Casila, J.C., Ali, I., Pasaribu, B.: Beyond surface: unveiling ecological and economic ramifications of microplastic pollution in the oceans. Water Environ. Res. **96**(7), e11070 (2024). 10.1002/wer.11070

21. Albert, S., et al.: Interactions between sea-level rise and wave exposure on reef island dynamics in the Solomon Islands. Environ. Res. Lett. **11**(5), 054011 (2016)
22. Camuffo, D.: A discussion on sea level rise, rate ad acceleration. Venice as a case study. Environ. Earth Sci. **81**(13), 10.1007/s12665-022-10482-x (2022)
23. Zanchettin, D., et al.: Sea-level rise in Venice: historic and future trends (review article). Nat. Hazards Earth Syst. Sci. **21**(8), 2643–2678 (2021). 10.5194/nhess-21-2643-2021
24. Casila, J.C., Andres, H.L., Haddout, S., Yokoyama, K.: Sediment transport modeling in the Pasig River, Philippines post Taal volcano eruption. Geosciences. **14**(2), 45 (2024). 10.3390/geosciences14020045
25. Casila, J.C., Dimapilis, D.A., Limbago, J.S., Delos Reyes, A.A., Casila, E.B., Haddout, S.: Characterization and quantification of surficial sediment microplastics and its correlation with heavy metals, soil texture, and flow velocity. Anal. Lett. **57**(7), 1092–1105 (2023). 10.1080/00032719.2023.2241090
26. Celis, N., Casallas, A., Lopez-Barrera, E.A., Felician, M., De Marchi, M., Pappalardo, S.E.: Climate change, forest fires, and territorial dynamics in the Amazon rainforest: an integrated analysis for mitigation strategies. ISPRS Int. J. Geo Inf. **12**(10), 436 (2023). 10.3390/ijgi12100436
27. Farid, A., et al.: A review of the occurrence and causes for wildfires and their impacts on the geoenvironment. Fire. **7**(8), 295 (2024). 10.3390/fire7080295

Expert Opinion

International Public Policies on Ocean Renewable: Effects of the Advisory Opinion of the International Law of the Sea Tribunal on Climate Change (and Other Cases) vis a vis Ocean Renewables

Sergio Peña-Neira(✉) and Patricio Araya-Meza

Universidad Mayor, Santiago de Chile 7500994, Chile
sergio.pena@umayor.cl

Abstract. The recent advisory opinion titled "Request for an Advisory Opinion Submitted by the Commission of Small Island States on Climate Change and International Law" marks a significant milestone in the fields of international relations, public policy, and international law. Ocean renewable energy is pivotal in crafting a comprehensive strategy to combat climate change, underscoring the need to thoughtfully consider innovative technologies in this area. Each sovereign state must develop its own policy approach, balancing maritime rights, the needs of its population, and a careful assessment of the environmental impacts on its natural resources. This analysis will explore the implications of this landmark advisory opinion on ocean renewables, particularly in the context of other international judgments and advisory opinions that shape global public policies on renewable energy. Ocean renewable energy is essential for an effective response to climate change, emphasizing the importance of integrating advanced technologies and environmental considerations into policy-making processes.

Keywords: Public policy · Ocean renewables · Climate Change

1 Ocean Renewables Policies

Policies, whether national or international, play a pivotal role in influencing the behavior of states, businesses, and individuals. The formulation of public policies is often grounded in formal documents and regulations, which can be either legal or non-legal in nature. These regulations may stem from a variety of sources, including international conventions, principles, customary practices, or rulings from international tribunals. The impact of such policies is far-reaching, often resulting in the establishment or modification of requirements that affect a wide array of activities, including those related to ocean renewable energy. This contribution focuses on an advisory opinion issued by a tribunal addressing maritime disputes. One of the key issues discussed is how new sources of energy, particularly in the context of ocean renewables, can mitigate the damaging

J. C. Casila et al. (Eds.): CCORE 2024, SPEES, pp. 17–22, 2025.
https://doi.org/10.1007/978-3-031-93887-0_2

effects of climate change. The challenge lies in harnessing these energy sources without exacerbating greenhouse gas emissions, which could further accelerate the already catastrophic shifts in global climate patterns.

1.1 International Public Policies on Ocean Renewable

The concept of "ocean renewables" serves as the starting point for this analysis. Operationally defined, *marine and hydrokinetic energy*, or *marine renewable energy*, refers to a sustainable energy source derived "from the natural movement of water, including waves, tides, and both river and ocean currents."[1] This form of energy harnesses the dynamic forces of water to generate power, offering a promising solution for renewable energy production. Conversely, "marine renewable energy" refers to a source of renewable power generated from the natural movement of water, encompassing ocean currents, tides, river currents, and waves. This form of energy leverages the dynamic forces of water to produce sustainable power, contributing to the broader landscape of renewable energy solutions[2]. The primary objective of international public policy is to transition away from coal and other non-renewable energy sources, such as petroleum, toward renewable energy alternatives by 2030, with many European countries targeting a shift by 2040–2050. To achieve this goal, marine renewable energy harnesses both thermal and kinetic energy. Heat is transferred to surface waters from deep, colder ocean layers, while movement is captured from natural forces such as wind, waves, ocean currents, and tides. This shift toward marine renewables plays a crucial role in the global effort to reduce reliance on fossil fuels and combat climate change[3]. The Advisory Opinion on Climate Change and International Law (AO)[4] is grounded in the United Nations Convention on the Law of the Sea (UNCLOS).[5]. Public policies aimed at addressing climate change and marine governance will be informed by the principles and legal frameworks outlined in this opinion. Ocean renewables includes any form of energy creation in the sea, therefore, in this research we will consider marine renewable energy, renewable energy from the sea wave or ocean currents energy sources as ocean renewables.

In this context, international public policy aims to mitigate the impacts of climate change by promoting environmentally sustainable energy sources. One such source, as discussed in this contribution, is ocean renewable energy. A key subgoal of international public policy is the transition from conventional electricity generation to renewable energy sources, with a significant portion of this shift expected to come from ocean renewables. This approach aligns with global efforts to reduce carbon emissions and transition to a more sustainable energy future[6].

[1] [US Department of Energy, 2024].
[2] [Chen, 2023: 298, Multon, 2013: xvii].
[3] [Chen, 2023: 298].
[4] [International Law of the Sea Tribunal, 2024].
[5] [United Nations Convention on the Law of the Sea, 1982].
[6] [International Energy Agency, 2020:13].

1.2 Advisory Opinion of the International Law of the Sea Tribunal on Climate Change

The Advisory Opinion of the International Tribunal for the Law of the Sea on climate change and international law, titled *"Request for an Advisory Opinion Submitted by the Commission of Small Island States on Climate Change and International Law"* (AO),[7] marks the first of its kind to address the intersection of climate change and international legal frameworks. This groundbreaking opinion represents a significant step in shaping global legal responses to the impacts of climate change, particularly for vulnerable regions such as small island states.

The Advisory Opinion on ocean renewables presents several noteworthy features, particularly in its comprehensive approach to climate change. The first part of the Opinion addresses the broader issue of climate change, with a specific focus on the greenhouse gases that drive these global shifts. From this analysis, several key principles can be derived, as outlined in the AO. These rules provide important guidance for shaping policies on ocean renewable energy and its role in mitigating climate change:

On Marine Pollution and a Solution: Ocean Renewables

The definition of marine pollution, as outlined in the Advisory Opinion, establishes that "anthropogenic greenhouse gas (GHG) emissions into the atmosphere constitute pollution of the marine environment.".[8] As well, it has included "specific sources of pollution, marine pollution from anthropogenic GHG emissions can be characterized as pollution from land-based sources, pollution from vessels, or pollution from or through the atmosphere".[9] Notably, these definitions exclude ocean renewable energy from being considered a source of contamination. Instead, it emphasizes the obligation of states to "take measures to prevent, reduce, and control marine pollution from any source.".[10] In this context, ocean renewables are placed as potential solutions mitigating marine pollution.

Permission for Ocean Renewables

The AO affirms that "States Parties to the Convention have specific obligations to take all necessary measures to prevent, reduce, and control marine pollution from anthropogenic GHG emissions and to endeavor to harmonize their policies in this regard.".[11] While the AO does not advocate for any single approach to mitigating pollution, it underscores the importance of implementing measures—such as the development of ocean renewables—that effectively address pollution without contributing to it.

On the Determination of Measures

Measures must be grounded in objectivity and scientific knowledge. The AO emphasizes that "measures should be determined objectively, taking into account, inter alia, the best available science and relevant international rules and standards contained in

[7] [International Law of the Sea Tribunal, 2024].

[8] [International Law of the Sea Tribunal, 2024: 179, 441 (3)(a)].

[9] [International Law of the Sea Tribunal, 2024: 441 (3) (e)].

[10] [International Law of the Sea Tribunal, 2024: 197].

[11] [International Law of the Sea Tribunal, 2024: 243, 441 3(b)(c)].

climate change treaties such as the UNFCCC and the Paris Agreement." Additionally, the obligation of due diligence requires States to adopt these measures as part of their commitments.[12]

Requirements for Public Policy Measures for Future Ocean Renewables Projects
States are obligated to adopt measures concerning ocean renewables, ensuring compliance with established requirements. These measures must reflect "due diligence"[13] encompassing the adoption of "appropriate rules and measures", "vigilance in their enforcement", and "the exercise of administrative control over public and private operators".[14] Public and private entities must adhere to these obligations not only in areas beyond national sovereignty or jurisdiction but also to prevent potential harm to the marine environment.

In the context of ocean renewables, this requirement is especially critical. The standard of due diligence is stringent due to the significant risks of serious and irreversible harm to the marine environment associated with such emissions. However, the implementation of due diligence obligations may vary depending on States' capabilities, technological development, available resources, and adherence to international rules and standards.[15]

Moreover, "available scientific knowledge" is a fundamental element that must inform public policies and guide the actions of both public and private operators. Its integration ensures that decisions are evidence-based and aligned with the goal of sustainable development.[16] Environmental Impact Assessment (EIA) represents a third critical requirement, carrying even greater significance than many other international standards.[17] Its primary objective is to evaluate activities with the potential to cause harm well before their implementation.[18] Consequently, EIA, alongside available scientific knowledge and due diligence, must be integral to the development of any public policy or project proposal related to ocean renewables.

Limits and Duties to International Obligations
The AO stipulates that greenhouse gas (GHG) emissions within a State's jurisdiction must not cause environmental harm to other States or to areas beyond the exercise of sovereign rights.[19]

National Laws
Sections 5 and 6 of Part XII of UNCLOS address the implementation of the Convention through national laws and regulations. States Parties have a *specific* obligation to adopt

[12] [International Law of the Sea Tribunal, 2024: 241].
[13] [International Law of the Sea Tribunal, 2024: 234].
[14] [International Law of the Sea Tribunal, 2024: 235].
[15] [International Law of the Sea Tribunal, 2024: 239, 243].
[16] [International Law of the Sea Tribunal, 2024: 321].
[17] [International Law of the Sea Tribunal, 2024: 345].
[18] [International Law of the Sea Tribunal, 2024: 352, 367].
[19] [International Law of the Sea Tribunal, 2024: 245].

laws and regulations that align with internationally agreed rules for the control, prevention, and reduction of greenhouse gas (GHG) emissions.[20] Consequently, any ocean renewables project must adhere to these rules, particularly when implemented in areas under national jurisdiction.

Cooperation (Ocean Renewables)
Cooperation in the area of ocean renewables is crucial in addressing the global challenge of climate change. This *specific* obligation, as outlined in international treaties, necessitates the formulation and adoption of "rules, standards, and recommended practices and procedures" that align with the provisions of UNCLOS.[21] These rules, along with other mandatory measures related to greenhouse gas (GHG) emissions, must be consistent with the Convention's framework.

This perspective is underscored by actions taken before the International Tribunal for the Law of the Sea (ITLOS) by a coalition of States, particularly small island nations. These States have highlighted the severe impacts of climate change on marine environments, including rising sea levels, marine contamination from GHG emissions, and the destruction of vital food sources.

2 Advisory Opinion of the International Court of Justice and of the Inter American Court on Human Rights

In 2025, the International Court of Justice[22] and the Inter-American Court of Human Rights[23] are set to issue Advisory Opinions on climate change and its impacts on human beings.

These opinions are expected to reflect similar reasoning and interpretation of international treaties, emphasizing their implications for public policies at both international and national levels. However, the scope of these opinions is likely to be broader, encompassing conventions, treaties, and other international directives. They are also expected to delve deeply into the relationship between climate change and the obligations arising from human activities that generate greenhouse gas (GHG) emissions.

Notably, these Advisory Opinions are likely to establish additional requirements for the development of ocean renewables projects, further shaping the legal and policy frameworks governing their implementation.

3 Conclusions and Effects on Ocean Renewables

While the AO does not explicitly focus on ocean renewables due to its emphasis on climate change, it indirectly supports their development as a clean and environmentally friendly energy source. This connection can be understood through several key points:

[20] [International Law of the Sea Tribunal, 2024: 260, 269, 270].
[21] [International Law of the Sea Tribunal, 2024: 321].
[22] [International Court of Justice, 2025].
[23] [Inter American Court of Human Rights, 2023].

1. **Support for Ocean Renewable Energy Development**: The AO implicitly encourages the exploration of ocean renewables as a viable source of sustainable energy.
2. **Compliance with Due Diligence and Assessments**: Meeting due diligence obligations, scientific and legal standards, and Environmental Impact Assessments is crucial. These requirements must be incorporated into national laws and adhered to in ocean renewable energy projects.
3. **Promotion of GHG-Free Solutions**: The AO highlights the importance of adopting solutions that do not generate greenhouse gas (GHG) emissions.
4. **Recognition of GHG Emissions as Pollution**: International policies and legal frameworks must explicitly address GHG emissions as a form of environmental contamination.

However, ocean renewables must also account for their own environmental impacts. This underscores the importance of compliance with due diligence and rigorous assessments in any project related to this energy source.

Finally, measures aimed at adapting ocean renewables to climate change must balance two objectives: reducing GHG emissions and maximizing the positive environmental benefits of ocean renewables. Ensuring that these projects do not inadvertently contribute to GHG emissions is essential for achieving sustainable outcomes.

References

Chen, F.: Marine energy: comparison of tidal and wave energy. Highligh. Sci. Eng. Technol. **29**, 298–307 (2023)

Inter American Court of Human Rights: Request for an Advisory Opinion submitted by Chile and Colombia before the Inter-American Court of Human Rights of January 9, 2023, San José de Costa Rica. https://www.corteidh.or.cr/docs/opiniones/soc_1_2023_en.pdf, last accessed 20 Jan 2025

International Court of Justice: Obligations of States in respect of Climate Change, The Hague. https://www.icj-cij.org/case/187, last accessed 22 Jan 2025

International Energy Agency: European Union 2020 Energy Policy Review, IEA, Paris, (2020). https://iea.blob.core.windows.net/assets/ec7cc7e5-f638-431b-ab6e-86f62aa5752b/European_Union_2020_Energy_Policy_Review.pdf last accessed 21 Jan 2025

International Law of the Sea Tribunal: Request for an advisory opinion submitted by the Commission of Small Island States on Climate Change and International Law", No 31, Hamburg (2024). https://www.itlos.org/fileadmin/itlos/documents/cases/31/Advisory_Opinion/C31_Adv_Op_21.05.2024_orig.pdf, last accessed 22 Jan 2025

Multon, B.: Preface. In: Multon, B. (ed.) Marine Renewable Energy Handbook. Wiley, Washington, DC (2013)

United Nations Convention on the Law of the Sea: United Nations Convention on the Law of the Sea (with annexes, final act and procès-verbaux of rectification of the final act dated 3 March 1986 and 26 July 1993). Concluded at Montego Bay on 10 December 1982, https://treaties.un.org/doc/publication/CTC/Ch_XXI_6_english_p.pdf, last accessed 22 Jan 2025

US Department of Energy: Marine Energy Basics, Washington (2024). https://www.energy.gov/eere/water/marine-energy-basics,

The Invention of Vacuum Insulated Wallpaper (VIW) for Energy Savings and Net-Zero Energy Sustainability Goals

Saim Memon[1,2,3](✉)

[1] Department of Industrial R&D in Vacuum Insulation Energy Technologies, Sanyou London Pvt Ltd, Level 18, 40 Bank Street, Canary Wharf, London E14 5NR, UK
Saim.Memon@sanyoulondon.com, Saim.Memon@bcu.ac.uk
[2] Department for Engineering, School of Engineering and the Built Environment, Birmingham City University, Millennium Point, Curzon Street, Birmingham B4 7XG, UK
[3] Jiangsu Sanyou Dior Energy-saving New Materials Co., Ltd (SANYOU DIOR), No.10 Guox-iang Road, West Tai Lake Science &Technology Industrial Park, Changzhou 213149, Jiangsu, China
https://saimmemon.com

Abstract. Buildings account for 34% of global energy demand and 37% of energy-related CO_2 emissions, underscoring the urgent need for innovative insulation solutions to address climate change, escalating energy costs, and sustainability goals. Vacuum Insulated Wallpaper (VIW) emerges as a transformative, ultra-thin (4 mm) insulation technology offering exceptional thermal conductivity of 5 mW/m·K. Designed for do-it-yourself (DIY) installation, VIW addresses traditional insulation challenges, such as high costs, complex installations, and limited retrofitting options, while achieving significant energy efficiency gains. This study highlights the superior performance of VIW compared to traditional materials. VIW shows a 93.2% reduction in thermal conductivity over 10 mm Thermal Insulating Wallpaper and a 90% reduction compared to 4 mm Wallrock Thermal Liner. Against bulk insulation materials, VIW achieves reductions of 88.6% for 200 mm Glass Mineral Wool and 85.3% for 250 mm Rock Mineral Wool. Even advanced materials like 180 mm PIR Insulation Board exhibit a 77.3% reduction in thermal conductivity when compared to VIW. VIW also excels across various wall types, reducing thermal conductivity by 71.74% in single brick walls, 41.98% in cavity walls, and 86.92% in concrete walls. Its unparalleled efficiency and space-saving design make it an ideal choice for both residential and commercial applications, while extending potential uses to electric vehicles and aviation. VIW bridges the gap between research and practice, providing a scalable, sustainable solution to improve energy efficiency, reduce carbon footprints, and reimagine spaces globally.

Keywords: Vacuum Insulated Wallpaper · Energy Efficiency · Net-Zero Buildings · Sustainable Insulation · Thermal Performance

J. C. Casila et al. (Eds.): CCORE 2024, SPEES, pp. 23–31, 2025.
https://doi.org/10.1007/978-3-031-93887-0_3

1 Introduction

The significance of insulating existing buildings is paramount, especially amid global challenges like climate change [1], escalating energy costs, and the imperative to reduce carbon footprints [2]. Buildings are substantial energy consumers, accounting for approximately 34% of global energy demand and 37% of energy-related carbon dioxide emissions in 2022 [3]. In the UK, nearly half of the housing stock remains below optimal energy efficiency standards, with 9% of dwellings rated in the least efficient categories (E to G) as of 2022 [4]. This inefficiency leads to significant energy waste, as poorly insulated buildings in cold regions, such as the UK and EU, require substantial energy for heating, while those in hot regions, like Qatar, Saudi Arabia, and the UAE, depend heavily on cooling systems to maintain comfort. The urgency of addressing these issues is underscored because 2024 is on track to be the warmest year recorded, with global temperatures expected to rise approximately 1.55 °C above pre-industrial levels [5]. Poor insulation exacerbates energy demands, leading to increased costs and environmental consequences. Addressing these challenges causes innovative and cost-effective solutions to enhance building energy efficiency and mitigate climate change.

The concept of vacuum insulation, first showed in 1913 by R. Burger and later commercialised by A. Zoller with the invention of the vacuum flask in Germany, provided an early example of the potential of vacuum technology for thermal insulation. This milestone marked the beginning of advancements in utilising vacuum between layers to achieve superior insulation performance [6, 7].

Decades after the invention of the vacuum flask, the concept of vacuum insulation evolved into vacuum glazing, which consists of two sheets of glass separated by an array of support pillars and a vacuum cavity [8]. The first successful vacuum glazing was constructed using high-temperature lead-based solder glass materials, notably Schott solder glass 8467, which hermetically sealed the edges at approximately 450 °C. This innovation, pioneered by Collins and Robinsons [9], became commercially significant under the trade name "Spacia" by Nippon Sheet Glass (NSG) Group in Japan [10]. However, the use of lead-based materials restricted its application due to environmental concerns, and the high processing temperature prevented the use of tempered glass and soft low-emissivity coatings such as silver [11].

Subsequent advancements were made in [12], who explored the use of semi-precious indium metal for edge sealing. This approach theoretically overcame the limitations of using tempered glass and silver coatings while achieving a low thermal transmittance value (U-value) of 0.86 $W/m^2 \cdot K$ for indium-sealed vacuum glazing [11]. Despite these technical improvements, indium's high cost and scarcity hindered its adoption for mass production. Recent innovations, such as composite hermetic edge seals combining Cerasolzer alloys and reinforced epoxy resin, were first invented by Professor Saim Memon have further enhanced the feasibility of vacuum glazing, but challenges related to manufacturing complexity and cost persist [13].

A key breakthrough came with the invention of lead-free vacuum glazing by Professor Saim Memon. This innovation employed glass-metallic vacuum materials and a fusion edge-sealing technique, cost-effective and eliminating the use of toxic lead while maintaining superior thermal performance [14]. Additionally, Professor Memon advanced the field by constructing first low-temperature edge-sealed triple vacuum gazing at laboratory scale [13]. Both of these recent innovations addressed the challenges of traditional high-temperature and low-temperature sealing methods, making it suitable for a broader range of applications. These technologies represented a leap forward in sustainable, high-performance insulation for windows [15]. However, the mass manufacturing and production and cost implications to consumers remain a challenge [16].

Subsequently, vacuum insulation panels (VIPs) and translucent vacuum insulation panels (TVIPs) were introduced compared to vacuum glazing, offering advanced thermal performance for building applications [17, 18].

Despite their promising thermal properties, vacuum-based insulation technologies have faced significant barriers to widespread adoption. High manufacturing costs, complexity in maintaining the vacuum integrity over time, and limited applicability in retrofitting existing buildings have constrained their scalability [19]. These challenges highlight the need for innovation to create accessible, cost-effective vacuum insulation solutions suitable for mass production and diverse building environments.

In response to these limitations, Professor Saim Memon and his team achieved the invention in 2023, progressing from a TRL 1 concept through laboratory R&D and prototyping to mass production. By early 2024, he successfully launched the revolutionary Vacuum Insulated Wallpaper (VIW) [20] as a physical consumer product, making it accessible to people worldwide. VIW is an ultra-thin, cost-effective, and easy-to-install insulation material, measuring just 4 mm in thickness and featuring an exceptional thermal conductivity of 5 mW/m·K, positioning it among the most efficient insulation materials available. Unlike traditional insulation methods [21], VIW is specifically designed for do-it-yourself (DIY) installation, eliminating the need for contractors and offering a practical solution for homeowners. This innovation effectively addresses key challenges in both cold and hot arid regions by reducing energy bills, minimising heating and cooling losses, and contributing to sustainability goals.

In this paper, the brief development, performance, and potential impact of VIW as a transformative solution for building insulation is explored. By addressing the barriers associated with traditional insulation methods, VIW offers a scalable and effective means of improving energy efficiency, reducing carbon footprints, and reimagining spaces across all climates.

2 Construction of Vacuum Insulated Wallpaper (VIW)

Figure 1 shows the construction of Vacuum Insulated Wallpaper (VIW), copyright-protected under Sanyou London Pvt. Ltd. [20], incorporates a multi-layered design engineered for superior thermal performance and aesthetic versatility. The front layer features a fully customisable design surface, allowing homeowners to choose from a wide range of colours and patterns to seamlessly integrate with any interior décor. The middle layer consists of a 3 mm thick vacuum insulation panel, delivering exceptional thermal resistance and significantly reducing heat transfer, making it one of the most efficient insulation solutions available. The back layer comprises a durable self-adhesive system, combining oil-based glue with a fiberglass mesh to provide a strong and lasting bond to wall surfaces. This advanced construction ensures that VIW not only enhances energy efficiency but also offers an easy-to-install, user-friendly solution suitable for diverse applications. All proprietary elements of VIW's design and construction are protected under copyright law, reinforcing Sanyou London Pvt. Ltd's commitment to innovation and intellectual property rights.

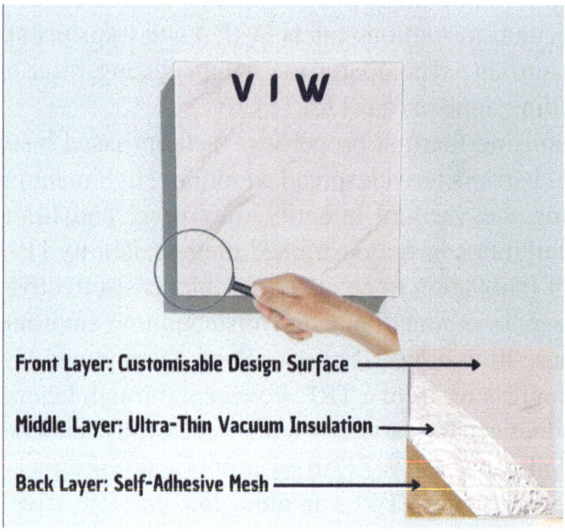

Fig. 1. .

3 Comparative Analysis of Thermal Conductivity and Thickness Efficiency in Insulation Materials

Figure 2 illustrates a comparative analysis of thermal conductivity across various insulation materials, highlighting the exceptional performance of Vacuum Insulated Wallpaper (VIW) with a thickness of approximately 4 mm. VIW demonstrates a thermal conductivity of only 5 mW/m·K, significantly outperforming other materials. Compared to traditional 10 mm thick Thermal Insulating Wallpaper with a conductivity of 74 mW/m·K, VIW achieves a dramatic 93.2% reduction in thermal conductivity. Similarly, when compared to the 4 mm thick Wallrock Thermal Liner, which has a conductivity of 50 mW/m·K, VIW reduces thermal conductivity by 90%, offering a much thinner yet more efficient alternative.

Traditional bulk insulation materials also fall short in efficiency despite their significantly greater thicknesses. For instance, 200 mm Glass Mineral Wool Insulation Board (44 mW/m·K) and 250 mm Rock Mineral Wool Insulation Board (34 mW/m·K) exhibit higher thermal conductivity, with VIW offering reductions of 88.6% and 85.3%, respectively. Even advanced options like the 180 mm thick PIR Insulation Board, with a conductivity of 22 mW/m·K, show a thermal conductivity reduction of 77.3% when compared to VIW.

These comparisons underscore the superior thermal efficiency of VIW relative to both thin and thick insulation materials, while simultaneously achieving significant space

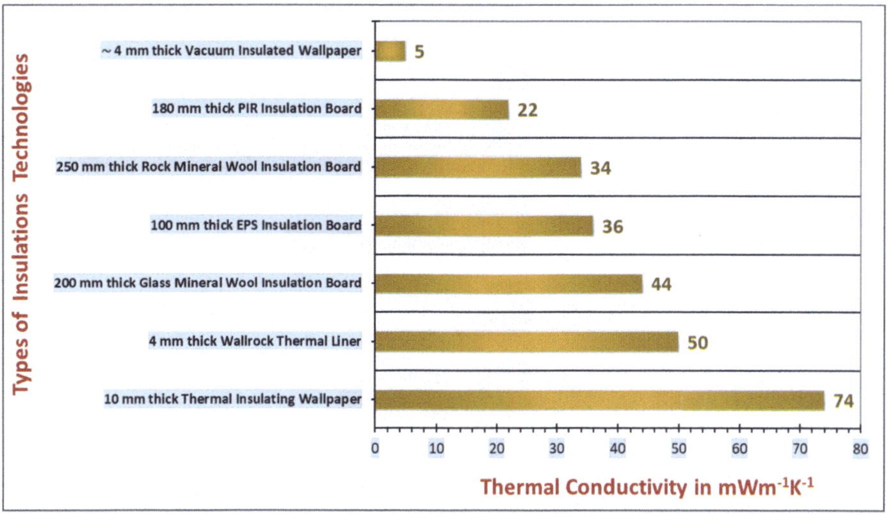

Fig. 2. Comparative Thermal Conductivity and Thickness Performance of Insulation Materials, highlighting the superior efficiency of Vacuum Insulated Wallpaper (VIW) at ~4 mm thickness with a thermal conductivity of 5 mW/m·K, compared to traditional and advanced insulation materials.

savings due to its ultra-thin construction. The combination of unparalleled thermal performance and reduced material thickness positions VIW as a transformative innovation in energy-efficient building insulation.

4 Impact of VIW on Thermal Conductivity Reduction Across Various Wall Types

Figure 3 shows a comparative analysis of thermal conductivity across various wall constructions, both with and without the application of Vacuum Insulated Wallpaper (VIW). Notably, the installation of VIW significantly reduces thermal conductivity in all tested wall types, thereby enhancing their insulation effectiveness. Specifically, the thermal conductivity of a single brick wall decreases from 700 mW/m·K to approximately 197.81 mW/m·K, achieving a reduction of about 71.74%. Similarly, a cavity wall shows a reduction from 280 mW/m·K to 162.44 mW/m·K, corresponding to a 41.98% decrease. Cob walls, originally at 900 mW/m·K, exhibit a reduction to 349.23 mW/m·K, marking a 61.20% decrease. Concrete or prefabricated walls, starting from 1700 mW/m·K, experience the most significant drop to 222.31 mW/m·K, an 86.92% reduction. Lastly, structural insulated panel systems (SIP) also benefit, with thermal conductivity decreasing from 140 mW/m·K to 68.68 mW/m·K, reflecting a 50.94% reduction. These results highlight VIW's potent impact on thermal performance, substantially decreasing heat transfer through these walls, which can lead to significant energy cost savings. Such data is crucial for homeowners and builders making informed decisions on insulation solutions that not only promote energy efficiency but also sustainability in building design.

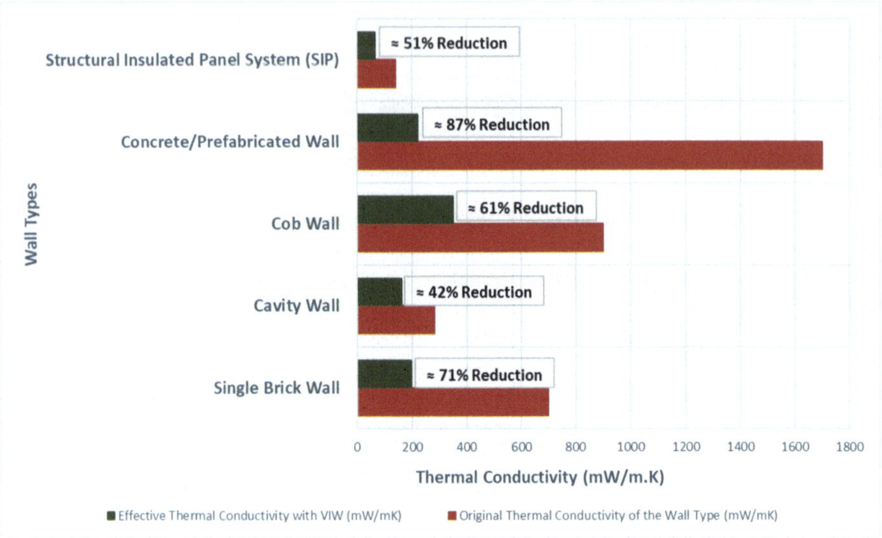

Fig. 3. Indicative impact of VIW on thermal conductivity reduction across various wall types

5 Applications of Vacuum Insulated Wallpapers (VIWs)

Vacuum Insulated Wallpapers (VIWs) combine cutting-edge insulation technology with aesthetic adaptability, offering exceptional thermal properties and lightweight construction that make them ideal for diverse applications. In residential settings, VIWs significantly enhance energy efficiency by improving insulation in living spaces such as bedrooms, kitchens, and bathrooms, without occupying much interior space. Their slim profile makes them particularly suitable for retrofits in older or heritage properties with solid walls, as well as for challenging applications like loft conversions. Moreover, their customisability allows seamless integration with various interior design styles, providing both functionality and aesthetic appeal.

In commercial environments, such as offices, retail stores, and restaurants, VIWs contribute to creating energy-efficient and comfortable indoor climates while optimising usable floor space. Their high insulation performance reduces heating and cooling demands, offering businesses the added advantage of cost savings and sustainability. Additionally, their modern and sleek design enhances the visual appeal of commercial spaces, aligning with brand aesthetics and improving customer experiences. Beyond buildings, VIWs find applications in electric vehicles (EVs), where they improve interior cabin insulation, reducing energy demands on climate control systems and contributing to extended driving ranges. They also serve as effective thermal management solutions for EV battery compartments, ensuring temperature stability, enhancing battery performance, and prolonging battery life under extreme weather conditions.

In high-end villas and luxury properties, VIWs cater to the dual need for superior energy efficiency and bespoke design, maintaining optimal indoor temperatures while complementing sophisticated interiors. Their potential extends to aviation, where their lightweight design and thermal efficiency could reduce cabin climate control requirements, saving energy and emissions. Additionally, their acoustic insulation properties promise a quieter cabin environment, enhancing passenger comfort, especially on long-haul flights. These diverse applications highlight the transformative potential of VIWs in reimagining spaces across residential, commercial, and specialised sectors.

6 Conclusion

In conclusion, Vacuum Insulated Wallpaper (VIW) proves to be a groundbreaking innovation in insulation technology, offering unparalleled thermal efficiency and versatility for diverse applications. The results implicate VIW as a superior alternative to traditional materials, achieving up to 93.2% reduction in thermal conductivity compared to 10 mm thick Thermal Insulating Wallpaper and 90% compared to the 4 mm Wallrock Thermal Liner. Even compared to bulkier insulation solutions like 200 mm Glass Mineral Wool (44 mW/m·K) and 180 mm PIR Insulation Board (22 mW/m·K), VIW achieves reductions of 88.6% and 77.3%, respectively, demonstrating its efficiency despite its ultra-thin 4 mm construction.

The results show that VIW dramatically enhances thermal performance in various wall constructions, achieving reductions in conductivity by up to 86.92% in prefabricated concrete walls, 71.74% in single brick walls, and 61.2% in cob walls. These reductions translate into significant energy savings and reduced heating and cooling demands, addressing key inefficiencies in both residential and commercial buildings. VIW's customisable design also enables seamless integration into modern and heritage properties, providing aesthetic flexibility alongside energy efficiency.

Beyond buildings, VIW's lightweight and high-performance properties make it a valuable solution for electric vehicles, enhancing cabin insulation and battery thermal management, contributing to extended driving ranges and improved battery safety. Its potential application in aviation could revolutionise aircraft insulation by reducing energy demands for cabin climate control and enhancing passenger comfort. The scalability, ease of installation, and adaptability of VIW make it a critical component in achieving global sustainability and net-zero energy goals by significantly reducing energy waste and carbon emissions across sectors.

Acknowledgement. This research was funded by Sanyou London Pvt. Ltd., under the leadership of Professor Saim Memon. Gratitude is also extended to Jiangsu Sanyou Dior Energy Saving New Materials Co. Ltd. for their invaluable support and collaboration in the development and evaluation of Vacuum Insulated Wallpaper (VIW). Their contributions have been instrumental to the success of this work.

References

1. Santamouris, M., Synnefa, A., Karlessi, T.: Building energy savings by green roofs and cool roofs in current and future climates. Nat. Sustain. **4** (2024). https://doi.org/10.1038/s42949-024-00159-8

2. Ürge-Vorsatz, D., Cabeza, L.F., Serrano, S., Barreneche, C., Broin, Ó., D.: Prioritize energy sufficiency to decarbonize our buildings. Nat. Human Behav. (2023). https://doi.org/10.1038/s41562-023-01752-0
3. UNEP: Global Status Report for Buildings and Construction. United Nations Environment Programme (2022) https://www.unep.org/resources/report/global-status-report-buildings-and-construction
4. UK Government: English housing survey 2022 to 2023: energy report. https://www.gov.uk/government/statistics/english-housing-survey-2022-to-2023-energy/english-housing-survey-2022-to-2023-energy-report (2022)
5. Financial Times: 2024 on track to be warmest year recorded, with temperature rise nearing 1.55°C. https://www.ft.com/content/ab1a51fa-a27f-4b19-b125-545b643a2ff1 (2024)
6. Burger, R.: Insulated container. US Patent 1,062,778. United States Patent and Trademark Office, Washington, DC (1913)
7. Zoller, A.: Commercialisation of the vacuum flask: Thermos (1913)
8. Collins, R.E., Simko, T.M.: Current status of the science and technology of vacuum glazing. Sol. Energy. **62**(3), 189–213 (1998). https://doi.org/10.1016/S0038-092X(98)00007-3
9. Collins, R.E., Robinson, S.J.: Evacuated glazing. Sol. Energy. **47**(1), 27–38 (1991). https://doi.org/10.1016/0038-092X(91)90060-A
10. Collins, R.E., Robinson, S.J.: Thermally insulating glass panel and method of construction. US Patent 5,657,607. University of Sydney (1997)
11. Eames, P.C.: Vacuum glazing: current performance and future prospects. Vacuum. **82**(7), 717–722 (2008). https://doi.org/10.1016/j.vacuum.2007.10.017
12. Griffiths, P.W., di Leo, M., Cartwright, P., Eames, P.C., Yianoulis, P., Leftheriotis, G., Norton, B.: Fabrication of evacuated glazing at low temperature. Sol. Energy. **63**(4), 243–249 (1998). https://doi.org/10.1016/S0038-092X(98)00019-X
13. Memon, S., Farukh, F., Eames, P.C., Silberschmidt, V.V.: A new low-temperature hermetic composite edge seal for the fabrication of triple vacuum glazing. Vacuum. **120**, 73–82 (2015). https://doi.org/10.1016/j.vacuum.2015.06.024
14. Memon, S., Eames, P.C.: Design and development of lead-free glass-metallic vacuum materials for the construction and thermal performance of smart fusion edge-sealed vacuum glazing. Energ. Buildings. **227**, 110430 (2020). https://doi.org/10.1016/j.enbuild.2020.110430
15. Memon, S., Katsura, T., Radwan, A., Zhang, S., Serageldin, A.A., et al.: Modern eminence and concise critique of solar thermal energy and vacuum insulation technologies for sustainable low-carbon infrastructure. Int. J. Solar Thermal Vacuum Eng. **1**(1), 52–71 (2020). https://doi.org/10.37934/stve.1.1.5271
16. Fang, Y., Memon, S., Peng, J., Tyrer, M.: Solar thermal performance of two innovative configurations of air vacuum layered triple glazed windows. Renew. Energy. **150**, 167–175 (2020). https://doi.org/10.1016/j.renene.2019.12.115
17. Radwan, A., Katsura, T., Memon, S., Serageldin, A.A., Nakamura, M., Nagano, K.: Thermal and electrical performances of semi-transparent photovoltaic glazing integrated with translucent vacuum insulation panel and vacuum glazing. Energy Convers. Manag. **215**, 112920 (2020). https://doi.org/10.1016/j.enconman.2020.112920
18. Katsura, T., Memon, S., Radwan, A., Nakamura, M., Nagano, K.: Thermal performance analysis of a new structured-core translucent vacuum insulation panel in comparison to vacuum glazing: experimental and theoretically validated analyses. Sol. Energy. **199**, 326–346 (2020). https://doi.org/10.1016/j.solener.2020.02.030
19. Katsura, T., Miyata, T., Memon, S., Radwan, A., Nagano, K.: Experimental analysis of evacuation pressure and gas flow rate in structured-core transparent vacuum insulation panels. Energy Rep. **9**(1), 1071–1078 (2023). https://doi.org/10.1016/j.egyr.2022.12.035

20. Sanyou London Pvt Ltd: Vacuum Insulated Wallpaper (VIW). https://sanyoulondon.com/department-of-products-and-sales/viw/ Accessed on 20 Nov 2024
21. Memon, S.: The scope of advanced smart vacuum insulation technologies for net-zero energy buildings. In: Sayigh, A. (ed.) Sustainable Energy Development and Innovation, pp. 1–20. Springer, Cham (2022). https://doi.org/10.1007/978-3-030-76221-6

Circular Economy Strategies for Reducing Marine Pollution: Implications for Climate Change and Ocean Renewable Energy

Anupam Khajuria(✉)

United Nations University Institute for the Advanced Study of Sustainability, Tokyo 150-8925, Japan
khajuria@unu.edu, a.khajuria@outlook.com

Abstract. Marine pollution and climate change are two of the most pressing environmental issues today, and they are closely interconnected. Marine ecosystems are vital for biodiversity, food security, and climate regulation. However, these ecosystems face severe threats due to pollution, particularly from plastics and other anthropogenic wastes, coupled with climate change impacts such as ocean acidification, rising sea levels, and habitat degradation. Ocean renewable energy technologies, such as offshore wind and tidal power, offer promising solutions for sustainable energy generation, but they depend on healthy marine environments for long-term success. The adoption of circular economy (CE) approach which focuses on sustainable resource use, waste minimization, and regenerative systems, provides a robust framework to mitigate marine pollution and address climate challenges. This article explores key circular economy strategies for addressing marine pollution and examines the broader implications for climate action and the development of ocean renewable energy.

Keywords: Circular economy · Marine pollution · Climate action · Sustainability

1 Introduction

The circular economy (CE) represents a transformative model that seeks to eliminate waste and extend the lifecycle of resources through reuse, recycling, and regeneration, as opposed to the traditional linear economy's "take, make, dispose" paradigm. CE principles offer solutions to reduce resource extraction, minimize environmental degradation, and foster a closed-loop system that maximizes resource efficiency and mitigates climate impacts. According to the Ellen MacArthur Foundation, CE approaches have demonstrated significant potential in reducing environmental footprints across industries by maintaining the value of resources through enhanced design, recycling, and waste recovery. The transition to CE is critical for achieving global sustainability goals, such as those outlined in the United Nations 2030 Agenda for Sustainable Development Goals (SDGs), and advancing towards a carbon-neutral society (Khajuria et al., 2022). This

J. C. Casila et al. (Eds.): CCORE 2024, SPEES, pp. 32–37, 2025.
https://doi.org/10.1007/978-3-031-93887-0_4

transition requires coordinated efforts through public-private partnerships, robust governmental policies, and active stakeholder engagement helps improve the use of natural resources and move closer to or achieve sustainability (Khajuria et al., 2022; Mohan et al., 2023).

The concept of planetary boundaries highlights the critical thresholds within which humanity can safely operate without causing irreversible environmental damage. CE strategies, such as the 3Rs (Reduce, Reuse, Recycle), offer practical pathways to stay within these boundaries by addressing critical stressors such as resource depletion, waste generation, and pollution (Khajuria, 2020). The adoption of circular practices can significantly reduce greenhouse gas (GHG) emissions (Ellen MacArthur Foundation, 2019). This paper explores how CE strategies can mitigate marine pollution and their interconnected benefits for climate change adaptation and renewable energy systems.

2 Key Circular Economy Strategies for Marine Pollution Mitigation

2.1 Reducing Plastic Waste at the Source

Plastic pollution is the largest contributor to marine pollution, causing widespread environmental and economic damage. Circular economy strategies aim to reduce plastic waste at its source through redesigning products and packaging to minimize plastic dependency. For example, increasing the use of recycled content and adopting biodegradable alternatives significantly reduces the volume of waste entering oceans. Policies such as Extended Producer Responsibility (EPR) incentivize manufacturers to design sustainable products and assume accountability for their lifecycle, ensuring proper disposal and recycling. Promoting refillable or reusable packaging systems can further reduce the demand for single-use plastics, contributing to a significant decline in ocean-bound waste.

2.2 Promoting Material Recovery and Recycling

Robust recycling systems play a crucial role in diverting plastics, metals, and textiles from marine environments. Circular economy practices enhance the recovery of materials from waste streams, reducing dependence on virgin resources and mitigating the environmental footprint of manufacturing processes. For example, the European Union's Circular Economy Action Plan (n.d.) promotes higher recycling targets for packaging materials, effectively reducing waste and preventing marine litter. Innovative programmes such as 'fishing-for-litter,' where fishing crews collect marine waste and bring it ashore for recycling, can significantly impact marine pollution mitigation if scaled and integrated with local infrastructure (Jambeck et al., 2015).

2.3 Innovative Materials and Eco-Design

Eco-design principles emphasize creating products that are durable, repairable, recyclable, and, where possible, biodegradable. Developing alternatives such as algae-based materials, biodegradable plastics, and other innovative substitutes for traditional plastics can significantly reduce marine waste. Eco-design also addresses waste generation at its source by extending the lifespan of products, reducing the need for new raw materials, and minimizing disposal rates. These approaches contribute directly to cleaner marine ecosystems by reducing pollution inflows and promoting resource efficiency.

2.4 Enhanced Waste Management Systems

Coastal areas often face unique waste management challenges due to their proximity to the sea. Strengthening waste collection, treatment, and disposal systems is essential for preventing marine pollution. Advanced waste management strategies such as waste-to-energy and circular bioprocessing of organic waste also offer avenues for cleaner energy production and resource recovery, contributing to both pollution reduction and energy generation. Such approaches not only help mitigate pollution but also reduce GHG (GHG) emissions from traditional landfill and incineration practices, aligning with broader climate goals.

2.5 Leveraging Digital Technologies for Waste Tracking and Management

Digital technologies, including blockchain, artificial intelligence (AI), and the Internet of Things (IoT), can revolutionize waste management by tracking materials through the supply chain. Blockchain systems, for instance, ensure transparency and traceability in recycling efforts, while AI can optimize waste sorting and processing operations. Digital monitoring can also help enforce regulations, hold industries accountable, and empower consumers to make more informed choices. For example, GPS tracking of waste exports could prevent illegal dumping at sea, and blockchain-based systems can facilitate transparency and traceability in plastic recycling.

3 Implications for Climate Change Mitigation

The intersection of circular economy principles with climate change mitigation is highly relevant, as waste management and resource extraction are major sources of GHG emissions. By reducing demand for virgin materials and minimizing waste, CE strategies help lower emissions and conserve energy.

3.1 GHG Reduction Through Waste Management

Improper waste management, particularly of plastics, generates methane and carbon dioxide emissions, both potent GHG emissions. Advanced waste management technologies like waste-to-energy systems capture and convert these emissions into energy, reducing pollution and contributing to the energy grid. Circular systems that minimize waste directly reduce the amount of GHG emissions associated with landfill use (Khajuria, 2024).

3.2 Efficient Use of Resources

Circular economy practices promote efficiency, reducing the carbon footprint associated with raw material extraction, processing, and disposal. By prioritizing recycled and renewable materials over virgin resource extraction, CE practices significantly reduce energy consumption and carbon footprints (Milhem et al., 2025). For example, producing plastic from recycled materials consumes significantly less energy than manufacturing from fossil fuels, thereby reducing emissions. For instance, using recycled plastic in new products can significantly cut energy consumption compared to producing plastic from virgin resources.

3.3 Climate-Resilient Marine Ecosystem

Reducing marine pollution supports the health of marine ecosystems, which act as carbon sinks. Healthy oceans play a key role in regulating atmospheric carbon dioxide and supporting biodiversity that maintains the ocean's ecological balance. Thus, CE strategies are directly linked to climate resilience and adaptation.

4 Conclusion

The adoption of circular economy principles provides an essential framework for addressing the interconnected challenges of marine pollution, climate change, and renewable energy development. Marine pollution and climate change, as interlinked issues, demand integrated solutions, and the circular economy offers a transformative framework that aligns economic development with ecological stewardship.

The implementation of circular economy principles addresses marine pollution at its roots. By reducing plastic waste, promoting material recovery, and designing eco-friendly alternatives, circular practices help maintain the health of marine ecosystems that are vital for biodiversity, climate regulation, and human livelihoods. These ecosystems serve as natural carbon sinks, absorbing atmospheric carbon dioxide and mitigating the impacts of global warming. Thus, reducing marine pollution not only safeguards biodiversity but also strengthens the resilience of oceanic systems to climate change impacts such as acidification and rising temperatures.

Moreover, the circular economy contributes significantly to climate change mitigation through its emphasis on resource efficiency and waste reduction. Advanced waste management systems, such as waste-to-energy technologies and bioprocessing, demonstrate how circular principles can simultaneously address pollution, conserve resources, and generate renewable energy. These systems reduce reliance on landfills and mitigate GHG emissions, which are major contributors to global warming.

The health of marine ecosystems also directly influences the success of ocean renewable energy systems, such as offshore wind, wave, and tidal power. These renewable energy sources are crucial for transitioning to a low-carbon economy, but they rely on stable and sustainable marine environments. Circular economy strategies, by reducing waste and pollution, support the long-term viability of these energy systems, fostering a mutually reinforcing relationship between clean energy development and environmental conservation.

In addition, the circular economy aligns with global climate goals, including the Paris Agreement and the United Nations Sustainable Development Goals (SDGs). Specifically, it contributes to SDG 12 (Responsible Consumption and Production), SDG 13 (Climate Action), and SDG 14 (Life Below Water). By fostering innovation, promoting collaboration, and encouraging public-private partnerships, circular practices pave the way for a holistic approach to addressing the planet's most pressing challenges.

The road to integrating circular economy practices into marine management requires coordinated action across governments, industries, academia, and civil society. Policymakers must prioritize frameworks that incentivize sustainable product design, enforce regulations to prevent illegal waste dumping, and invest in advanced technologies for waste tracking and recycling. Collaboration at local, regional, and global levels is essential to scale these efforts and ensure their effectiveness.

In conclusion, the circular economy is not a one-dimensional solution but a comprehensive pathway that addresses marine pollution, mitigates climate change, and supports sustainable energy systems. It offers a blueprint for harmonizing economic activities with ecological preservation, ultimately leading to a healthier planet and a more equitable future. As nations work towards carbon neutrality and climate resilience, integrating circular economy principles into marine and climate policies is not only feasible but indispensable for securing a sustainable future for all (Fig. 1).

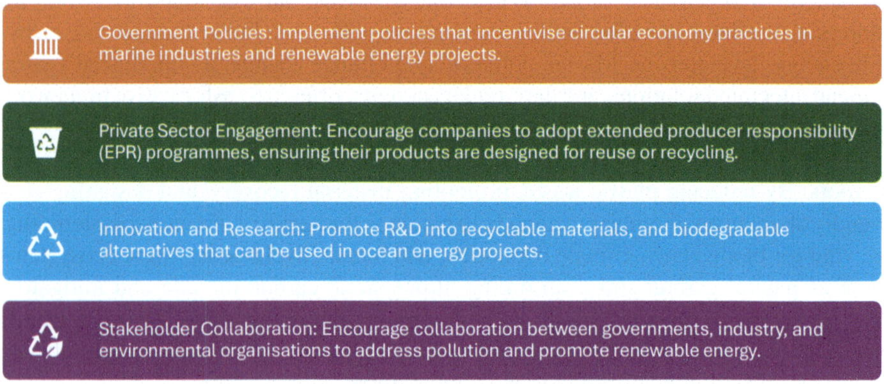

Fig. 1. Potential solutions and recommendations

References

Ellen MacArthur Foundation (2019). Completing the Picture: How the Circular Economy Tackles Climate Change. https://www.ellenmacarthurfoundation.org/

European Union: Circular Economy Action Plan (n.d.) https://eur-lex.europa.eu/legal-content/EN/TXT/?qid=1583933814386&uri=COM:2020:98:FIN

Jambeck, J.R., et al.: Plastic waste inputs from land into the ocean. Science. **347**(6223), 768–771 (2015) https://www.science.org/doi/10.1126/science.1260352

Khajuria, A.: Integrated approach between DPSIR, planetary boundaries and sustainable development goals ~ towards 3Rs and resource efficiency world. Environment. **10**(2), 52–56 (2020). https://doi.org/10.5923/j.env.20201002.03

Khajuria, A.: Diplomatic Courier. Circular economy education is a critical tool for NDCs (2024). Published on 11 November

Khajuria, A., Atienza, V.A., Chavanich, S., Henning, W., Islam, I., Kral, U., Liu, M., Liu, X., Murthy, I.K., Timothy, T.D., Oyedotun, P.V., Guochang, X., Zeng, X., Li, J.: Accelerating circular economy solutions to achieve the 2030 agenda for sustainable development goals. Circ. Econ. **1**(1), 100001 (2022). https://doi.org/10.1016/j.cec.2022.100001

Milhem, M., et al.: Advancing sustainability: embracing the circular economy for environmental and economic resilience. In: AlDhaen, E., Braganza, A., Hamdan, A., Chen, W. (eds.) Business Sustainability with Artificial Intelligence (AI): Challenges and Opportunities. Studies in Systems, Decision and Control, vol. 566. Springer, Cham (2025). https://doi.org/10.1007/978-3-031-71318-7_6

Mohan, J., Mohanty, C.R.C., Cross, J.S., Khajuria, A.: Mapping 3R and circular economy policy implementation in Asia and the Pacific. Circ. Econ. Sustain. **4**, 671–692 (2023). https://doi.org/10.1007/s43615-023-00302-0

Transforming Groundwater Research with Drones and Machine Learning

Md Munir Hayet Khan(✉)

Faculty of Engineering and Quantity Surveying (FEQS), INTI International University (INTI-IU), Persiaran Perdana BBN, Nilai 71800, Negeri Sembilan, Malaysia
munir.hayetkhan@newinti.edu.com

Abstract. Groundwater is a critical yet often overlooked resource that sustains billions of people worldwide, providing nearly half of the global drinking water supply, supporting agriculture, and driving industrial growth. However, excessive extraction, climate-induced disruptions, and contamination from human activities threaten its availability and quality. Conventional monitoring techniques face challenges in spatial coverage, accessibility, and real-time data acquisition. To address these issues, the integration of drones and machine learning (ML) is emerging as a transformative approach to groundwater management. Drones equipped with advanced sensors such as LiDAR, thermal imaging, and multispectral cameras offer high-resolution data on recharge zones, land-use changes, and potential contamination sources. Simultaneously, ML algorithms process vast datasets to predict groundwater levels, assess contamination risks, and model surface-groundwater interactions with improved accuracy. The synergy between drones and ML enables real-time monitoring, cost-effective analysis, and data-driven decision-making, making groundwater management more sustainable and adaptive to climate challenges. Despite its potential, barriers such as regulatory constraints, data integration complexities, and financial costs remain. Open-source platforms and public-private partnerships offer promising solutions to enhance the accessibility and scalability of these technologies. As groundwater sustainability becomes an urgent priority, integrating drone-based data collection with ML-driven analysis presents a promising pathway toward informed resource management and climate resilience. By embracing these advancements, policymakers, researchers, and communities can work collaboratively to safeguard groundwater resources for future generations.

Keywords: Machine Learning (ML) · Groundwater Management · Drone

1 Introduction

1.1 The Hidden Value of Groundwater

Groundwater is a hidden yet vital resource that sustains billions of lives worldwide. It provides nearly half of the global drinking water supply, irrigates agricultural fields in arid and semi-arid regions, and supports industries critical to economic growth. Despite

J. C. Casila et al. (Eds.): CCORE 2024, SPEES, pp. 38–41, 2025.
https://doi.org/10.1007/978-3-031-93887-0_5

its importance, groundwater is facing unprecedented challenges. Over-extraction driven by population growth and agricultural demands is depleting aquifers faster than they can recharge. Climate change has further intensified this problem by disrupting rainfall patterns, increasing drought frequencies, and reducing natural recharge rates. In addition, contamination from agricultural runoff, industrial effluents, and urban wastewater poses a significant threat to groundwater quality. The need for innovative, efficient, and scalable solutions to manage and protect this precious resource has never been greater.

Groundwater is often referred to as a hidden resource, as it exists beneath the surface and is less visible compared to rivers, lakes, and reservoirs. Yet, its significance cannot be overstated. Globally, groundwater supports drinking water, agriculture, and industrial processes. However, unsustainable extraction practices and contamination continue to threaten its availability and quality. As pressure mounts on this resource, it becomes imperative to rethink how we manage and sustain it.

2 Drones and Machine Learning: A New Frontier in Groundwater Monitoring

Drones, or unmanned aerial vehicles, have fundamentally changed how we collect environmental data. They offer a cost-effective, flexible, and rapid means of surveyinglarge areas, even in remote or difficult-to-access terrains. Equipped with advanced sensors such as LiDAR, thermal cameras, and multispectral imagers, drones can generate detailed maps of groundwater recharge zones, track land-use changes, and identify pollution sources [1]. These applications have redefined the possibilities for groundwater monitoring and opened new avenues for efficient data collection.

Machine learning excels in analyzing complex datasets and identifying patterns that traditional methods might miss. For groundwater research, these models are particularly valuable in predicting groundwater levels, recharge rates, and contamination risks. From forecasting the impacts of climate variability to simulating groundwater-surface water interactions, ML provides actionable insights that help manage groundwater resources more sustainably [2]. Its predictive power has become indispensable in addressing the growing complexities of water resource management.

3 The Synergy Between Drones and Machine Learning

The combination of drones and ML represents a paradigm shift in groundwater management. Drones serve as efficient data collectors, capturing high-resolution spatial and temporal data, while ML processes this data to generate meaningful insights. Together, they enable real-time monitoring, enhanced predictive models, and cost-effective solutions. These technologies complement each other to create a dynamic framework for understanding and managing groundwater systems at unprecedented scales and resolutions [2].

Drones and machine learning are proving indispensable in adapting to the impacts of climate change on groundwater resources. By providing real-time data and predictive insights, these technologies help communities anticipate and respond to climate-induced challenges such as prolonged droughts, erratic rainfall, and groundwater depletion. For instance, drones can identify areas where artificial recharge structures, like check dams, are most needed, while ML models optimize their placement to maximize water retention. These tools not only enhance climate resilience but also support long-term resource sustainability.

4 Challenges, Opportunities and Future Directions

While the potential of drones and ML is immense, several challenges remain. Regulatory barriers, such as restrictions on drone usage and data sharing, can limit their deployment. Additionally, the high cost of advanced drones and ML tools, coupled with the need for skilled operators and analysts, poses a hurdle for widespread adoption. Data integration remains another challenge, as processing large volumes of drone data requires robust computational infrastructure. However, technological advancements, such as autonomous drones and user-friendly platforms, are gradually addressing these barriers.

One of the most promising developments in recent years is the rise of open-source platforms for drone data processing and machine learning. Platforms like Google Earth Engine and QGIS enable researchers and resource managers to analyze spatial data without requiring costly software licenses. Integrating drone data with these platforms democratizes access to advanced analytics, allowing smaller organizations and developing nations to benefit from cutting-edge technologies. By leveraging open-source tools, we can accelerate the adoption of drones and ML in groundwater management on a global scale [3].

Public-private partnerships are emerging as vital enablers in scaling the application of drones and machine learning for groundwater research. Governments, research institutions, and private technology firms can pool resources and expertise to develop cost-effective solutions tailored to local needs. For example, technology companies specializing in drones and AI can collaborate with water management agencies to deploy these tools in regions facing acute groundwater stress. These partnerships can also address challenges related to funding and capacity-building by providing financial resources, technical training, and operational support.

The future of groundwater research will be shaped by the integration of emerging technologies [3]. Autonomous drones capable of self-directed monitoring will enable continuous, large-scale data collection with minimal human intervention. The Internet of Things (IoT) can complement drones by creating networks of sensors that provide real-time updates on groundwater conditions. Meanwhile, advances in deep learning will further enhance the analytical capabilities of ML models, allowing them to handle increasingly complex datasets and scenarios. Collaborative efforts will be key to realizing the full potential of these innovations.

5 The Ethical Imperative of Equitable Technology Deployment

As we embrace advanced technologies like drones and ML, it is imperative to consider their equitable deployment. Regions with limited access to these tools, particularly in low-income countries, may face a widening gap in groundwater management capabilities. Ensuring affordable access, capacity-building, and knowledge-sharing across borders can prevent disparities in resource management. Moreover, involving local communities in technology adoption ensures that solutions are context-sensitive and address the real needs of stakeholders. Groundwater is a shared resource, and its sustainable management requires shared responsibility.

6 Conclusion

The integration of drones and machine learning is transforming how we study, monitor, and manage groundwater resources. By providing faster, more accurate, and cost-effective solutions, these tools address critical challenges in groundwater sustainability. From mapping recharge zones to predicting contamination risks, drones and ML empower researchers and policymakers with the insights needed to safeguard groundwater for future generations. As someone who has dedicated many years to the study of hydrology and water resources, I am confident that these technologies will play a pivotal role in shaping a sustainable water future. Therefore, by fostering collaboration and innovation, we can secure a sustainable water future for generations to come.

References

1. García-López, S., Vélez-Nicolás, M., Zarandona-Palacio, P., Curcio, A.C., Ruiz-Ortiz, V., Barbero, L.: UAV-borne LiDAR revolutionizing groundwater level mapping. Sci. Total Environ. **859**(1), 160272 (2023). 10.1016/j.scitotenv.2023.160272
2. Boo, K.B.W., El-Shafie, A., Othman, F., Khan, M.M.H., Birima, A.H., Ahmed, A.N.: Groundwater level forecasting with machine learning models: a review. Water Res. **252**, 121249 (2024). 10.1016/j.watres.2024.121249
3. Davamani, V., John, J.E., Poornachandhra, C., Gopalakrishnan, B., Arulmani, S., Parameswari, E., Santhosh, A., Srinivasulu, A., Lal, A., Naidu, R.: A critical review of climate change impacts on groundwater resources: a focus on the current status, future possibilities, and role of simulation models. Atmos. **15**(1), 122 (2024). 10.3390/atmos1501

Climate Change and River Basins: A Global Perspective on Water Resources

Soufiane Haddout[1(✉)], Joan Cecilia C. Casila[2], K. L. Priya[3], and Mariusz Ptak[4]

[1] Department of Physics, Ibn Tofail University, Kenitra, Morocco
soufian.haddout@gmail.com

[2] Land and Water Resources Engineering Division, IABE, CEAT, University of the Philippines Los Baños, Los Baños, Laguna, Philippines

[3] Department of Civil Engineering, TKM College of Engineering, Kollam, Kerala, India

[4] Department of Hydrology and Water Management, Adam Mickiewicz University, Poznań, Poland

Abstract. Climate change is profoundly reshaping the dynamics of river basins worldwide, significantly impacting water resources. This study provides a comprehensive global perspective on how rising temperatures, altered precipitation patterns, and increased frequency of extreme weather events are affecting river basins and their ecosystems. We explore the interplay between climate variability and water availability, focusing on critical challenges such as changing river flows, diminishing water quality, and heightened risks of floods and droughts. Particular attention is given to the socio-economic and environmental consequences for regions dependent on these vital water systems. The paper also examines adaptation strategies, including integrated water resource management, sustainable infrastructure development, and policy interventions to mitigate climate-induced disruptions. By synthesizing case studies, this work underscores the urgent need for collaborative and innovative approaches to safeguard river basins and ensure the sustainable management of water resources in a rapidly changing climate.

Keywords: Climate Change Adaptation · River Basin Hydrology · Water Resource Management · Extreme Weather Events

1 Introduction

Rivers serve as the lifeblood of human civilization, supporting diverse ecosystems, providing essential freshwater resources, and facilitating economic and social development. River basins, which encompass entire catchment areas drained by river networks, play a critical role in sustaining global water supplies for agriculture, industry, and domestic use. However, the impacts of climate change are increasingly disrupting the natural balance of these hydrological systems, altering river flow patterns, reducing water availability, and exacerbating extreme weather events such as floods and droughts [1].

The warming of the Earth's atmosphere has been linked to significant shifts in precipitation patterns, glacier melt, and evapotranspiration rates, all of which influence river

J. C. Casila et al. (Eds.): CCORE 2024, SPEES, pp. 42–48, 2025.
https://doi.org/10.1007/978-3-031-93887-0_6

basin hydrology [2]. In some regions, excessive rainfall is leading to catastrophic flooding, while other areas are experiencing prolonged droughts that threaten water security and food production [3]. The combined effects of climate change, land use modifications, and increasing water demand pose unprecedented challenges to the sustainable management of river basins worldwide.

The consequences of these hydrological changes extend beyond environmental concerns. Water scarcity and extreme climate events have profound socio-economic implications, particularly in water-stressed regions. Declining river flows and deteriorating water quality affect drinking water supplies, agriculture, hydroelectric power generation, and biodiversity conservation [4]. Moreover, transboundary river basins, shared by multiple countries, present geopolitical challenges as nations compete for limited water resources, increasing the potential for conflicts [5].

Given these pressing challenges, effective adaptation and mitigation strategies are essential to enhance the resilience of river basins against climate-induced disruptions. This paper provides a global perspective on the impacts of climate change on river basins, examining the key hydrological, ecological, and socio-economic consequences. Furthermore, it explores strategies for sustainable water management, including integrated water resource management (IWRM), ecosystem-based adaptation (EbA), and policy interventions aimed at mitigating the adverse effects of climate variability. By synthesizing global case studies and current research, this study underscores the need for a collaborative and science-based approach to safeguard the world's river basins in the face of a changing climate.

2 Impacts of Climate Change on River Basins

- *Altered Hydrological Cycles*

Rising global temperatures have led to significant changes in precipitation patterns, resulting in altered hydrological cycles within river basins. Some regions experience increased rainfall and flooding, while others face prolonged droughts due to decreased precipitation. These shifts disrupt the natural flow regimes of rivers, affecting water availability and quality.

- *Glacial Melt and Reduced Snowpack*

In mountainous regions, higher temperatures contribute to accelerated glacial melt and reduced snowpack. This phenomenon leads to increased river flows in the short term but poses long-term risks of diminished water supplies as glaciers recede. Communities dependent on glacial-fed rivers are particularly at risk of future water shortages.

- *Increased Frequency of Extreme Weather Events*

Climate change has been linked to a rise in the frequency and intensity of extreme weather events, including heavy rainfall and prolonged droughts. These events exacerbate the variability of river flows, leading to challenges in water resource management and increased risks of flooding and water scarcity.

3 Socio-Economic and Environmental Consequences

- *Water Scarcity and Agriculture*

Altered river flows and reduced water availability directly impact agricultural productivity. Irrigation-dependent farming systems face challenges in maintaining crop yields, leading to economic losses and food security concerns. Regions such as the Mediterranean are projected to experience significant reductions in natural water resources, exacerbating water scarcity issues.

- *Ecosystem Degradation*

Changes in river flow regimes and water quality adversely affect aquatic ecosystems. Species that rely on specific flow conditions may decline, leading to reduced biodiversity. Additionally, increased water temperatures and altered sediment loads can harm fish populations and other aquatic life.

- *Human Health and Infrastructure*

Flooding and water scarcity pose significant risks to human health and infrastructure. Flood events can lead to water contamination and the spread of waterborne diseases, while water shortages may compromise sanitation and access to clean drinking water. Infrastructure such as dams and levees may also be strained under changing hydrological conditions.

4 Adaptation Strategies

- *Integrated Water Resource Management (IWRM)*

Implementing IWRM approaches allows for the coordinated development and management of water, land, and related resources. This strategy promotes sustainable water use and enhances resilience to climate-induced changes in river basins.

- *Ecosystem-Based Adaptation (EbA)*

EbA measures, such as restoring wetlands and riparian buffers, leverage natural systems to mitigate the impacts of climate change. These approaches are cost-effective and provide co-benefits, including biodiversity conservation and improved water quality.

- *Policy and Governance Interventions*

Developing and enforcing policies that promote sustainable water use and protect water resources are essential. This includes setting water quality standards, regulating water withdrawals, and encouraging water-saving technologies. Transboundary cooperation among nations sharing river basins is also crucial for effective water management.

5 Case Studies

To understand the global implications of climate change on river basins, it is essential to examine specific cases where hydrological changes have led to significant environmental, social, and economic consequences. The following case studies highlight the diverse challenges faced by river basins across different climatic and geographical regions, as well as the adaptive measures taken to mitigate these impacts.

(a) *The Danube River Basin: Integrating Climate Change Adaptation into Water Management*

The **Danube River Basin**, the second-largest river basin in Europe, spans 19 countries and serves as a vital water source for over 80 million people. Climate change has increased the frequency and intensity of extreme weather events, including floods and droughts, disrupting water availability and quality [6].

- *Key Climate Change Impacts*

Increased rainfall variability leading to severe floods, particularly in 2002, 2006, and 2013, causing billions of euros in damage.

Rising temperatures contributing to more frequent droughts, especially in the basin's southern and eastern regions.

Changing precipitation patterns affecting hydropower generation and irrigation systems.

- *Adaptation Strategies*

The **Danube River Basin Climate Change Adaptation Strategy**, developed by the International Commission for the Protection of the Danube River (ICPDR), focuses on integrating climate change considerations into water management.

Floodplain restoration projects have been implemented to absorb excess water during floods and maintain biodiversity.

Cross-border cooperation agreements have been established among riparian states to improve data sharing and emergency preparedness for extreme hydrological events.

This case study underscores the importance of international collaboration in transboundary river basin management and the need for proactive adaptation planning.

(b) *The Colorado River Basin: Addressing Water Scarcity in a Changing Climate*

The **Colorado River Basin**, one of the most heavily managed river systems in the United States, supplies water to over 40 million people and supports agriculture, hydropower, and ecosystems across seven U.S. states and Mexico. Climate change has intensified the region's longstanding water scarcity issues, threatening its long-term sustainability [7].

- *Key Climate Change Impacts*

Rising temperatures have contributed to a **20% decline in the river's flow** over the past century, largely due to increased evapotranspiration.

Reduced snowpack in the Rocky Mountains, the river's primary water source, has led to lower runoff and decreased reservoir levels.

Prolonged droughts, such as the **2000–2023 megadrought**, have pushed the basin's two largest reservoirs—Lake Mead and Lake Powell—to record-low levels.

- *Adaptation Strategies*

The **Colorado River Drought Contingency Plan (DCP)**, signed in 2019, allocates water cuts to reduce overuse and protect critical reservoir levels.

Large-scale investment in **water conservation programs**, such as urban water reuse and agricultural efficiency measures.

Exploration of **desalination and groundwater banking** as alternative water sources to supplement declining river flows.

The Colorado River Basin case highlights the urgent need for demand management and innovative water-saving technologies to adapt to long-term climate-induced water shortages.

(c) *The Ganges River Basin: Balancing Climate Change, Population Growth, and Water Demand*

The **Ganges River Basin**, stretching across India, Nepal, Bangladesh, and China, supports a population of over 650 million people. However, climate change, combined with rapid population growth and industrialization, is exacerbating water stress and environmental degradation in the region [8].

- *Key Climate Change Impacts*

Increased temperatures are accelerating **glacial melt in the Himalayas**, leading to temporary increases in river flow followed by long-term reductions in water availability.

Unpredictable monsoon patterns are causing **more frequent flooding** in Bangladesh and northern India, affecting millions of people annually.

Declining water quality due to **rising pollution levels**, exacerbated by lower flows and higher temperatures.

- *Adaptation Strategies*

Himalayan hydrological monitoring programs have been established to track glacier melt and predict water availability changes.

The **Ganga River Basin Management Plan**, launched by the Indian government, focuses on pollution control, ecosystem conservation, and improved water governance.

Investments in **sustainable agriculture and irrigation efficiency** to reduce excessive water withdrawals from the river.

This case study highlights the complex interplay between climate change, human activity, and policy interventions in ensuring water security for millions of people.

(d) *The Nile River Basin: Transboundary Water Conflicts and Climate Resilience*

The **Nile River Basin**, shared by 11 countries in Africa, is a lifeline for over 300 million people. Climate change and rising water demands have intensified geopolitical tensions, particularly between Egypt, Sudan, and Ethiopia [9].

- *Key Climate Change Impacts*

Increasing temperatures and declining rainfall are **reducing water availability** in key tributaries such as the Blue Nile.

Extreme drought events are threatening food security and hydropower production in Ethiopia and Sudan.

The construction of the **Grand Ethiopian Renaissance Dam (GERD)** has raised concerns in downstream countries about reduced water flows to Egypt's Aswan High Dam.

- *Adaptation Strategies*

Transboundary water agreements, such as the Nile Basin Initiative (NBI), aim to promote cooperative water sharing among riparian nations.

Investments in **drought-resistant crops and water-efficient irrigation systems** to improve agricultural resilience.

Exploration of **alternative water sources**, including desalination and groundwater management, to reduce reliance on the Nile.

This case study illustrates the intersection of climate change and geopolitics, emphasizing the need for diplomatic solutions and regional cooperation in water governance.

(e) *The Murray-Darling Basin: Climate Variability and Water Rights Management*

The **Murray-Darling Basin**, covering southeastern Australia, is one of the most important agricultural regions in the country. Climate variability, prolonged droughts, and water over-extraction have placed immense pressure on this river system [10].

- *Key Climate Change Impacts*

Increased drought frequency, particularly during the 1997–2009 Millennium Drought, resulted in severe water shortages.

Higher temperatures have reduced **runoff efficiency**, decreasing the volume of water entering the river system.

Declining river health and **loss of wetlands**, affecting biodiversity and ecosystem services.

- *Adaptation Strategies*

The **Murray-Darling Basin Plan**, implemented in 2012, establishes sustainable water diversion limits to protect the environment.

Water trading markets allow farmers and industries to buy and sell water rights, improving allocation efficiency.

Large-scale **river restoration projects** aim to revive wetland ecosystems and maintain environmental flow requirements.

This case study demonstrates the role of policy-driven water management strategies in adapting to climate-induced water challenges.

6 Conclusion

Climate change presents significant challenges to river basins and their associated water resources. Addressing these challenges requires a multifaceted approach that combines scientific understanding, adaptive management practices, and robust policy frameworks. By implementing integrated and ecosystem-based strategies, societies can enhance the resilience of river basins and ensure the sustainable availability of water resources in the face of a changing climate.

Declaration of Generative AI in Scientific Writing. During the preparation of this work, the corresponding Author used ChatGPT to collect some information. After using this tool, the corresponding author reviewed and edited the content as needed and takes full responsibility for the publication's content.

References

1. Kundzewicz, Z.W., Krysanova, V., Dankers, R., Hirabayashi, Y., Kanae, S., Hattermann, F.F., et al.: Differences in flood hazard projections in Europe–their causes and consequences for decision making. Hydrol. Sci. J. **63**(1), 1–14 (2018)
2. IPCC: Climate Change 2021: The Physical Science Basis. Contribution of Working Group I to the Sixth Assessment Report of the Intergovernmental Panel on Climate Change. Cambridge University Press (2021)
3. Pahl-Wostl, C., Palmer, M., Richards, K.: Enhancing water security for the benefits of humans and nature–the role of governance. Curr. Opin. Environ. Sustain. **5**(6), 676–684 (2013)
4. Grafton, R.Q., Williams, J., Perry, C.J., Molle, F., Ringler, C., Steduto, P., et al.: The paradox of irrigation efficiency. Science. **361**(6404), 748–750 (2018)
5. Zeitoun, M., Mirumachi, N.: Transboundary water interaction I: reconsidering conflict and cooperation. Int. Environ. Agreem.: Politics Law Econ. **8**(4), 297–316 (2008)
6. ICPDR: Danube River Basin Climate Adaptation Strategy (2018)
7. Udall, B., Overpeck, J.: The 21st century Colorado River hot drought and implications for the future. Water Resour. Res. **53**(3), 2404–2418 (2017)
8. Immerzeel, W.W., van Beek, L.P., Bierkens, M.F.: Climate change will affect the Asian water towers. Science. **328**(5984), 1382–1385 (2010)
9. Conway, D., Hulme, M.: The impacts of climate variability and future climate change in the Nile Basin. Clim. Chang. **34**(2), 145–156 (1996)
10. CSIRO: Murray-Darling Basin Plan: Water Management in a Changing Climate (2012)

Precipitation and Temperature Trends on the Southern Croatian Coast and Islands (NE Mediterranean)

Igor Ljubenkov(✉)

Water Development Ltd., Split, Croatia
iljubenkov@gmail.com

Abstract. This study analyzes the series of annual air temperatures and annual precipitations at three meteorological stations located on the southern Croatian coast and islands, which belongs to the northeastern Mediterranean. Basic statistical values of annual precipitation and temperature were compared for two consecutive 30-year periods: 1961–1990 and 1991–2020. The results show a decrease in precipitation and an increase in temperatures, which is characteristic of climate change in the Mediterranean. Linear trends in precipitation and temperature were calculated, based on available measurements from the period 1949–2024. Calculated trends show that the decrease in precipitation is more pronounced on islands (up to −23.4 mm/10 years) than on the coast (−2.2 mm/10 years), while warming has different gradients across stations (from +0.06 °C/10 years to + 0.24 °C/10 years). The identified climate trends, which will continue or even intensify, will have a strong impact on numerous natural processes and economic activities. A particular impact is expected on water resources, which are already very limited in the Mediterranean coast and on the islands.

Keywords: Precipitation · Air Temperature · Trend

1 Introduction

The Croatian coast is the eastern coast of the Adriatic Sea and together with its islands, belongs to the Dalmatia region and northeastern Mediterranean. Analysis of climate parameters (precipitation, temperature) was conducted for three meteorological stations: Split, Vela Luka and Korčula. The Split (Marjan) station is located in the city of the same name, which is also the largest city on the eastern coast of the Adriatic Sea. The other two stations are located on islands, Vela Luka on the island of Korčula, and Komiža on the island of Vis. The islands of Korčula (area 276 km^2) and Vis (89 km^2) are considered as the small and very small islands of the Mediterranean (Diaz Arenas and Febrillet Huertas 1986; Ljubenkov 2013; Bonacci, 2019).

According to Köppen's classification, the Dalmatian coast and islands have a Mediterranean climate (Csa), characterized by dry and hot summers and mild winters, mostly with rainfall. Climate change is a current topic in professional and scientific literature, and the Mediterranean is recognized as a particularly sensitive area (MedCC 2019;

J. C. Casila et al. (Eds.): CCORE 2024, SPEES, pp. 49–55, 2025.
https://doi.org/10.1007/978-3-031-93887-0_7

Lazoglou et al. 2024; Sakalis 2024). In general, annual temperature in the Mediterranean region has an increasing trend in the last 100 years. Precipitation in most of this region has shown a marked decreasing trend over the last 50 years (Longobardi and Villani 2010; Bonacci et al. 2019; Varlas et al. 2022). Climate trends are not uniform across the Mediterranean and depend on a number of factors such as the choice of measuring stations, their location, the length of available measurements, measurement interruptions, etc. Bonacci et al. (2021) showed that even on the same island there can be significant differences in trends of climate indicators, due to the influence of relief, proximity to the sea, influence of vegetation, etc. Therefore, each regional and local climate analysis is useful, as it provides the best assessment of climate change in that area. In this context, the present study contributes to the understanding and quantification of changes in mean annual temperature and total annual precipitation on the eastern Adriatic coast and nearby islands.

2 Climate Data

This study analyzed the series of air temperatures and precipitation measured at the three main meteorological stations of the DHMZ (Croatian Meteorological and Hydrological Service): Split, Vela Luka and Komiža (Fig. 1), which are monitored daily at 7, 14 and 21 o' clock in local time. The basic characteristics of these stations are given in Table 1. The Split station began operating in the mid-nineteenth century (1865), while the remaining two stations were founded in the mid-twentieth century (Vela Luka 1949 and Komiža 1956). This analysis includes all available observations from 1949 to 2024. In certain years, there is an interruption in the operation of the measuring stations, so they do not have an annual value. Therefore, the number of data points in the analyzed series varies by stations and by meteorological indicators (P or T) (Table 1).

The station Komiža had the lowest mean annual precipitation (776 mm), because it is further from the mainland (approximately 50 km) than Vela Luka (approximately 40 km). The highest mean annual precipitation was recorded by Vela Luka (836 mm). The highest average mean annual temperature was recorded by Komiža (16.7 °C), and the lowest by Vela Luka (15.7 °C) (Table 2).

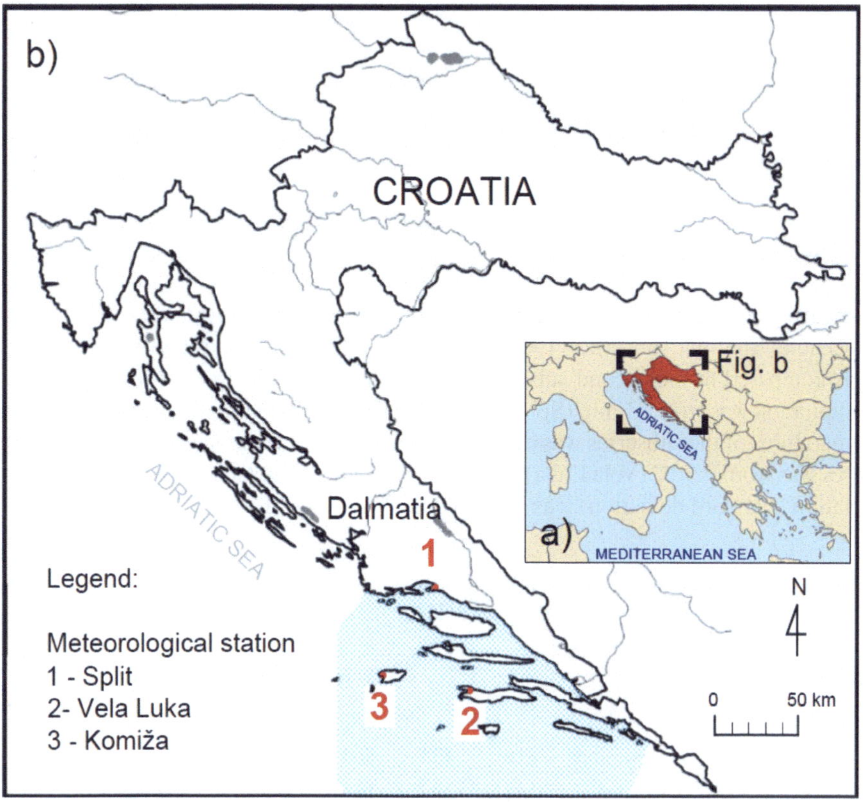

Fig. 1. Map of the study area with marked station location.

Table 1. The main characteristics of analyzed meteorological stations.

Station name	Split	Vela Luka	Komiža
Geographical coordinates	43° 30′ 30″N, 16° 25′ 35″E	42° 57′ 33″N, 16° 43′ 14″E	43° 2′ 55″N, 16° 5′ 14″E
Elevation (m a.s.l.)	122	14	20
Distance from the sea (m)	560	365	95
Analyzed period	1949–2024	1949–2023	1956–2024
No. of data for P	70	71	65
No. of data for T	65	63	61

Table 2. Mean annual precipitation (mm) and temperature (°C) (1949–2024).

Station name	Split	Vela Luka	Komiža
Precipitation	798	836	776
Temperature	16.4	15.7	16.7

3 Results

Comparing the basic statistical values for the two 30-year subseries, differences are noticeable. The first subseries (1961–1990) had higher average precipitation at all stations compared to the second subseries (1991–2020). The decrease in average annual precipitation was from 31 mm (Split) to 82 mm (Vela Luka). Also, an increase in average temperatures of the subseries was evident at all stations. The increase in average temperatures was from 0.5 °C (Vela Luka) to 0.9 °C (Komiža). Also, an increase in temperature variation (standard deviation) was observed at all stations (Table 3).

Table 3. Mean annual precipitation (P, mm) and temperature (T, °C) with standard deviations for two subperiods: a) 1961–1990 and b) 1991–2020.

Station name	Split		Vela Luka		Komiža	
1961–1990						
Mean precipitation/St. deviation	816	136	864	201	809	210
Max. P/Min. P	1041	487	1223	496	1214	426
Mean temperature/St. deviation	16.0	0.37	15.4	0.37	16.3	0.40
Max. T/Min. T	16.8	15.1	16.2	14.5	17.1	15.5
1991–2020						
Mean precipitation/St. deviation	785	188	782	182	768	192
Max. P/Min. P	1209	522	1163	464	1269	479
Mean temperature/St. deviation	16.8	0.59	15.9	0.50	17.2	0.49
Max. T/Min. T	17.5	15.5	16.8	14.7	18.0	16.2

Figure 2 shows the series of annual precipitations observed at the Split, Vela Luka and Komiža meteorological stations over a 76-year period (1949–2024). The trend lines are plotted with the associated linear functions and corresponding coefficients of determination (R^2). A decrease in precipitation is evident at all three stations, with different gradients of decrease. The largest trend is recorded at Vela Luka station, and the smallest at Split (Table 4).

Figure 3 shows the series of mean annual air temperatures observed at the Split, Vela Luka and Komiža, with associated trend lines. All stations have a trend of increasing temperatures, with the gradient being the smallest in Vela Luka and the largest in Komiža (Table 4).

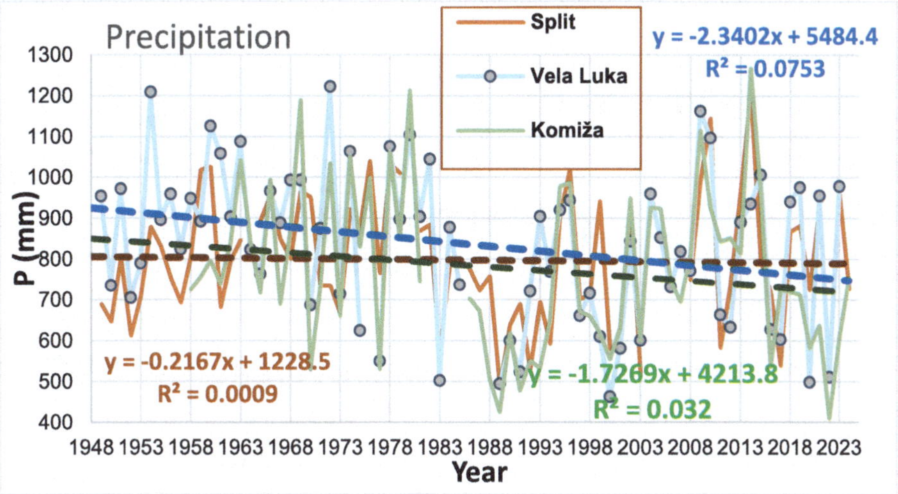

Fig. 2. Annual precipitation series measured at meteorological stations Split, Vela Luka and Komiža in the period 1949–2024.

Table 4. Precipitation and temperature trend (1949–2024).

Station name	Split	Vela Luka	Komiža
Precipitation	−2.2 mm/10 years	−23.4 mm/10 years	−17.3 mm/10 years
Temperature	+ 0.18 °C/10 years	+ 0.06 °C/10 years	+ 0.24 °C/10 years

Bonacci (2019) analyzed a series of annual, monthly, and daily air temperatures, and annual and monthly precipitation at the Lastovo meteorological station on the island of Lastovo (Croatia) during the 1948–2018 period. This island is located approximately 25 km southeast of Vela Luka. The linear trend of the mean annual temperature for Lastovo was +0.19 °C/10 years, approximately corresponding to the calculated trends in Split and Vela Luka. Total annual precipitation on the island of Lastovo had a decreasing trend (−5.1 mm/10 years). It is significantly less than trends in Vela Luka and Komiža.

Precipitation and temperature trends in Vela Luka for a shorter period (1952–2018, N = 55) have gradients of −25.0 mm/10 years and + 0.08 °C/10 years (Bonacci et al. 2021). The trends roughly match the results of this study (1949–2024, N = 63), which is logical since the input series are similar.

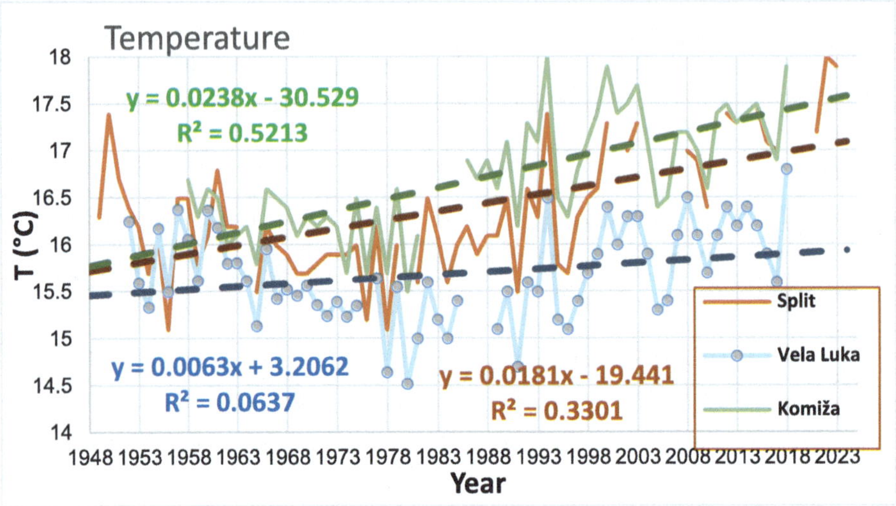

Fig. 3. Series of mean annual temperatures measured at meteorological stations Split, Vela Luka and Komiža in the period 1949–2024.

4 Conclusion

The mean annual temperatures and total annual precipitation of three meteorological stations from the Croatian coast and islands show trends of increase or decrease, respectively, which has generally been established in the Mediterranean. However, although these are relatively close stations at a distance of up to 65 km, there are evident differences in climatic parameters. Therefore, it can be emphasized that local conditions such as the location of the station, relief, proximity to the sea, etc. have an important influence on climatic conditions. The conducted analysis can only serve for a qualitative assessment of climatic processes and could be a first step for future research. Since the climate is a very complex system, a detailed analysis should consider other climatic parameters (e.g. wind, humidity, etc.) on various time scales (day, month, year). However, the aforementioned climate changes are already being felt in many activities, including in the water management, particularly on the Croatian coast and islands.

References

Bonacci, O.: Air temperature and precipitation analyses on a small Mediterranean island: the case of the remote island of Lastovo (Adriatic Sea, Croatia). Acta Hydrotech. **32**(57), 135–150 (2019)

Bonacci, O., Ljubenkov, I., Roje-Bonacci, T.: Different climate changes at two locations on a small karst island Korčula (Adriatic Sea, Croatia). Naše more. **68**(1), 1–13 (2021)

Diaz Arenas, A.A., Febrillet Huertas, J.: Hydrology and Water Balance of Small Islands. A Review of Existing Knowledge. Technical Documents in Hydrology. UNESCO, Paris (1986)

Lazoglou, G., Papadopoulos-Zacho, S.A., Georgiades, P., Zittis, G., Velikou, K., Manios, E.M., Anagnostopoulou, C.: Identification of climate change hotspots in the Mediterranean. Sci. Rep. **14**(29817), 1–11 (2024)

Longobardi, A., Villani, P.: Trend analysis of annual and seasonal rainfall time series in the Mediterranean area. Int. J. Climatol. **30**, 1538–1546 (2010)

Ljubenkov, I.: Hydro climatic ranking of small Mediterranean islands. J. Environ. Sci. Water Res. **2**(4), 112–121 (2013)

MedCC: Risks Associated to Climate and Environmental Changes in the Mediterranean Region. UN Environment Programme (2019)

Sakalis, V.D.: Trend analysis and forecast of annual precipitation and temperature series in the Eastern Mediterranean region. Atmosfera. **38**, 1–27 (2024)

Varlas, G., Stefanidis, K., Papaioannou, G., Panagopoulos, Y., Pytharoulis, I., Katsafados, P., Papadopoulo, S.A., Dimitriou, E.: Unravelling precipitation trends in Greece since 1950s using ERA5 climate reanalysis data. Climate. **10**(2), 1–19 (2022)

Green Energy Horizons

Enhancing Offshore Wind Resource Projection in India: A Multi-Model, Multi-Method Ensemble Framework for Climate Change Impact Assessment

Garlapati Nagababu(✉)

School of Technology, Pandit Deendayal Energy University, Gandhinagar, Gujarat, India
garlapatinagu@gmail.com

Abstract. The rising global energy demand and the need to mitigate climate change have driven the search for sustainable energy solutions, with offshore wind energy emerging as a key component. With its vast coastline, India recognizes offshore wind energy as essential to its renewable energy strategy. However, the availability of wind resources is highly sensitive to climate change, making it critical to assess the potential impacts of future climatic shifts on offshore wind energy. This study employs the Multi-Model and Multi-Method Ensemble (4ME) framework to refine climate model projections using statistical downscaling techniques, providing high-resolution wind resource assessments for India's offshore regions. A total of 13 Global Climate Models (GCMs) from the CMIP6 dataset were selected, and five downscaling methods were applied to generate refined wind speed projections from 1990–2014. The performance of individual model-method combinations was evaluated using statistical metrics, and the most accurate combinations were integrated into the 4ME ensemble. The results demonstrate the superiority of the 4ME ensemble over individual GCM projections, offering more reliable and consistent wind resource estimates. Spatial bias analysis reveals the effectiveness of the 4ME framework in reducing model discrepancies and improving alignment with ERA5 reanalysis data, particularly in offshore regions. This study provides valuable insights into future wind energy potential and underscores the importance of ensemble-based frameworks for more accurate climate impact assessments in energy planning.

Keywords: CMIP6 · Climate Change · Statistical Downscaling Techniques · Global Climate Models · Multi-Model Multi-Method Ensemble

1 Introduction

Recently, the global surge in energy-driven globalization, industrialization, and population growth has contributed to rising global temperatures. In response, the 2015 Paris Agreement set an ambitious goal to limit the increase in global temperature to below 1.5 °C [1]. Achieving this goal requires significant reductions in fossil fuel consumption, making renewable energy like solar, wind and wave energy sources as a more sustainable alternative.

J. C. Casila et al. (Eds.): CCORE 2024, SPEES, pp. 59–67, 2025.
https://doi.org/10.1007/978-3-031-93887-0_8

India, recognizing the importance of renewable energy, has set an ambitious target to achieve 500 gigawatts (GW) of renewable capacity by 2030 and aims for net-zero emissions by 2070. Offshore wind energy, in particular, plays a critical role in this strategy due to India's extensive coastline of approximately 7600 km [2]. The estimated wind energy potential on the coasts of Tamil Nadu and Gujarat at around 70 GW. However, the availability of wind resources is intrinsically linked to climate conditions, making it crucial to study the impact of climate change on offshore wind potential [3]. Assessing these impacts is vital for ensuring that future developments can effectively adapt to and mitigate the challenges posed by climate change [4].

The sensitivity of wind energy production to climate variability highlights the limitations of traditional resource methods, which predominantly rely on historical observations. These methods often fail to account for future climate changes. In contrast, climate models provide a pathway for analyzing potential shifts in wind patterns and resource availability under varying climate scenarios, thus integrating long-term resilience into energy planning [5]. Recent studies [4, 6–8] have highlighted the regional variability of wind energy potential under changing climatic conditions. For instance, an assessment of wind energy potential in China using Coupled Model Intercomparison Project (CMIP6) models revealed significant spatial variations in wind intensity Northern regions experienced notable increases in wind speeds, whereas areas such as the Bohai Sea and the Yellow Sea exhibited declining trends. Similarly, the southern South China Sea showed a decrease in wind speed, underscoring the heterogeneous impacts of climate change on wind resource availability [9].

Global climate models (GCMs) are indispensable for large-scale climate assessments; however, their coarse spatial resolution and limited capacity to accurately capture regional and localized climate dynamics pose significant challenges [10]. These limitations are particularly pronounced in areas with complex atmospheric patterns, leading to discrepancies between model outputs and observed data [11]. To address these challenges, multi-model ensemble frameworks have gained prominence. By integrating outputs from multiple GCMs, these ensembles reduce biases and uncertainties, providing more robust and reliable projections of wind resource variability [9, 12].

The present study leverages the 4ME framework, which refines coarse-resolution GCM outputs into high-resolution data using statistical downscaling techniques. By capturing localized atmospheric patterns, the 4ME methodology addresses the inherent limitations of traditional models, providing a detailed assessment of wind energy resources in the Indian offshore region. The novelty of this work lies in the application of 4ME to generate ensemble data specifically for India's offshore region, which has not been studied using this advanced framework. Furthermore, the study systematically compares ensemble outputs and individual GCM datasets with ERA5 reanalysis data to evaluate their performance and ensure the reliability of the results.

2 Data and Method

2.1 Data

In this study, simulations from GCMs under the latest phase of the CMIP6 framework were used to assess future wind energy resources under different emission scenarios. A total of 13 GCMs (listed in Table 1) were selected for analysis, with daily wind data from 1990 to 2014. For comparison, the ERA5 dataset was used. ERA5 is a fifth-generation reanalysis product developed by ECMWF. It includes daily averaged wind speeds for the same period. These models represent diverse geographical regions and spatial resolution, providing a broad spectrum of climate projections for the study area.

Table 1: List of CMIP6 GCMs used in this study, including their respective regions and grid resolutions.

S. No	Model Name	Region	Grid
1	ACCESS-CM2	Australia	1.25° × 1.875°
2	AWI_ESM_1_REcoM	Germany	0.9375° × 0.9375
3	BCC-CSM2-MR	China	1.12° × 1.125°
4	CMCC-CM2-SR5	Italy	0.182° × 0.215°
5	CMCC-ESM2–0	Italy	1.25° × 0.9375°
6	EC-EARTH3-VEG	European countries	0.5625° × 1.125°
7	GFDL-ESM4	America	1° × 1.25°
8	IPSL-CM6A-LR	France	1.25° × 1.25°
9	MPI-ESMI	Germany	1.875° × 1.875°
10	MIROC 6	Japan	1.40625° × 2.8125°
11	CanESM5	USA	1.25° × 1.875°
12	BCC_CSM2-MR	China	1.125° × 2.25°
13	NESM1	China	1.875° × 1.875°

2.2 Method

In this study, the 4ME framework is applied to create an ensemble of selected GCMs from CMIP6. This framework enhances the reliability of wind energy resource projections by refining the raw outputs of GCMs using statistical downscaling techniques. The process begins with the application of five statistical downscaling methods, each designed to adjust large-scale wind speed projections from GCMs, making them more representative of observed data at regional scales. The downscaling methods include Additive Correction (AC), Additive Scaling (AS), Variance Correction (VC), and Quantile Mapping (QM) using the Weibull distribution (QMW) and Kappa distributions (QMK) [13].

Each downscaling method is applied to the output of each GCM, aligning the modeled wind speeds more closely with the observed or reference wind data. The resulting downscaled (or predicted) wind speeds are then compared against ERA5 reanalysis wind speed data from 1990–2014. Various statistical parameters (SP), such as bias, mean absolute error (MAE), root mean square error (RMSE), standard deviation (STD), and correlation coefficient (CC) are used to assess the performance of downscaling methods. The aim is to minimize bias, MAE, RMSE, and STD while maximizing CC. The inverse of their absolute values is calculated for all metrics except CC, where higher values correspond to better performance. This framework ensures that model-method combinations with fewer errors are prioritized. Given 13 GCMs and 5 downscaling methods, a total of 65 combinations are evaluated.

To ensure all metrics are equally weighted, each metric value is normalized by dividing it by the mean value of the same metric across all 65 combinations. This step standardizes the scale of the metrics. The normalized performance metrics are then summed, and their values adjusted to total one. An overall performance measure for each combination is calculated as the average of the standardized values across all five metrics. This integrated measure provides a relative performance score for each combination.

The above metrics are calculated for a specific period of data at a single grid cell. To identify the most relevant climate change patterns for a given region, the performance metrics are spatially averaged across all grid cells within that region. This spatial averaging process is then used to compute the final skill score (SS), as represented in Eq. 1 [14–16].

$$SS_{ia,ib} = \frac{1}{m_a m_b m_c} \sum_{j_c=1}^{m_c} \sum_{j_b=1}^{m_b} \sum_{j_a=1}^{m_a} \frac{\frac{n_a n_b}{|SP_{ia,ib,ja,jb,jc}|}}{\sum_{i_b=1}^{n_b} \sum_{i_a=1}^{n_a} \left(\frac{1}{|SP_{ia,ib,ja,jb,jc}|} \right)} \tag{1}$$

Here, $SS_{ia,\,ib}$ represents the skill score for the combination of the $i_a - th$ model and the $i_b - th$ method. $SP_{ia,\,ib,\,ja,\,jb,\,jc}$ denotes the $j_a - th$ statistical parameter for the same model-method combination, evaluated in the $j_b - th$ cross-validation iteration. A higher SS value indicates better output of the model-method combination. These scores are then used to derive the weights for the 4ME using Eq. 2.

$$W_{ia,ib} = \frac{SS_{ia,ib}}{\sum_{i_b=1}^{n_b} \sum_{i_a=1}^{n_a} SS_{ia,ib}} \tag{2}$$

3 Results and Discussions

3.1 Skill Scores of Model-Method Combinations

As discussed in Sect. 2.2, the SS for various combinations of model-method were estimated using Eq.1, and their corresponding values are depicted in Fig. 1. In this figure, the x-axis denotes the statistical downscaling techniques, while the y-axis corresponds to the GCMs. Figure 1 reveals significant disparities in the effectiveness of various model-method combinations in simulating wind resources. This variation highlights the heterogeneity in performance across combinations, emphasizing the importance of the interaction between statistical methods and models in accurately replicating observed wind dynamics.

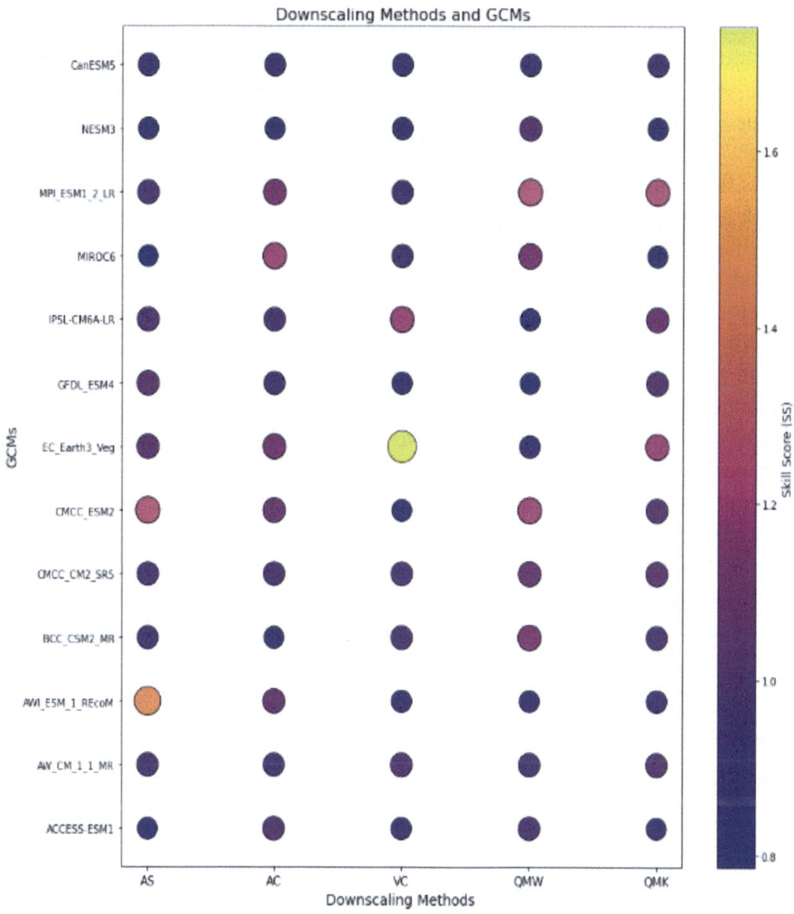

Fig. 1. Skill score for various model-method combinations

Among the evaluated combinations, the application of the VC method to the EC-Earth3_Veg model achieves the highest skill score of 1.74, indicating superior performance in capturing localized wind dynamics. In contrast, most combinations yield skill scores ranging between 0.8 and 1.2, indicating varying degrees of agreement with observed wind resource patterns.

The observed variability arises from two primary factors. First, inherent uncertainties in GCMs result from their imperfect representation of large-scale climatic and atmospheric processes, which directly impact their accuracy in simulating wind characteristics. Second, the statistical downscaling methods vary in their ability to effectively translate large-scale climatic signals into localized wind projections. For example, while the VC method performs well with the EC-Earth3_Veg model, its performance is less consistent with other GCMs, highlighting the sensitivity of results to the choice of both model and method.

These results highlight the limitations of depending on a single model or method for wind resource assessment. This finding emphasizes the need for ensemble frameworks, such as the 4ME framework, to enhance the reliability of projections. Variability in model performance is partly attributed to uncertainties caused by imperfect representations of climatic and atmospheric dynamics, as well as limitations of downscaling methods. To address these uncertainties, the 4ME framework is implemented in this study. This framework dynamically assigns the weights for different model-method combinations based on their skill scores, calculated using Eq. 2. The estimated weights for these models and methods range from 0.012 to 0.027, showing slight variations among the model-method combinations. The maximum weight of 0.027 is obtained from the VC method with the EC_Earth3_Veg model, while the minimum weight of 0.012 is associated with the QMW method with the GFDL_ESM4 model. In the study area, most model-method combinations exhibit slight variations in weights, as models with similar predictive accuracy contribute equally to the final ensemble. This is likely due to a large number of combinations, many of which perform similarly within the region. As a result, models with comparable performance often receive similar weights, reflecting the ensemble's balanced approach to aggregating their contributions. Additionally, by integrating a diverse set of models and methods, the framework reduces individual biases and enhances overall predictive accuracy. The effectiveness of the 4ME framework is further evaluated in the following section, demonstrating its capacity to provide more reliable wind resource projections.

3.2 Comparison of Performance of 4ME with ERA-5 Reanalysis Data

The performance of the 4ME ensemble is assessed against ERA 5 reanalysis data for the historical period of 1990–2014. Figure 2 presents a comparative analysis of the performance between the individual GCM downscaling method and the 4ME ensemble framework, revealing notable differences in their predictive capabilities across various performance metrics. The individual GCM method demonstrates considerable variability in its performance, exhibiting a wider range of bias values, including an outlier with an unusually low bias. Furthermore, the individual GCM method shows higher values for RMSE (~2.8–3.0) and MAE (~2.3–2.4), indicating relatively poor accuracy and precision in capturing wind resource dynamics. The presence of outliers and the broader spread of results further underscore the instability and less reliable performance of this method. In contrast, the 4ME ensemble framework shows a more consistent and reliable performance across all metrics, with lower RMSE and MAE values, reflecting improved accuracy. Additionally, the correlation coefficient for the GCM method is relatively low (~0.35–0.45), signifying a weak alignment with observed wind resources. The standard deviation for the individual GCM method is also higher (~2.4–2.8), indicating greater variability in model predictions. These results highlight the superiority of the 4ME framework in reducing error and enhancing the reliability of wind resource projections, offering a more stable and robust alternative to individual GCM downscaling methods.

In contrast, 4ME consistently outperforms GCM across all metrics. It shows minimal bias with a narrow range close to zero, demonstrating high accuracy. The RMSE and MAE values for 4ME are significantly lower (~2.8 for RMSE and ~2.2 for MAE), indicating better precision and reliability. Furthermore, 4ME achieves a higher correlation

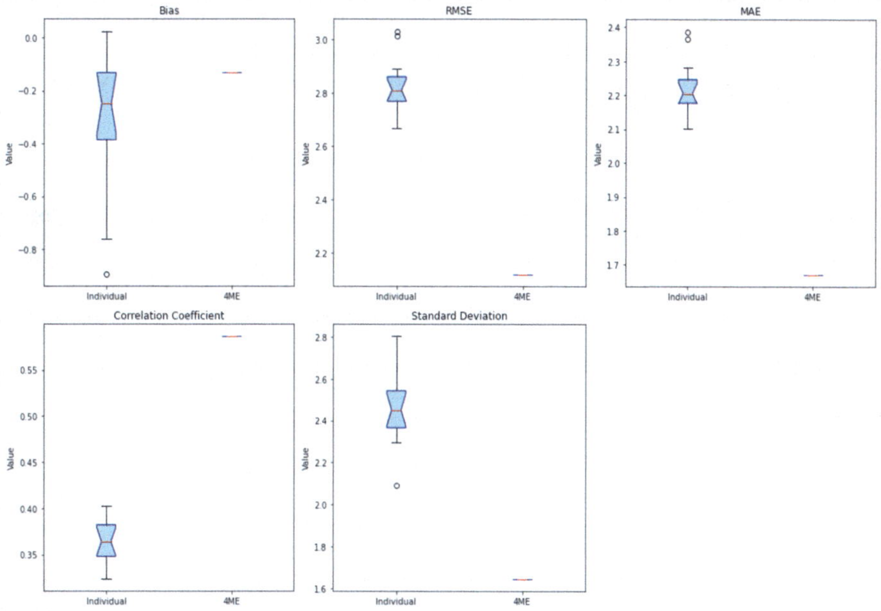

Fig. 2. Comparison of performance metrics between individual GCM downscaling methods and the 4ME ensemble using ERA5 data (1990–2014).

coefficient (~0.55), signifying stronger agreement with observations, and its standard deviation is smaller (~2.2), reflecting consistent performance. Unlike GCM, 4ME does not exhibit any outliers, highlighting its stability. These results confirm that 4ME is a more accurate, reliable, and consistent method for downscaling compared to GCM.

3.3 Spatial Bias Analysis

Figure 3 illustrates the spatial bias of wind speed between 13 individual climate models, and the 4ME ensemble, compared to ERA5 reanalysis data as the reference. Bias values range from approximately −2.5 m/s (underestimation) to +2.5 m/s (overestimation) for individual models, while the 4ME ensemble demonstrates a narrower range of −1.5 m/s to +1.5 m/s, indicating reduced variability and improved alignment with the reference dataset. Most models underestimate wind speeds in the eastern region of the study area, except for models such as GFDL-ESM4, ACCESS-ESM1, CMCC-CM2-SR5, and CMCC-ESM2, which show overestimation near the coastline. Models like BCC-CSM2-MR, CMCC-CM2-SR5, and CMCC-ESM2 exhibit smaller biases in coastal areas. Across the southern tip, all models significantly underestimate wind speeds by more than 2 m/s. Notably, AWI-ESM-1-REcoM overestimates wind speeds in the western and southern regions of the study area. Offshore regions show pronounced extremes in bias for individual models, but the 4ME ensemble effectively smooths these biases, producing more spatially coherent and moderate patterns. By compensating for individual model biases and reducing variability, the ensemble demonstrates superior performance, particularly in offshore regions critical for wind energy development. These findings emphasize the value of multi-model ensembles in minimizing uncertainties, offering robust, reliable wind speed estimates for renewable energy planning under future climate scenarios.

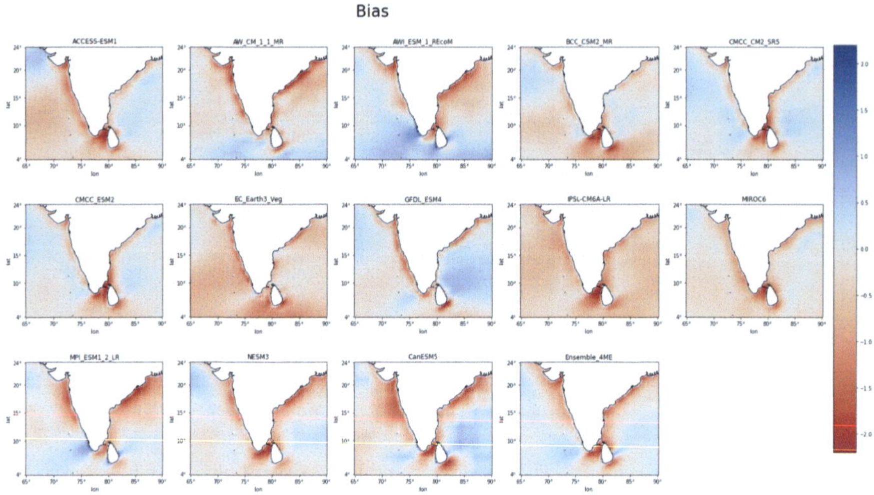

Fig. 3. Bias of wind speed (m/s) between individual climate models, the 4ME ensemble, and ERA5 reanalysis data

4 Conclusion

This study demonstrates the effectiveness of the 4ME framework in refining wind resource projections for India's offshore regions, accounting for the impacts of climate change on wind energy potential. The integration of multiple GCMs with statistical downscaling methods results in a more robust and accurate representation of future wind patterns compared to traditional methods that rely on individual models. The 4ME ensemble not only mitigates the uncertainties associated with individual model biases but also provides a more consistent and reliable forecast for offshore wind energy development. The spatial bias analysis further highlights the advantages of using the ensemble method, with the 4ME framework producing more coherent and moderate bias patterns, particularly in critical offshore areas. These findings emphasize the importance of using ensemble frameworks in climate change impact assessments to guide renewable energy planning, ensuring long-term resilience and sustainability. The results provide a solid foundation for future offshore wind energy projects and offer valuable insights into the evolving wind resource potential in India under changing climate conditions.

References

1. IPCC: Global Warming of 1.5°C. Cambridge University Press (2022)

2. Government of India: India's Wind Potential Atlas at 120 m agl National Institute of Wind Energy Chennai Under Ministry of New and Renewable Energy (2019)
3. Siram, O., Sahoo, N., Saha, U.K.: Changing landscape of India's renewable energy and the contribution of wind energy. Clean Eng. Technol. **8**, 100506 (2022). https://doi.org/10.1016/j.clet.2022.100506
4. He, J.Y., Li, Q.S., Chan, P.W., Zhao, X.D.: Assessment of future wind resources under climate change using a multi-model and multi-method ensemble approach. Appl. Energy. **329**, 120290 (2023). https://doi.org/10.1016/j.apenergy.2022.120290
5. Ji, L., Li, J., Sun, L., Wang, S., Guo, J., Xie, Y., Wang, X.: China's onshore wind energy potential in the context of climate change. Renew. Sust. Energ. Rev. **203**, 114778 (2024). https://doi.org/10.1016/j.rser.2024.114778
6. Sillmann, J., Kharin, V.V., Zwiers, F.W., Zhang, X., Bronaugh, D.: Climate extremes indices in the CMIP5 multimodel ensemble: part 2. Future climate projections. J. Geophys. Res. Atmos. **118**, 2473–2493 (2013). https://doi.org/10.1002/jgrd.50188
7. Kim, Y.T., Yu, J.U., Kim, T.W., Kwon, H.H.: A novel approach to a multi-model ensemble for climate change models: perspectives on the representation of natural variability and historical and future climate. Weather Clim. Extrem. **44**, 100688 (2024). https://doi.org/10.1016/j.wace.2024.100688
8. Hemri, S., Bhend, J., Liniger, M.A., Manzanas, R., Siegert, S., Stephenson, D.B., Gutiérrez, J.M., Brookshaw, A., Doblas-Reyes, F.J.: How to create an operational multi-model of seasonal forecasts? Clim. Dyn. **55**, 1141–1157 (2020). https://doi.org/10.1007/s00382-020-05314-2
9. Xu, J., Li, J., Pan, S., Yao, Y., Chen, L., Wu, Z.: Assessment of wind and wave energy in China seas under climate change based on CMIP6 climate model. Energy. **310**, 133207 (2024). https://doi.org/10.1016/j.energy.2024.133207
10. Nagababu, G., Srinivas, B.A., Kachhwaha, S.S., Puppala, H., Kumar, S.V.V.A.: Can offshore wind energy help to attain carbon neutrality amid climate change? A GIS-MCDM based analysis to unravel the facts using CORDEX-SA. Renew. Energy. **219**, 119400 (2023). https://doi.org/10.1016/j.renene.2023.119400
11. Lupo, A., Kininmonth, W.: Global Climate Models and their Limitations (2013)
12. Zhang, S., Li, X.: Future projections of offshore wind energy resources in China using CMIP6 simulations and a deep learning-based downscaling method. Energy. **217**, 119321 (2021). https://doi.org/10.1016/j.energy.2020.119321
13. Zhang, T., Yan, P., Li, Z., Wang, Y., Li, Y.: Bias-correction method for wind-speed forecasting. Meteorol. Z. **28**, 293–304 (2019). https://doi.org/10.1127/metz/2019/0950
14. Chang, T.P.: Estimation of wind energy potential using different probability density functions. Appl. Energy. **88**, 1848–1856 (2011). https://doi.org/10.1016/j.apenergy.2010.11.010
15. Li, Y., Wu, X.P., Li, Q.S., Tee, K.F.: Assessment of onshore wind energy potential under different geographical climate conditions in China. Energy. **152**, 498–511 (2018). https://doi.org/10.1016/j.energy.2018.03.172
16. Han, Z., Shi, Y., Wu, J.I.A., Xu, Y., Zhou, B.: Combined dynamical and statistical downscaling for high-resolution projections of multiple climate variables in the Beijing–Tianjin–Hebei region of China. J. Appl. Meteorol. Climatol. **58**, 2387–2403 (2019). https://doi.org/10.1175/JAMC-D-19-0050.1

Ranking Candidate Offshore Wind Power Sites Using SMART Technique: A Case Study on Türkiye

Samet Öztürk(✉)

Bursa Technical University, Environmental Engineering Department, Bursa 16310, Turkey
samet.ozturk@btu.edu.tr

Abstract. Offshore wind energy becomes one of the promising energy sources around the globe due to its high energy yield and reducing investment cost. Türkiye is surrounded by coasts with great wind energy potential. Although there are more than 12 GW installations of onshore wind in the country, there is not a single plant of offshore wind as of 2024 in the country. Along this line, the Turkish Ministry of Energy and Natural Resources (MENR) set a 5 GW offshore wind installation goal for 2035. There are candidate areas for offshore wind installations, however the priorities among these areas are not discussed in detail in the literature. This study utilizes Simple Multi-Attribute Rating Technique (SMART) to rank candidate suitable regions for offshore power installations for Türkiye. Eight criteria, namely wind potential, water depth, shipping density, bird migration density, port availability, closeness to relevant industry, number of touristic regions and number of electrical substations nearby are selected to rank the candidate sites. The results show that Karabiga region comes first followed by Bozcaada and Bandırma regions. Hoşköy coast has the least scores among all four candidate areas.

Keywords: Offshore Wind · SMART · Türkiye

1 Introduction

Global climate change and increasing need for energy directs governments to invest more on renewable energy systems. Wind energy systems are among the most invested ones among renewable energies globally. Although wind turbines inland occupy more of the wind energy systems as of 2024, offshore wind turbines are gaining more deployment for several reasons. First, offshore wind yields greater and more stable energy comparing onshore counterparts. Second, offshore wind energy systems do not require land whereas onshore wind turbines occupy land which is a problem for countries with scarce regions. Third, offshore wind systems require less turbines than onshore counterparts, because the turbines offshore have greater capacities than onshore counterparts due to higher wind potential offshore. Because of above mentioned reasons, offshore wind deployed rapidly in last 10 years and reached 75 GW capacity as of 2023, globally [1] (see Fig. 1).

Many countries which have coastal shores started deploying offshore wind energy recently. China, United Kingdom, Germany, Netherlands, Denmark and United States

J. C. Casila et al. (Eds.): CCORE 2024, SPEES, pp. 68–77, 2025.
https://doi.org/10.1007/978-3-031-93887-0_9

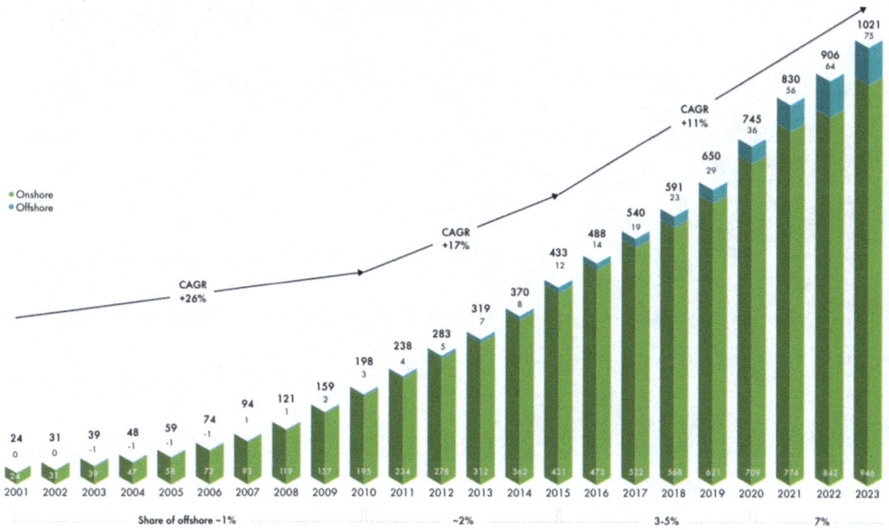

Fig. 1. Global historic development of total wind energy installations [1].

of America are some of the countries which collect energy from offshore wind. Türkiye, on the other hand, has no offshore wind farm on their coasts, yet. Nevertheless, the Turkish Ministry of Energy and Natural Resources (MENR) set a 5 GW offshore wind installation goal for 2035. Therefore, studies related to offshore wind site suitability and prioritization of the sites are becoming important. There are many studies investigated candidate site locations on Turkish territorial waters in the literature. Satir et al. (2018) investigated offshore wind potential site locations on Aegan Sea, while Caceoglu et al. (2022) searched for best locations on Northwest Türkiye [2, 3]. While Karipoglu and Ozturk (2021) investigated potential of a local region, Argin et al. (2019) researched all coasts of Türkiye to determine suitable offshore wind locations [4, 5]. On the other hand, although are proposed by MENR proposed candidate sites for the country no studies yet, to the author's knowledge, investigated these regions specifically for their prioritization.

This study utilizes one of the multi-criteria decision making method to prioritize candidate offshore wind sites which were determined by MENR. The importance of this study is being as the first research on the determined offshore wind locations and their prioritization.

2 Material and Method

In this study, candidate offshore wind sites which are determined by Turkish Ministry of Energy and Natural Resources (MENR) are analyzed and ranked by using several open public data and Simple Multi Attribute Ranking Technique (SMART). Candidate offshore sites which were determined by MENR, namely Bandırma, Karabiga, Hoşköy and Bozcaada regions are shown in Fig. 2.

Fig. 2. Four candidate sites which are determined by Turkish Ministry of Energy and Natural Resources

2.1 Criteria Selection and Data

The criteria selection is important for an accurate ranking of candidate offshore wind sites. Therefore, a thorough selection of criteria which have impact on offshore wind site establishment is considered. Eight criteria, i.e. wind potential, water depth, shipping density, bird migration density, port availability, closeness to relevant industry, number of touristic points, electrical infrastructure nearby are selected and analyzed in this paper after a careful literature review. Following subsections explain the criteria and corresponding data acquisition.

Wind Potential

Wind energy yield is one of the most prominent aspects for an offshore wind farm. Wind potential of a region directly impacts the energy yield, thus higher the wind potential greater the outcome of energy yield. In this study Global Wind Atlas is utilized to determine wind potential for candidate regions [6]. As it can be seen from Fig. 3 average wind speed at 100 m height is around 7.75 m/s for Bandırma, 6.75 m/s for Karabiga, 7.5 m/s for Hoşköy, and 9 m/s for Bozcaada regions.

Water Depth

The depth of water is important for offshore wind site development because wind turbines on deeper seas need more complicated and expensive foundation. That means more shallow coasts are advantageous for offshore wind farm deployment. Water depth for Marmara Sea are found by European Marine Observation and Data Network (EMODNet) data [7]. According to the data, average depth of seas are the least for Bozcaada region whereas deepest seas are at Bandırma coast among four alternatives. Water depth on average of Bandırma region is around 55 m, Karabiga region is 40 m, Hoşköy region is 45 m, and Bozcaada region is 20 m as can be seen from Fig. 4.

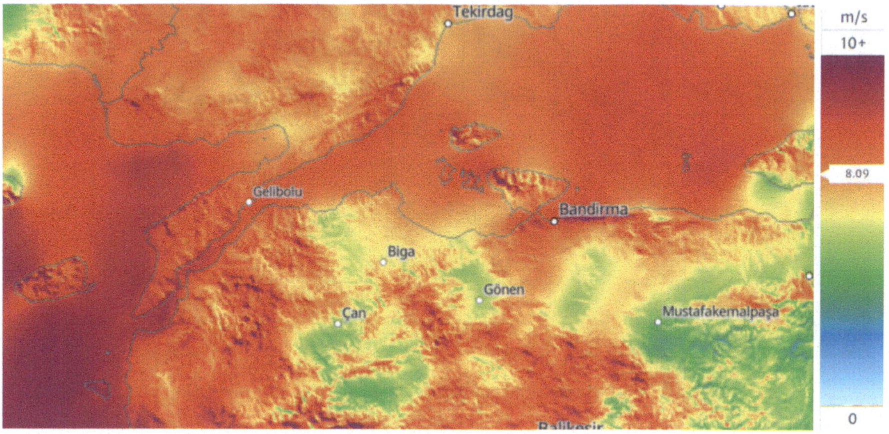

Fig. 3. Average wind speed map on candidate regions [6].

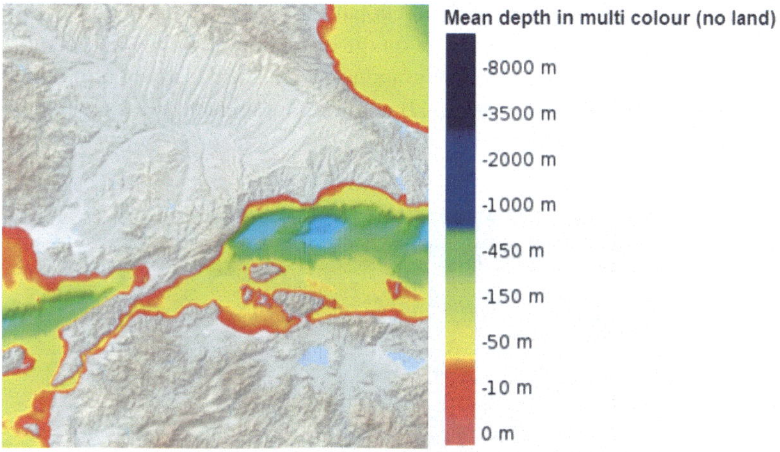

Fig. 4. Bathymetry map for candidate regions [7].

Shipping Density

Offshore wind power plants are generally a conflict for shipping operations. Therefore, shipping routes must be considered for any candidate regions before any investment realizes. Figure 5 shows the shipping density map on South Marmara and Northern Aegean seas which include candidate offshore regions [8]. These data show that on average Bandırma region has around 100 routes per sq. km per year, Karabiga has around 5 routes per sq. km per year, Hoşköy has around 75 routes per sq. km per year, and Bozcaada has around 200 routes per sq. km per year (Fig. 4).

Bird Migration Density

Renewable energy systems are well-known with its clean and environmentally friendly aspects. However, one of the critiques on wind energy systems is that bird killings

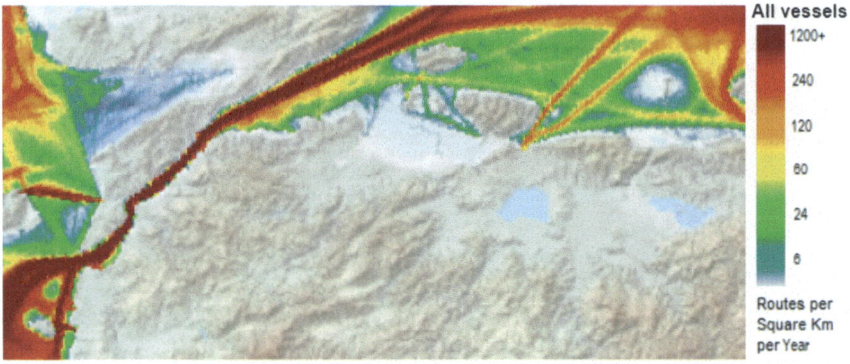

Fig. 5. Shipping route densities on candidate regions [8].

of wind turbines. In order to avoid this problem, bird migration routes are avoided to construct wind turbines. Therefore, bird migration route density is taken as a criterion for this study. The data are obtained from Global Biodiversity Information Facility [9]. Figure 6 shows the bird migration routes on the candidate offshore wind plant regions for spring and fall seasons when only migrations occur. The density of bird migration is classified as high, mid and low densities. Bandırma, Karabiga and Hoşköy regions have high densities while Bozcaada has low density.

Fig. 6. Bird immigration routes on candidate regions at spring seasons (top), and fall seasons (bottom) [9].

Port Availability

Offshore wind power plants require installation and operations and maintenance activities on the seas. Therefore, port operations are required for offshore wind energy systems.

Port availability for candidate offshore wind power sites is an important factor to establish a power plant. In this study, port availability is determined by numerical values for four candidate regions [10]. Bandırma regions has two container ports namely Bandırma and Gemlik ports, Karabiga region has Karabiga container port, Hoşköy has Tekirdağ container port and Bozcaada has no nearby ports but Karabiga container port is on the 200 km sphere of the region that is given only 0.5 as numerical value (see Fig. 7).

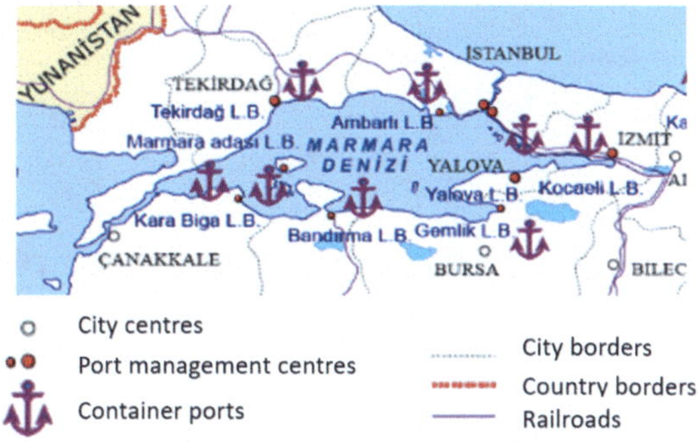

Fig. 7. Port availabilities of candidate regions [10].

Closeness to Relevant Industry

Industrial readiness is an important factor to develop an offshore wind farm. Wind energy equipment producers, R&D and O&M companies and Industrial zones help sustain the offshore wind farms nearby and can be clamied as relevant industry. This study takes closeness to relevant industry in three stages such as far, mid and close. Figure 8 shows the relevant industry nearby the candidate regions. Bandırma region is determined as close while Karabiga region is found as mid. Hoşköy and Bozcaada regions are determined as far from the relevant industry [11].

Number of Touristic Points

Touristic points such as beaches and historical landmarks can be an obstacle to deploy offshore wind farms. In this study, candidate regions are investigated based on the number of touristic points nearby to the candidate regions. For Bandırma region, Armutlu, Eğerce, Malkara and Kurşunlu beaches and Karacabey floodplain forests, for Karabiga region, Kadınlar and Saraylar beaches, for Hoşköy region, Mürefte beach, for Bozcaada region, Ayazma, Sulubahce, Habbele, Tuzburnu and Çayır beaches are found as touristic points.

Electrical Infrastructure Nearby

Offshore wind power plants require power lines and substations for electricity transmission as other power plants do. Therefore, electrical infrastructure readiness is a significant factor for offshore wind power plant installations. There are four substations nearby Bandırma region, namely Erdek, Bandırma, Hasköy and Demirtaş substations, three for Karabiga region, namely Erdek, Bandırma and Biga substations, one for Hoşköy region, namely Malkara substation, one for Bozcaada region, namely Ezine substation.

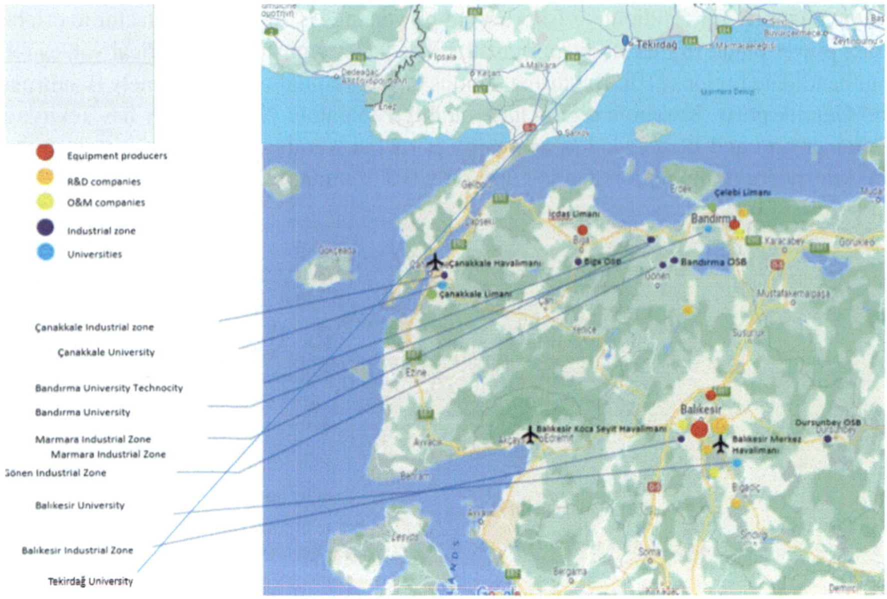

Fig. 8. Relevant industry nearby candidate regions [11].

2.2 Method

In this study, four determined regions for offshore wind deployment are ranked using a multi-criteria decision making (MCDM) method, namely Simple Multi-Attribute Rating Technique (SMART). MCDMs are utilized to decide the best option or rank several options. There are many MCDMs used in the literature such as Analytical Hierarchy Process (AHP), Technique for Order of Preference by Similarity to Ideal Solution (TOPSIS), Elimination and Choice Translating Reality (ELECTRE) and SMART [12–15]. This study utilizes SMART due to its applicability and transparency about the analysis [15]. SMART uses a linear additive model to determine the rank of the alternatives. First, criteria for the decision are settled, then, weights for criteria are assigned by experts or decision makers. Second, weights based on criteria are given for each alternative. Third, the weights of criteria and alternatives are multiplied to reach total score which determine the ranks of alternatives. Equation 1 is used to apply SMART for four offshore wind candidate regions [15].

$$\sum\nolimits_{k=1}^{n} w_k . a_{ik} \tag{1}$$

where n is the number of criteria, w_k is weight of kth criterion, a_{ik} is the alternative weight for ith alternative of kth crietierion. The result of Eq. 1 shows the score of ith alternatives' total score. Ranking is done when all alternatives' scores are calculated.

3 Results

The results of this study show that wind potential criterion has the highest weight among all alternatives while shipping density, port availability and number of touristic points have the least weights (Table 1.). Weights are given between 0 and 100 to the relevant criteria.

Table 1. Criteria weights for offshore wind candidate regions

Criteria	Weights	Standardized weights
Wind potential	95	14.8
Water depth	90	14.1
Shipping density	70	10.9
Bird migration density	90	14.1
Port availability	70	10.9
Closeness to relevant industry	75	11.7
Number of touristic points	70	10.9
Electrical infrastructure nearby	80	12.5
TOTAL	640	

Furthermore, each criterion has different weights on each alternative. These values are gathered from abovementioned data sources. Table 2. gives the real values from the data sources and weights are given between 0 and 100 to the relevant criteria- alternative matches.

Table 2. Criteria data and scores on each alternative

Alternatives								
	Port availability - score		Closeness to relevant indusry - score		Number of touristic regions - score		Electrical infrastructure - score	
Bandırma	2	85	Close	90	5	70	4	85
Karabiga	1	80	Mid	85	2	80	3	80
Hoşköy	1	80	Far	70	1	85	1	70
Bozcaada	0.5	70	Far	70	5	70	1	70
	Wind potential (m/s) - score		Water depth (m) - score		Shipping density (route/sq. km) - score		Bird migration density - score	
Bandırma	7.75	80	55	65	100	75	High	70
Karabiga	6.75	70	40	80	5	90	High	70
Hoşköy	7.5	75	45	70	75	70	High	70
Bozcaada	9	90	20	90	200	65	Low	85

Table 3. shows the total scores and ranks of candidate regions. Accordingly, Karabiga region is found to get the highest score, followed by Bozcaada and Bandırma regions, respectively. It should be noted that the score difference is very little among these three candidate regions. On the other hand, Hoşköy region has the least score among four candidate regions by a great score difference.

Table 3. Total scores and ranks of candidate offshore wind energy regions.

Alternatives	Total Score	Rank
Bandırma	7719	3
Karabiga	7879	1
Hoşköy	7348	4
Bozcaada	7734	2

When it comes to the criteria that candidate regions show strength, Bandırma region has advantages with port availability closeness to industry and electrical infrastructure. Karabiga region has advantages with low shipping density and closeness to the industry. Hoşköy region has an advantage with the low touristic points nearby. Bozcaada has advantages with high wind potential and the least water depth and bird migration density.

4 Conclusions

This study investigated ranking of four offshore wind farm candidate regions on Türkiye's seas by applying Simple Multi-Attribute Rating Technique (SMART). Several data sources are utilized to weight alternatives based on predetermined criteria. Results show that three candidate regions are stepping forward while Karabiga region has the highest score. One of the regions, namely, Hoşköy region has the least score with having generally low scores on criteria. This study is the first type of study ranking candidate offshore wind farms using SMART, to the author's knowledge.

References

1. Global Wind Energy Council. Global Wind Report (2024)
2. Satir, M., Murphy, F., McDonnell, K.: Feasibility study of an offshore wind farm in the Aegean Sea, Turkey. Renew. Sustain. Ener. Rev. **81**, 2552–2562 (2018)
3. Caceoğlu, E., Yildiz, H.K., Oğuz, E., Huvaj, N., Guerrero, J.M.: Offshore wind power plant site selection using analytical hierarchy process for Northwest Turkey. Ocean Eng. **252**, 111178 (2022)
4. Karipoğlu, F., Öztürk, S., Genç, M.S.: Determining suitable regions for potential offshore wind farms in Bandırma Bay using multi-criteria-decision-making method. Mühendislik Bilimleri ve Araştırmaları Dergisi. **3**(1), 123–132 (2021)
5. Argin, M., Yerci, V., Erdogan, N., Kucuksari, S., Cali, U.: Exploring the offshore wind energy potential of Turkey based on multi-criteria site selection. Energ. Strat. Rev. **23**, 33–46 (2019)

6. Global Wind Atlas, https://globalwindatlas.info/en, last accessed 18 Oct 2024
7. European Marine Observation and Data Network, https://emodnet.ec.europa.eu/en/bathymetry, last accessed 18 Oct 2024
8. European Marine Observation and Data Network, https://emodnet.ec.europa.eu/en/human-activities, last accessed 18 Oct 2024
9. Global Biodiversity Information Facility, https://bbecquet.github.io/bird-tracking/, last accessed 18 Oct 2024
10. Coğrafya Harita, http://cografyaharita.com/, last accessed 18 Oct 2024
11. South Marmara Development Agency. TR 22 Region Offshore Wind Power Plant and Port Pre-Feasibility Report, https://www.gmka.gov.tr/dokumanlar/yayinlar/TR22-Bolgesi-Deniz-Ustu-Ruzgar-Santrali-Limani-On-Fizibilite-Raporu.pdf, last accessed 18 Oct 2024
12. Ozturk, S., Karipoglu, F.: Investigation of the best possible methods for wind turbine blade waste management by using GIS and FAHP: Turkey case. Environ. Sci. Pollut. Res. **30**(6), 15020–15033 (2023)
13. Karipoğlu, F., Ozturk, S., Efe, B.: A GIS-based FAHP and FEDAS analysis framework for suitable site selection of a hybrid offshore wind and solar power plant. Energy Sustain. Dev. **77**, 101349 (2023)
14. Öztürk, S., Atalay, K.: Bulanık TOPSIS Yöntemiyle Rüzgar Türbin Kanat Atıkları İçin En Uygun Bertaraf Yönteminin Belirlenmesi: Bandırma Örneği. Mühendislik Bilimleri ve Araştırmaları Dergisi. **5**(2), 293–303 (2023)
15. Siregar, D., Arisandi, D., Usman, A., Irwan, D., Rahim, R.: Research of simple multi-attribute rating technique for decision support. J. Phys. Conf. Ser. **930**(1), 012015 (2017)

Ocean Current Effects on Modified SPAR Wind Floaters: Implications for Motion and Load Management

Fuad Mahfud Assidiq[1](✉), Daeng Paroka[1], Syerly Klara[2], Un-Ryong Rim[3], Hema Carla Rodrigues Moreira[4], Jamhari Hidayat Mustofa[5], Dwiki Timur Pratama Bakri[6], Muhammad Mustafa Algifari[1], and Sulkifli[1]

[1] Department of Ocean Engineering, Universitas Hasanuddin, Poros Malino 92171, Gowa, Indonesia
assidiqfm@unhas.ac.id

[2] Department of Marine Engineering, Universitas Hasanuddin, Poros Malino 92171, Gowa, Indonesia

[3] Institute of Ocean Engineering, Kim Chaek University of Technology, Kyogu-dong, 950003 Pyongyang, Democratic People's Republic of Korea

[4] Department of Ocean Engineering, Federal University of Rio de Janeiro, Av. Athos da Silviera Ramos, 21945-970 Rio de Janeiro, Brazil

[5] Qatar Navigation QPSC, Umm Al Saneem Area, 153 Doha, Qatar

[6] Hakata Shipbuilding Co. Ltd., Imabari 794-2305, Ehime, Japan

Abstract. This research examines the impact of flow-induced motion (FIM) on SPAR-type floating offshore wind turbine substructures under various uniform current conditions. Motivated by concerns over increased structural stress, reduced efficiency, and imbalanced mooring tensions, the study aims to optimize lower SPAR substructures for improved stability and efficiency. A combination of numerical simulations and experimental validation was used to analyze in-line and cross-flow motion characteristics, hydrodynamic properties, and mooring tension distribution. The configurations studied include the base, 3VP, 4VP, and 5VP models. Results show that the 4VP configuration offers superior stability, reduced oscillation, effective resonance suppression, and balanced mooring loads, particularly at low to medium current angles (0°–45°). In contrast, the Base model was highly susceptible to resonance and exhibited significant oscillation and uneven load distribution. While the 5VP configuration shows potential under high reduced velocities and steep angles, further optimization is needed. This research highlights the importance of enhancing vertical plates (VP) to mitigate FIM effects and ensure structural integrity in evolving SPAR designs for offshore renewable energy. Future studies will explore multi-directional and extreme current scenarios.

Keywords: Current Loads · Cross-flow motion · Flow-induced Motion · In-line motion · Vertical Plates

J. C. Casila et al. (Eds.): CCORE 2024, SPEES, pp. 78–90, 2025.
https://doi.org/10.1007/978-3-031-93887-0_10

1 Introduction

A Floating Offshore Wind Turbine (FOWT), especially the stable single point anchor point-type (SPAR) design, is crucial for deep-water renewable energy [1–3]. However, flow-induced motion (FIM) from ocean currents can cause structural stress, reduce efficiency, and increase mooring tension. To improve performance and longevity, SPAR designs must tackle these issues [4–8]. FIM creates complex motion patterns as currents interact with structures. Dahl [9] identified resonant conditions that heighten vibrations, while, Duan et al. [10] described cross-flow *CF* attraction, where larger sways lead to smaller surges, impacting mooring stability. Zheng et al. [11] found that low-frequency currents affect SPAR motions, and Li et al. [12] noted that cross-flow *CF* is influenced by lift forces and in-line *IL* by drag forces. These findings underscore the need for effective control measures to reduce structural fatigue and maintain efficiency. The effects of varying currents on SPAR substructures are unclear, as most research has focused on steady-state conditions. Leal et al. [13] observed displacement peaks at certain current velocities, but their study was limited to multi-directional currents. Gonçalves et al. [14] showed that motion-suppressing devices are effective for mono-column platforms, which do not apply to SPAR configurations. Innovative strategies are needed to enhance SPAR performance in realistic conditions.

This study proposes a new substructure SPAR wind floater design to mitigate FIM effects and improve efficiency. It will use boundary element methods (BEM) and additional experiments to analyze *IL* and *CF* motions, hydrodynamic properties, and mooring tensions under various current scenarios. The findings will guide location-specific design improvements, focusing on optimized motion responses for the lower SPAR substructure, excluding the wind turbine to eliminate hydrodynamic influences. Hybrid solutions should utilize insights from past research on active controls for drag reduction [15] and predicted vertical plate (VP) configurations [16]. Integrating active control mechanisms with structural enhancements is expected to reduce FIM and improve stability.

These advancements aim to create SPAR FOWT designs that minimize FIM impacts, extend operational lifespans, and enhance platform efficiency, promoting offshore renewable energy adoption [17]. Additionally, recommended design considerations may lead to better outcomes in various marine environments, addressing challenges in the transition to sustainable energy.

2 Numerical Method

The present study employs a hybrid experimental and numerical analysis approach to investigate flow-induced motion (FIM) in a modified SPAR 6 MW floating offshore wind turbine (FOWT). This combination of experimental and numerical techniques is consistent with previous work on SPAR platforms, as referenced in the studies conducted by [18, 19], and ensures a comprehensive understanding of the dynamics of FIM in offshore renewable energy systems. This model was employed to replicate the inline and cross-flow natural frequencies of a full-scale platform. A comprehensive numerical matrix, presented in Table 1, was employed to study a wide range of reduced velocities V_r and uniform current flow directions, thereby identifying crucial FIM conditions. In

Fig. 1, all the principal physical and hydrodynamic parameters are defined that were inputted into the model simulations, in addition to the characteristics of the cable for the catenary section at a depth of 100 m. The total length of the analysed catenary section is 400 m, and its mass per unit length is 370 kg/m. The equivalent cross-sectional area of the section in question is 0.09 m^2, with a stiffness of 80 MN and maximum tensile capacity of 11.98 MN. The hydrodynamic drag coefficients are 2.4 for transverse drag and 1.15 for longitudinal drag, with an equivalent diameter of 0.13 m. Furthermore, the added mass coefficient was set to 1.0, in accordance with the environmental fluid assumptions.

Table 1. Load case matrix.

Cases	C1	C2	C3	C4	C5	C6	C7	C8	C9	C10
V_r	2.0	2.5	3.0	3.5	4.0	4.5	5.0	5.5	6.0	6.5
Cases	C11	C12	C13	C14	C15	C16	C17	C18	C19	C20
V_r	7.0	7.5	8.0	8.5	9.0	9.5	10	11	12	15

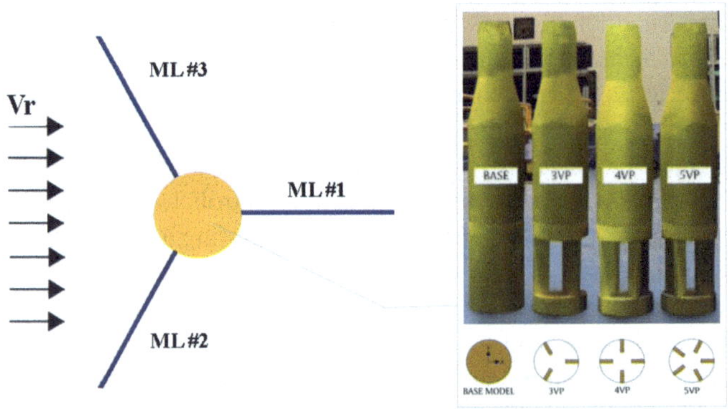

Fig. 1. Layout of SPAR FOWT simulation.

This study employed the boundary element method (BEM), a well-established numerical technique that accurately represents hydrodynamic forces [20–22] including effects at the far-field boundary [23, 24]. A total of 240 simulations were conducted, with each lasting for 4,000 seconds (excluding the initial 400 seconds, which were identified as being transient according to [25]. Viscous correction damping was applied to enhance the precision of the numerical results [26]. A statistical and spectral analysis of the data identified the dominant vibration frequencies and quantified the variability of the FIM responses. Ethical considerations were carefully addressed in this research, with all hybrid procedures following safety protocols and guidelines. Data privacy and confidentiality were protected, and potential environmental impacts were thoroughly assessed.

3 Results and Discussion

3.1 Numerical Validation

The numerical validation of SPAR models in surge and sway decay was performed under a variety of configurations and incidence angles, namely 0°, 45°, and 90°. A summary of the findings can be found in Table 2, which presents the calculated natural frequencies for both the surge and sway modes, as applicable to SPAR configurations: The Base, 3VP, 4VP and 5VP configurations were considered.

Table 2. Validation of Surge and Sway Decay on All Proposed SPARs.

Incidence angle (°)	SPAR model	Surge natural frequencies f_{xn} (Hz)	Sway natural frequencies f_{yn} (Hz)	Validation range [27] (Hz)	Assessment
0°	Base	0.0118	0.0113	0.0050–0.0250	Validated
	3VP	0.0095	0.0095		Validated
	4VP	0.0098	0.0098		Validated
	5VP	0.0098	0.0098		Validated
45°	Base	0.0118	0.0113	0.0050–0.0250	Validated
	3VP	0.0095	0.0095		Validated
	4VP	0.0098	0.0098		Validated
	5VP	0.0098	0.0098		Validated
90°	Base	0.0118	0.0118	0.0050–0.0250	Validated
	3VP	0.0095	0.0095		Validated
	4VP	0.0098	0.0098		Validated
	5VP	0.0098	0.0098		Validated

It can be seen from the analysis that all these constellations satisfy the specified frequency range as outlined by [27]. This range is widely regarded as a reference for evaluating the stability and dynamic performance of floating offshore structures subjected to load excitation. The findings indicate that the SPAR models, including the baseline and those enhanced through the utilisation of vortex suppression devices, exhibited an adequate level of stability with regard to their natural frequency behaviour. This is a notable outcome given that models whose natural frequencies fall outside the acceptable range are susceptible to resonance phenomena, which could potentially lead to structural damage. Consequently, maintaining frequency values within the established range provides a crucial resilience mechanism against resonance-induced instability.

3.2 Motion Trajectories

Figure 2 presents the motion trajectories for the four SPAR configurations (Base, 3VP, 4VP, and 5VP) analysed at three current incidence angles: 0°, 45°, and 90°. The results demonstrate that the Base configuration exhibited the most pronounced oscillatory behaviour, particularly in the cross-flow direction (*CF/D*). This indicates a greater variability and instability in its performance. The addition of vertical plates (VPs) has demonstrated the most significant reduction in these oscillations at 0°-incidence. Furthermore, the maximum stability and confinement of trajectories have been observed in the 4VP and 5VP configurations.

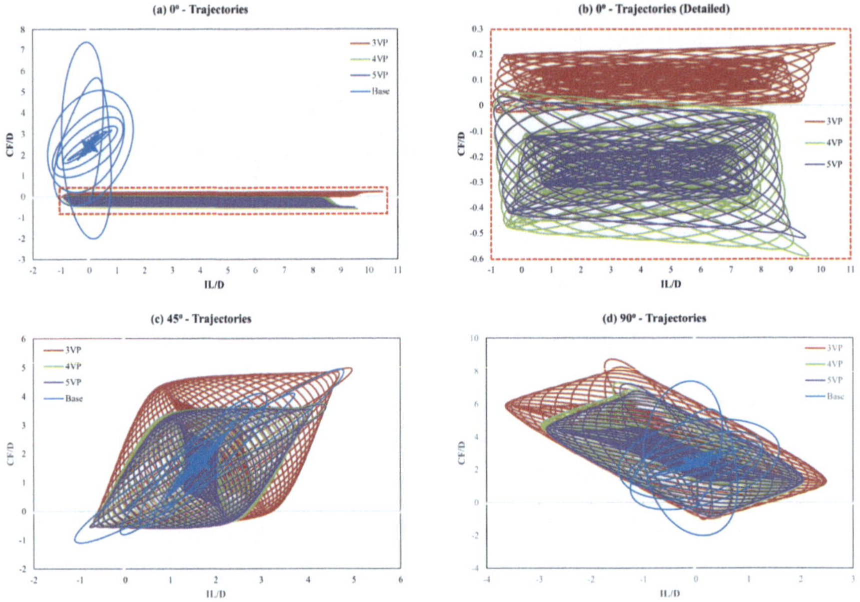

Fig. 2. Motion Trajectory on All Proposed SPARs.

As the incidence angle increases from 45° to 90°-incidence, a corresponding expansion in spread is evident for all configurations in both the in-line direction *IL/D* and cross-flow direction *CF/D* directions. At 45°-incidence, the largest loop-shaped trajectory was observed in the 3VP configuration, whereas the 4VP and 5VP designs demonstrated superior control over their oscillatory amplitudes. However, from an angle of 90°-incidence, the dispersion of trajectories is most extensive across all configurations. In comparison to the 3VP and Base designs, the 4VP and 5VP configurations demonstrate a more favourable outcome regarding their dispersion amplitude.

This finding is in accordance with those reported by [28], who observed an increase in the mean current load due to *CF* motions. Similarly, Shen and Liu [29] highlighted the destabilising effects of current-induced flow asymmetry. It can be concluded that the addition of vertical plates was successful in enhancing the stability of low tilt angles. However, further measures are necessary to address the oscillatory behaviour at steeper angles. Configurations such as 4VP and 5VP have proven to be highly effective in reducing damage due to instability, making them suitable for SPAR FOWT applications that require precise motion control.

3.3 Response Amplitudes

An investigation of the response amplitude indicates that the in-line amplitude $A_{IL\,RMS}$ and the cross-flow amplitude $A_{CF\,RMS}$ are notably influenced by the configurations and reduced velocities V_r across cases. These observations are represented in Fig. 3. The base configuration demonstrates a pronounced rise in both the $A_{IL\,RMS}$ and the $A_{CF\,RMS}$ when V_r exceeds 7 at 0°-incidence, indicating a high sensitivity to oscillatory forces. However, the 3VP, 4VP and 5VP configurations all demonstrated low RMS, indicating a minimal response to in-line and cross-flow oscillations at this angle.

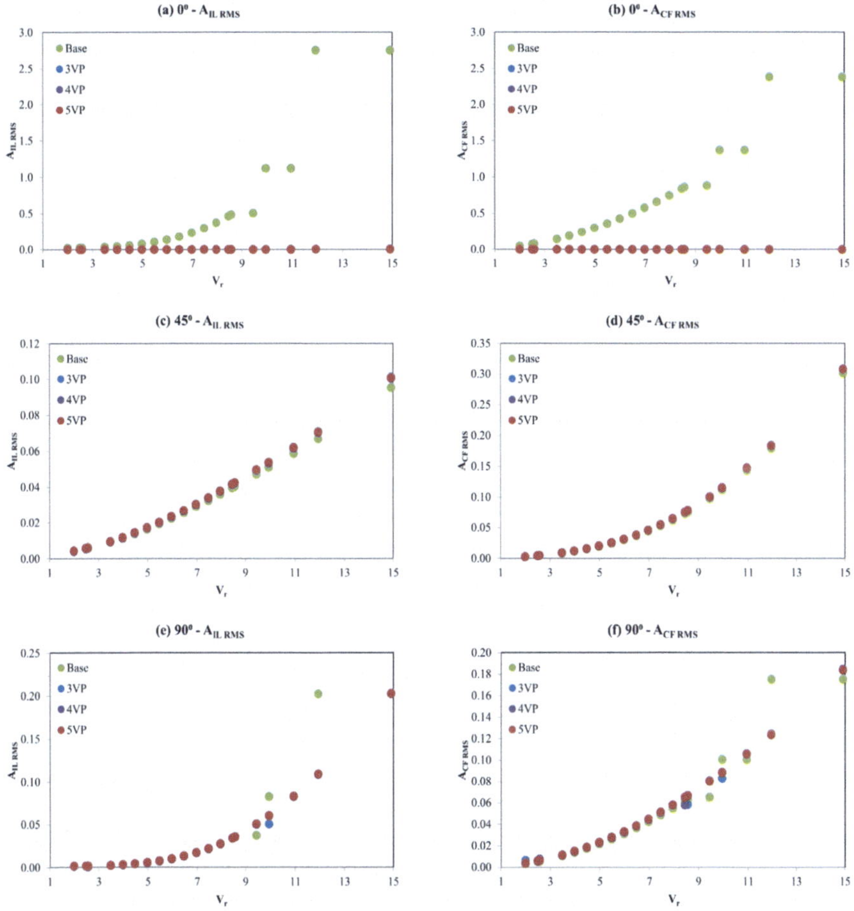

Fig. 3. Response Frequencies on All Proposed SPARs.

Both the in-line and cross-flow response amplitudes demonstrate an increase with increasing reduced velocity V_r at an angle of 45°. Within a 45 -incidence, $A_{IL\,RMS}$ values demonstrate a notable degree of consistency across configurations at lower V_r levels. However, at increased V_r levels, 5VP model exhibited higher amplitudes. With regard to $A_{CF\,RMS}$, the highest values, particularly at elevated V_r, are consistently attributed to 5VP. The response at 90°-incidence becomes increasingly dependent on the configuration, with the 5VP exhibiting the highest $A_{IL\,RMS}$ at high V_r when in-line sensitivity is enhanced. Conversely, the 3VP configuration demonstrates a dominant influence on the A_{CF} response, which strongly indicates cross-flow oscillations. It can be observed that other configurations, such as the Base and 4VP, display relatively stable, lower amplitudes in both directions.

It is anticipated that the majority of movements will demonstrate a pronounced influence of structural characteristics and flow dynamics on the oscillatory response. The significant oscillation observed in the Base configuration indicates the presence of specific forces or resonance effects that are unique to that particular location. Conversely, the enhanced dynamic sensitivity of the 5VP configuration at higher V_r suggests that the designer may have faced certain risks. In contrast, the 3VP and 4VP configurations exhibit a notable degree of stability, demonstrating effective damping, if not optimal flow conditions. These findings align with those of prior studies [30–32] reinforcing the necessity for incorporating configuration-specific responses as a pivotal measure for enhancing structural stability and performance.

3.4 Response Frequencies

The frequency of responses observed in the SPAR configurations were subjected to analysis, with comparisons drawn between Base, 3VP, 4VP and 5VP configurations for both in-line *IL* and cross-flow *CF* motions, across a range of current incidence angles (0°, 45° and 90°). These observations are represented in Fig. 4. With regard to in-line motion, the ratio of frequencies ($f^*_x = f_x/f_n$) remained essentially unchanged across the various configurations and reduced velocity V_r. This indicates a lack of significant increases or resonances. The Base configuration exhibited consistently higher frequency ratios, reaching a value of 0.7. In contrast, the 3VP, 4VP, and 5VP configurations exhibited lower frequency ratios, clustering between 0.5 and 0.6. The frequency ratios for all angles remained below the Strouhal number threshold (<0.1 and <0.2), indicating that no synchronisation was observed to vortex-shedding frequencies. With regard to *CF* motion, the frequency ratios ($f^*_y = f_y/f_n$) are found to be uniformly low across all angles and configurations. The Base model exhibited a frequency ratio below 0.8 initially, which then showed a slight increase with V_r; however, it did not appear to generate resonance. In contrast, the 3VP, 4VP, and 5VP configurations demonstrated lower ratios (0.2–0.4) with minimal variability throughout the testing period. Similarly to the case of *IL* motion, the *CF* ratios remain below the Strouhal lines, indicating that the frequency of shedding vortices does not have a significant effect on the excitation of the system.

It can thus be concluded that the implementation of vortex suppression plates has resulted in a notable reduction in the occurrence of peak frequency ratios and resonance under the influence of these motions themselves. The results align with those of [30], which indicate that the design of structures may offer a viable approach to countering phenomena associated with lock-in and responses from motion-induced by flow. Moreover, the findings indicate that structural designs can be an efficacious strategy for mitigating lock-in and flow-induced motion responses. The three configurations and the full range of angles exhibited a uniform damped response, which is also stable, thereby

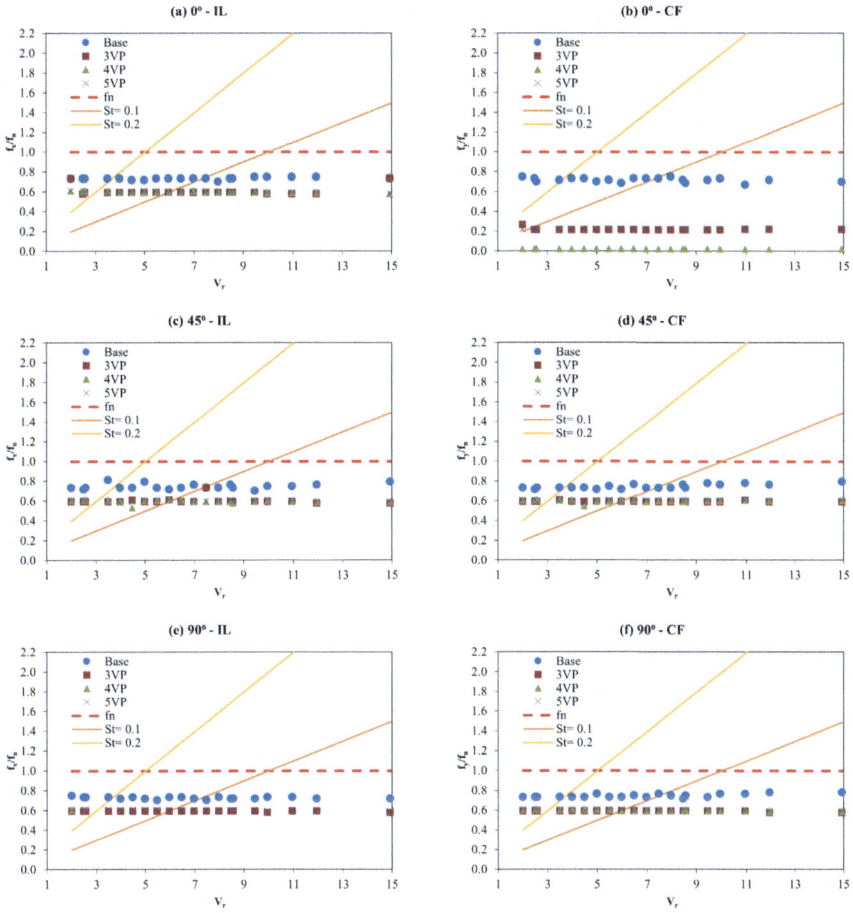

Fig. 4. Response Frequencies on All Proposed SPARs.

demonstrating the resilience of SPAR designs with regard to stability in varying flow conditions.

3.5 Mooring Tensions

The tension in the mooring over the three mooring lines (ML #1, ML #2, and ML #3) was analysed at an angle of current incidence of 0°, 45°, and 90° in accordance with the methodology illustrated in Fig. 5. In the 0°-incidence, ML #2 experiences the highest tensile force, reaching a level of nearly 2.0 MN at the 5VP configuration. Conversely, ML #1 experiences a moderate tensile force, and ML #3 exhibits the lowest tensile force, registering below 1.0 MN, with the exception of a minor spike at the 5VP configuration. At 45°-incidence, the distribution of tension loads becomes relatively balanced. However, ML #2 continues to record the highest values, followed by ML #1, which exhibits moderate tension levels approaching 1.0 MN, and finally ML #3, for which the recorded values remain consistently lower. At 90°-incidence, the effect of the angle increases stress levels. Nevertheless, the configuration remains unchanged: ML #2 has the highest tension, approximately 2.5 MN; ML #1 has moderate levels; and ML #3 has the lowest tension readings, averaging around 1.0 MN.

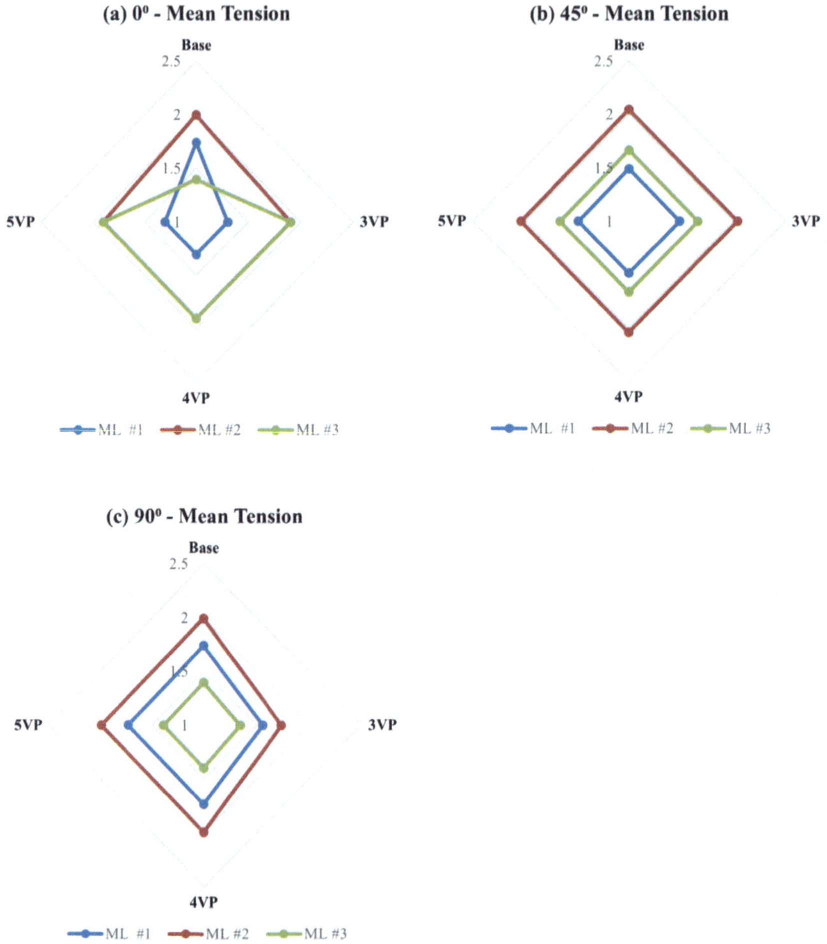

Fig. 5. Mooring Tensions on All Proposed SPARs; units: MN.

From these results, it can be inferred that ML #2 is the most critical element, bearing the main load across all configurations and angles. This highlights the necessity for careful monitoring to avoid overload. Conversely, the moderate tension of ML #1 indicates that it is performing well, with a relatively balanced structural position. ML #3 regularly experiences very little stress, which suggests that it occupies a more favourable structural position. The uniformity of the data across all angles provides evidence that the moorings exhibit a consistent tension distribution, which aligns with the hypothesis of [33] concerning flow-induced motion and its impact on mooring design.

3.6 Modified SPAR Design Factor Recommendations

Table 3 enables the determination of the specific requirements for the design factor of SPAR FOWT substructures, with particular attention to the following aspects: motion stability, oscillation reduction, prevention of resonances, and distribution of mooring loads. The 4VP configuration offers superior stability, reduced response amplitude, and consistent mooring stress distribution, indicating enhanced reliability in varying ocean currents. In contrast, the Base Model showed low amplitude, high resonance susceptibility, and poor mooring stress distribution. While the 3VP and 5VP models faced challenges with sensitivity at high angles and uneven load distribution, they still demonstrated notable performance improvements.

Table 3. Summary of key parameters of All Proposed SPARs.

Aspect	Base model	3VP model	4VP model	5VP model
Motion trajectory	Highly unstable	Moderate stability	Superior stability	Initial stability, but sensitive at high incidences
Response amplitude	High amplitudes	Moderate reduction	Lowest amplitudes	Low amplitudes at low V_r, high at large V_r
Response frequency	High likehood of resonance	Moderate suppression	Most effective suppression	Variability at high V_r
Mooring tension	Imbalanced, peak on ML #2	Moderate improvement	Most uniform load	Higher peak on ML #2 variability

In order to ensure the stability and durability of a modified SPAR-type floating offshore wind turbine sub-structure, which is designed to resist the impact of ocean current loads, the following principal recommendations must be taken into account:

1. The optimal solution in this context is one that offers the greatest stability in motion, the lowest amplitude response, and the most uniform mooring stresses. The 4VP configuration is capable of preventing resonance, which makes it a suitable choice for offshore structures that are required to operate in uniform currents or with low-to-moderate incidence angles (between 0° and 45°).
2. It is possible to apply a 5VP-shaped approach to sites where the incidence angle is substantially larger than 45° or the velocity is higher. Localised gain has the potential to address the 5VP's acute sensitivity to higher speeds, especially at critical nodes, for example ML#2.
3. Mitigating the effects of resonance is a significant challenge in the design of offshore structures. 4VP configurations incorporating active vortex plates serve to maintain a frequency ratio that is distant from resonance. This results in a notable reduction in damage due to vortex-induced vibrations. This assertion is corroborated by data which demonstrate the capacity of 4VP to maintain the response frequency while simultaneously reducing lock-in effects.

4. It is unavoidable that the configuration, the Base Model, will ultimately fail, in the same way as any other model, due to the significant and uncontrolled resonance with high amplitudes and the unequal load distribution that could potentially diminish the service life of the structure or induce an unsafe condition with regard to the fastener.

4 Conclusion

The study investigated the hydrodynamic-induced motion with load distribution of an enhanced SPAR-type floating offshore wind turbine (FOWT) in the context of diverse ocean current conditions. The study employed both numerical and experimental methods for the investigation of four distinct SPAR configurations. This study assessed the motion trajectories, response amplitudes and frequencies, and mooring tensions of the base, 3VP, 4VP and 5VP models in order to evaluate their lower substructure performance.

The data indicated that the 4VP configuration exhibited enhanced performance relative to the other comparable parameters. The configuration exhibited enhanced motion stability, diminished oscillatory amplitudes, effective damping of resonance frequencies, and optimal mooring tension distribution. Consequently, it is the most dependable and efficient design for low-to-moderate operational states with incidence angles between 0° and 45°. In contrast, the Base configuration was found to be markedly unstable, exhibiting considerable oscillations, a pronounced tendency to resonance, and an unequal load distribution, making it unsuitable for deployment in offshore wind turbines. Modelling of the 5VP demonstrated efficiency at elevated velocities and high incursion angles, although further optimisation may be needed to accommodate dynamic loads in cross-flow directions. The 3VP configuration exhibited marginal enhancements in performance in comparison to the remaining models. However, in terms of overall stability and load management, it demonstrated inferiority to the 4VP.

The present study contributes to the growing body of knowledge on the topic of vertical plate (VP) designs and their role in mitigating FIM effects and structural fatigue. The findings highlight the necessity for the design of bespoke SPAR configurations in order to reinforce offshore wind energy infrastructures and achieve broader sustainability objectives. It is acknowledged that, despite the robust findings, several limitations have been identified. These relate to the fact that the study is based on uniform current scenarios and is therefore unable to take into account multi-directional or extreme oceanic conditions. Furthermore, the focus on SPAR configurations may lead to limitations in the application of the study results to every type of offshore substructure. Finally, all limitations will be further researched in order to address them.

Acknowledgments. The authors wish to express their gratitude to the Laboratory Based Education Program 2024 at the Faculty of Engineering, Universitas Hasanuddin for providing financial support through Grant No. 12680/UN4.7.2/PM.01.01/2024.

References

1. Uzunoglu, E., Karmakar, D., Guedes Soares, C.: Floating Offshore Wind Platforms. In: Floating Offshore Wind Farms. Green Energy and Technology, pp. 53–76. Springer, Cham (2016). https://doi.org/10.1007/978-3-319-27972-5_4

2. Chen, J., Kim, M.-H.: Review of recent offshore wind turbine research and optimization methodologies in their design. J. Mar. Sci. Eng. **10**(1), 28 (2021). https://doi.org/10.3390/jmse10010028
3. Bashetty, S., Ozcelik, S.: Review on dynamics of offshore floating wind turbine platforms. Energies (Basel). **14**(19), 6026 (2021). https://doi.org/10.3390/en14196026
4. Lee, I., Kim, M., Jin, C.: Impact of hull flexibility on the global performance of a 15 MW concrete-spar floating offshore wind turbine. Mar. Struct. **100**, 103724 (2025). https://doi.org/10.1016/j.marstruc.2024.103724
5. Peng, Y., Chang, T.: Reliability-based design optimization of tuned mass-damper-inerter for vibration mitigation of spar-type floating offshore wind turbines. Ocean Eng. **310**, 118630 (2024). https://doi.org/10.1016/j.oceaneng.2024.118630
6. Zhang, S.Y., Tan, G., Deng, Y., Zhang, Y., Wang, Y., Wang, Z.: Vibration mitigation of spar-buoy floating wind turbines using a nonlinear energy sink. Ocean Eng. **304**, 117883 (2024). https://doi.org/10.1016/j.oceaneng.2024.117883
7. Zuo, H., Zhang, J., Bi, K., Zhu, S., Hao, H., Ma, R.: Structural vibration control of spar-buoy floating offshore wind turbines. Eng. Struct. **294**, 116732 (2023). https://doi.org/10.1016/j.engstruct.2023.116732
8. Sant, T., Buhagiar, D., Farrugia, R.N.: Evaluating a new concept to integrate compressed air energy storage in spar-type floating offshore wind turbine structures. Ocean Eng. **166**, 232–241 (2018). https://doi.org/10.1016/j.oceaneng.2018.08.017
9. Dahl, J.: Vortex-induced vibration of a circular cylinder with combined in-line and cross-flow motion. Ph.D dissertation, Massachusetts Institute of Technology (2008)
10. Duan, F., Hu, Z., Wang, J.: Investigation of the VIMs of a spar-type FOWT using a model test method. J. Renew. Sustain. Ener. **8**(6) (2016). https://doi.org/10.1063/1.4966968
11. Zheng, Z., Chen, J., Liang, H., Zhao, Y., Shao, Y.: Hydrodynamic responses of a 6 MW spar-type floating offshore wind turbine in regular waves and uniform current. Fluids. **5**(4), 187 (2020). https://doi.org/10.3390/fluids5040187
12. Li, Y., Liu, L., Zhu, Q., Guo, Y., Hu, Z., Tang, Y.: Influence of vortex-induced loads on the motion of SPAR-type wind turbine: a coupled aero-hydro-vortex-mooring investigation. J. Offshore Mech. Arct. Eng. **140**(5) (2018). https://doi.org/10.1115/1.4040048
13. Peres Leal, A., Luís Condino Fujarra, A., Matos Carnier, R., Trentin Gonçalves, R., Suzuki, H.: Flow-induced vibration analysis using a low-cost inertial measurement unit. In: Volume 5: Ocean Engineering. American Society of Mechanical Engineers (2023). https://doi.org/10.1115/OMAE2023-105138
14. Gonçalves, R.T., Rosetti, G.F., Fujarra, A.L.C., Nishimoto, K.: An overview of relevant aspects on VIM of spar and monocolumn platforms. J. Offshore Mech. Arct. Eng. **134**(1) (2012). https://doi.org/10.1115/1.4003698
15. Korpus, R.A., Liapis, S.: Active and passive control of spar vortex-induced motions. J. Offshore Mech. Arct. Eng. **129**(4), 290–299 (2007). https://doi.org/10.1115/1.2746400
16. Assidiq, F.M., Paroka, D., Hidayatullah, Ramadan, M.F.F., Tjuti, T.T.: Flow-induced motion investigation on novel SPAR wind floater on damping plate configurations. In: Proceedings of the 2nd International Conference on Climate Change and Ocean Renewable Energy. CCORE 2022 Springer Proceedings in Earth and Environmental Sciences, pp. 60–76. Springer, Cham (2024). https://doi.org/10.1007/978-3-031-71555-6_8
17. Kaptan, M., Skaare, B., Jiang, Z., Ong, M.C.: Analysis of spar and semi-submersible floating wind concepts with respect to human exposure to motion during maintenance operations. Mar. Struct. **83**, 103145 (2022). https://doi.org/10.1016/j.marstruc.2021.103145
18. Assidiq, F.M., Paroka, D., Palippui, H., Hidayatullah, H., Ramadan, M.F.F.: Experimental investigation on the characteristics of pitch motion for a novel SPAR–type FOWT in regular waves. Kapal: Jurnal Ilmu Pengetahuan dan Teknologi Kelautan. **20**(2), 163–174 (2023). https://doi.org/10.14710/kapal.v20i2.52191

19. Jie, Y., Ping, H.Y., Sheng, Yang, Z., Lin, S.Y., Chong, H.: Current effect on the hydrodynamic responses of spar type floating offshore wind turbine. In: Proceedings of the Thirtieth (2020) International Ocean and Polar Engineering Conference, pp. 398–405 (2020)
20. Zabala, I., Henriques, J.C.C., Kelly, T.E., Ricci, P.P., Blanco, J.M.: Post-processing techniques to improve the results of hydrodynamic boundary element method solvers. Ocean Eng. **295**, 116913 (2024). https://doi.org/10.1016/j.oceaneng.2024.116913
21. Li, B., Liu, Z., Liang, H., Zheng, M., Qiao, D.: BEM modeling for the hydrodynamic analysis of the perforated fish farming vessel. Ocean Eng. **285**, 115225 (2023). https://doi.org/10.1016/j.oceaneng.2023.115225
22. Nguyen, H.P., Liang, H., Luong, V.H.: Finite element - dual boundary element method for hydroelastic analysis of very large floating structures protected by perforated barrier. Ocean Eng. **268**, 113511 (2023). https://doi.org/10.1016/j.oceaneng.2022.113511
23. Jamshidi, S., Haghighi Poshtiri, A., Maali, M.E.: Numerical simulation of flow-induced vibration of the two-degree-of-freedom circular cylinder by random vortex-boundary element method. Ocean Eng. **262**, 112276 (2022). https://doi.org/10.1016/j.oceaneng.2022.112276
24. Nguyen, H.P., Wang, C.M.: Finite element - multi-domain boundary element method for hydroelastic analysis of large floating pontoons with perforated plates. Ocean Eng. **246**, 110659 (2022). https://doi.org/10.1016/j.oceaneng.2022.110659
25. Guidance Notes on Global Performance Analysis for Floating Offshore Wind Turbine. American Bureau of Shipping Incorporated by Act of Legislature of the State of New York 1862, 2020. [Online]. Available: https://ww2.eagle.org/content/dam/eagle/rules-and-guides/archives/design_and_analysis/206_globalperanalyfowti1/fowt-gpa-gn-july20.pdf
26. Assidiq, F.M., Paroka, D., Habibi, Hidayatullah, Ramadan, M.F.: Influence of vertical plates on the pitching motion of a SPAR wind floater in waves. In: Climate Change and Ocean Renewable Energy, pp. 30–42. Springer Nature, Cham (2023). https://doi.org/10.1007/978-3-031-26967-7_3
27. Moan, T., Gao, Z., Bachynski, E.E., Nejad, A.R.: Recent advances in integrated response analysis of floating wind turbines in a reliability perspective. J. Offshore Mech. Arct. Eng. **142**(5) (2020). https://doi.org/10.1115/1.4046196
28. O. J. Waals, A. C. Phadke, and S. Bultema, "Flow induced motions on multi column floaters," in Volume 1: Offshore Technology; Special Symposium on Ocean Measurements and Their Influence on Design, ASMEDC, 2007, pp. 669–678. https://doi.org/10.1115/OMAE2007-29539
29. Shen, M., Liu, Y.: Current effects on global motions of a floating platform in waves. Ocean Syst. Eng. **7**(2), 121–141 (2017). https://doi.org/10.12989/ose.2017.7.2.121
30. Liu, Y., Ge, D., Bai, X., Li, L.: A CFD study of vortex-induced motions of a semi-submersible floating offshore wind turbine. Energies (Basel). **16**(2), 698 (2023). https://doi.org/10.3390/en16020698
31. Li, X., Xiao, Q., Gonçalves, R.T., Peyrard, C.: A coupled wave-current-structure study for a floating offshore wind turbine platform. In: Volume 7: CFD and FSI. American Society of Mechanical Engineers (2022). https://doi.org/10.1115/OMAE2022-79016
32. Vengatesan, V.: Hydrodynamic Force Cofficients for Rectangular Cylinders in Waves and Current. University of Glasglow, 2002. [Online]. Available: https://theses.gla.ac.uk/6351/1/2002venugopalphd.pdf
33. Gonçalves, R.T., Chame, M.E.F., Silva, L.S.P., Koop, A., Hirabayashi, S., Suzuki, H.: Experimental flow-induced motions of a FOWT semi-submersible type (OC4 phase II floater). J. Offshore Mech. Arct. Eng. **143**(1) (2021). https://doi.org/10.1115/1.4048149

Blue Horizons: Innovations in Sustainability

Influence of Climatic Actions on the Dynamic Response for Insulating Glass Units

Enrico Zacchei[1](✉) and Reyolando Brasil[2,3]

[1] Department of Mechanical Engineering, Higher Polytechnic School of Avila, University of Salamanca (USAL), Salamanca, Spain
enricozacchei@gmail.com

[2] Polytechnic School, University of São Paulo (USP), 380 Prof. Luciano Gualberto, São Paulo, SP, Brazil

[3] Center for Engineering, Modeling and Applied Social Sciences, Federal University of ABC (UFABC), Alameda da Unversidade, São Bernardo do Campo, SP, Brazil

Abstract. Insulating glass units (IGUs) are composite structures typically consisting of glass plates, spacers, and internal gas. The edge spacers enclose the gas, which plays a key role in providing effective thermal and acoustic insulation, as well as contributing to energy efficiency. However, most research has treated these structures without adequately accounting for the real influence of the internal gas, which can undergo density variations due to climatic factors, temperature changes, and, more broadly, climate change. For this, in many cases IGUs are also used by combining with photovoltaic panels. In this paper, an accurate analytical and numerical analyses have been carried out to quantify the dynamic behaviour of IGUs, as a function of the glass and cavity thickness, glass size, temperature variations, gas density variation, gas type, etc. Some scenarios and benchmark examples have been analysed. Results shown that the gas presence can produce a system frequency greater of ~5.0 than the fundamental frequency of the glass. More significant effects were noted for higher vibration modes. Additionally, results shown that, due to daily variations of solar radiation and environmental temperature, the gas is subjected to temperature variations in the range about ±20.0 °C, leading to density variations that significantly impact the frequency response of IGUs.

Keywords: Insulating glass units · modal analyses · gas influence · temperature change

1 Introduction

Insulating glass units (IGUs) are composed by glass plates entrapping a cavity filled with a gas sealed by perimeter spacers. The structural function of the glass plates is to withstand the external loadings, whereas the internal gas must reduce the thermal conductivity from an external environment to an internal one as well as to improve the acoustic insulation, and blast resistances as studied in [1, 4].

IGUs are widely used for building facades, and due to the brittle nature of glass, specific safety concepts are required to design glass structural members (e.g., transparent

J. C. Casila et al. (Eds.): CCORE 2024, SPEES, pp. 93–102, 2025.
https://doi.org/10.1007/978-3-031-93887-0_11

roofs) in buildings in a safe way [5]. The effects of the gas and spacers on the glass plates are called "load sharing" and they have been studied in [6, 7]. The influence of the sealants (i.e., usually made by silicone place between spacers and glasses) has been also treated by numerical and experimental tests in [13, 17]. Due to its relatively simple interaction in mechanical terms, some studies focused on simple models studying the influence of the glass type/size [14].

In general, there are not many papers regarding IGUs as also mentioned in [1, 8]. In the last decades several studies on IGUs have been conducted mainly for the evaluation of their thermal performance, whereas a limited attention has been dedicated to the assessment of their structural response [9]. Various methods are available to evaluate the influence between the glasses, under external uniformly distributed loads, and changes in barometric pressure/temperature [10]. Very recently, the more efficient Betti's analytical method (BAM), that can be readily used for double-GUs (DGUs) of any size and shape, under the most various boundary and loading conditions, has been proposed in [11, 20] and validated in [12].

Regarding dynamic analyses, some studies have treated the behaviour of laminated glass beams, in particular the frequencies of PVB-glass systems in function of the temperature [15]. In [16], dynamic analyses for monolithic glass panels under different restraint configurations have been also studied. These studies combine dynamic analyses with IGU performances.

The static evaluation of the gas temperature in an IGU is an important aspect, in fact, in addition to the heat exchanges with the surrounding environment, by means of convection and radiation, also the heat exchange between the different plates and the gas must be considered for the temperature evaluation [24]. According to [25], the gas temperature can be assumed to be uniform in different regions in the IGU: the central portion of the plate can be considered fully irradiated by the sun and/or extreme temperature caused by climate change, whereas the peripheral regions are not.

Figure 1 represent a creative and futuristic building generated purely to emphasize the model under extreme high temperature.

The temperature change has also an effect on the dynamic behaviour of IGUs, however, the vibration frequencies have not yet been fully explored for this purpose [18]. More analyses are necessary not only to quantify the structural behaviour of IGUs, but also to improve their performance under extreme high temperature. In this paper, the dynamic vibrations of IGUs have been studied mainly focusing on the influence of the density and temperature (correlated to the climate change [2, 3]) of the gas. To the best of the authors' knowledge, there are no modern studies on this issue.

2 Model

2.1 Mechanical Model

The mechanical model is formed by two rectangular plates, $a \times b$ ($b > a$), with thickness h_j (where $j = 1.0$ for the top plate, and $j = 2.0$ for the bottom plate), as shown in the cross-sectional in the plane, x, z, in Fig. 1. The top and bottom glasses, parallel each other, are separated by a gas chamber of width, s, sealed by perimeter rigid spacers. The top and bottom glass are subjected to external time-varying (t-varying) vertical distributed

Fig. 1. Creative and futuristic image (created by [26] indicating "low building made of insulating glass units, subject to high temperatures").

loads q_1(x,y,t) and q_2(x,y,t), respectively, as well as to the gas pressure, which produce a time-dependent vertical deflections w_j(x,y,t) of the glass plates.

As treated in this paper, the gas pressures are directly correlated to the external temperature variations due to, for instance, the extreme climate change.

According to the Kirchhoff's model, the partial differential equation (PDE), and its boundary conditions, governing the response of each simply supported plate, are respectively [18]:

$$D_j\left(\frac{\partial^4 w}{\partial x^4} + 2\frac{\partial^4 w}{\partial x^2 \partial y^2} + \frac{\partial^4 w}{\partial y^4}\right) + m_j\frac{\partial^4 w}{\partial t^4} = q_j(x, y, t) + p_j(x, y, t) \tag{1}$$

$$w_j = \frac{\partial^2 w}{\partial x^2} = 0, \text{ for } x = 0, x = a \text{ (analogue in other side, b)} \tag{2}$$

where m_j is the mass of the j-th plate, p_j(x,y,t) is the internal dynamic gas pressure, $D_j = \frac{E_g h_j^3}{12\left(1-\upsilon_g^2\right)}$ is the flexural stiffness of the j-th glass, with E_g and υ_g denoting the elastic modulus and Poisson's ratio of the glass, respectively.

To consider the influence of the gas on the dynamic response of the glass plate, it is necessary to evaluate their interactions, in terms of gas pressures exerted on the upper plate (i.e., p_j, correspondent to the gas pressure at $z = -s/2$), and on the lower plate (at $z = s/2$). By assuming the gas between two plates to be irrotational, inviscid, compressible and homogeneous, the motion equation of the gas can be expressed by a potential ϕ(x,y,z,t) described by its velocity components of any gas particle [18].

The internal dynamic pressures, p_j, and the velocity potential of the gas are correlated, to account for the gas-glass interaction. In this way it is possible the dependence of p_j on the gas density, ρ_0, thus on the temperature T (discussed in Sect. 2.2). The correlation of these three parameters (i.e., $p_j \leftrightarrow \rho_0 \leftrightarrow T$) is a key aspect of this study, allowing to estimate the temperature during the dynamic behaviour of the gas-glasses system.

The solution is divided in two portions (for brevity only the solution of the top bottom plate, w_1, is here shown). By considering $q_1 = 0$ and only p_1, the solution of Eqs. (1) and (2) is:

$$w_1(x, y, t) = \sum_{m=1}^{M}\sum_{n=1}^{N} C_{mn}\sin\frac{m\,\pi\,x}{a}\sin\frac{n\,\pi\,y}{b}e^{i\omega_s t} \tag{3}$$

where C_{mn} are unknown constants that define the amplitude of w_1(x,y,t), which provide its physical meaning, and ω_s is the circular frequency of the only one plate where the gas act, and i is the imaginary unit.

Therefore, the key problem is to find these unknown constants C_{mn} (plus the coefficients in Eq. (5)) for each pairs (m, n), which are correlated each other by:

$$C_{mn}\left(\frac{D_1}{m_1}\gamma_{mn}^4 - \omega_s^2\right) + \frac{i\,\omega_s\,\rho_0}{m_1}\sum_{k=0}^{K}\sum_{l=0}^{L}\Phi_{kl}\left(-\frac{s}{2}\right)\alpha_m^k\beta_n^l = 0 \tag{4}$$

where γ_{mn} is a parameter defined in function of geometry, a, b, Φ_{kl} is the modal shape of the system with $\{K, L\} \in \mathbb{N}$, and

$$\begin{cases} \alpha_m^k = \frac{2}{\pi}\frac{m\left[1-(-1)^{m+k}\right]}{m^2-k^2}, \ \alpha_k^k = 0 \\ \beta_n^l = \frac{2}{\pi}\frac{n\left[1-(-1)^{n+l}\right]}{n^2-l^2}, \ \beta_l^l = 0 \end{cases} \tag{5}$$

By adequate substitutions (for more details see [27]) at z = -s/2, we obtain:

$$-\sum_{k=0}^{K}\sum_{l=0}^{L}\left[\frac{\partial\Phi_{kl}}{\partial z}\right]_{-\frac{s}{2}}\alpha_m^k\beta_n^l = i\omega_s C_{mn} \tag{6}$$

where, by extrapolating C_{mn}, substituting it in Eq. (4) and multiplying all terms for ($-i\,\omega_s$), the following set of equations is obtained:

$$\sum_{k=0}^{K}\sum_{l=0}^{L}\left[\left(\frac{D_1}{m_1}\gamma_{mn}^4 - \omega_s^2\right)\frac{\partial\Phi_{kl}\left(-\frac{s}{2}\right)}{\partial z} + \frac{\omega_s^2\,\rho_0}{m_1}\Phi_{kl}\left(-\frac{s}{2}\right)\right]\alpha_m^k\beta_n^l = 0 \tag{7}$$

Equation (7) is called "eigenvalue or characteristic value problem" [19], which estimates ω_s for all mode vibrations obtained as a determinant of a matrix 2 × 2 (for each pair m, n) formed by the expressions in the brackets and coefficients α_m^k and β_n^l. Thus, it is possible to estimate the modal shapes. In a similar way, it is possible to define the dynamic analysis for the bottom plate.

2.2 Climatic Model: Temperature Variation

The temperature distribution in a glass panel is strongly affected by the variation of external temperature along the day. Changes in the temperature of the gas in the cavity produces an increase of internal pressure and density deforming the glasses. By considering p_0, T_0, and V_0, which are the atmospheric pressure, uniform temperature, and initial volume (i.e., $V_0 = a \times b \times s$) at the time of sealing, respectively, the variation of the cavity volume ΔV, correlated to a pressure variation, Δp, and a change in temperature, ΔT, is [20]:

$$\Delta V \approx k' \overline{\psi} \left(\frac{p_0 \, v_0}{k' \, p_0 \, \overline{\psi} + V_0} \times \frac{\Delta T}{T_0} \right), \text{ with } k' = A^3 \left(\frac{1}{D_1} + \frac{1}{D_2} \right) \tag{8}$$

where A is the glass area (i.e., $A = a \times b$), and $\overline{\psi}$ is a non-dimensional parameter, related to the mean value of the shape function for the plate deflection (for more details see [11]). Note that Eq. (8) is valid only under the ideal gas law.

Thus, by considering that ΔV is inversely proportional to gas density, ρ_0, its variation may be evaluated as:

$$\Delta \rho \approx -\rho_0 \left(\frac{k' \, \overline{\psi} \, p_0 \Delta T}{T_0 \left(k' \, \overline{\psi} \, p_0 + V_0 \right) + \left(k' \, \overline{\psi} \, p_0 \Delta T \right)} \right) \tag{9}$$

which allows to evaluate the gas density variation, for a given ΔT, as a function of the insulating glass geometry.

3 Analyses and Results

3.1 Mechanical and Thermal Data

Two examples of applications have been considered. The first, is a triple-GU (TGU) formed by a top monolithic glass and a bottom two-layered laminated glass plate that contain the gas. The second example is a DGU formed by two identical monolithic glasses (top and bottom) confining the gas (like Fig. 2). The laminated glass of the TGU is composed by two glass plates 12.0 mm thick, bonded by a polyvinyl butyryl (PVB) interlayer of 1.52 mm (= 4 × 0.38 mm). Table 1 lists the geometrical and mechanical properties of both examples. The values in common are the glass density, ρ_g (= 2600.0 kg/m^3), the elastic modulus, E_g (= 70.0 GPa), and Poisson's ratio, υ_g (= 0.22).

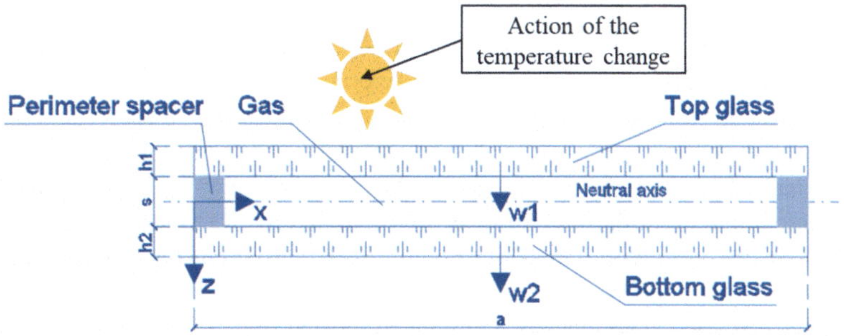

Fig. 2. Cross-section of the mechanical model for DGU.

Table 1. Geometrical, mechanical and thermal parameters

Parameter	TGU glass	DGU glass
Plate length, a × b (m)	3.0 × 4.0	2.0 × 3.0
Top layer thickness, h_1 (mm)	12.0	6.0
Bottom layer thickness, h_2 (mm)	12.0 (glass) + 1.52 (PVB) + 12.0 (glass)	6.0
Flexural rigidity, D_j (kN m) [a]	11.04–66.96	1.38
Gas cavity thickness, s (mm)	20.0	14.0
k' (mm^5/N) (Eq. (8))	1.82×10^{14}	3.13×10^{14}
$\overline{\psi}$ (−) [11]	1.56×10^{-3}	1.46×10^{-3}
$\Delta\rho/\rho_0$ (%), for $\Delta T = \pm$ 20.0 °C (Eq. (9))	−6.34 to +7.26	−6.38 to +7.31
Atmospheric pressure, p_0 (MPa) [20]	0.101	
ρ_0, for T_0 = 20.0 °C (kg/m^3)	1.204 (air), 1.641 (argon), 3.485 (krypton)	

[a]Calculated values by D_j relation shown for Eq. (1). For DGU $D_1 = D_2$, whereas for TGU it is necessary to estimate the effective flexural rigidity of the bottom layered glass (in accordance with [21], D_2 = 66.96 kN m).

Regarding the chosen values for the parameter s, as shown in [1], a range $14.0 < s < 20.0$ mm provides good performances in terms of thermal transmittances.

The considered gases (air, argon and krypton) are usually more common for this purpose. The values show that the quasi-linear $\Delta\rho/\rho_0$ ratio varies between approximatively −6.50% and +7.50% altering its real contribution, which is usually neglected when the dynamic response of insulating glasses is studied. Thus, the gas influence must be considered to carry out an accurate analysis to estimate the real interactions between gas-glasses system.

3.2 Results

Analytical and numerical analyses have been carried out. The analytical Eq. (7) provides all frequency (i.e., k = 0, 1, 2, …, K; l = 0, 1, 2, …, L) of the gas-glass system (here K = L = 5). The analysis should provide more complete results by accounting for the influence of the gas on the glass dynamic response. In this sense, it represents the exact solution. By considering $\rho_0 = 0$, it is also possible to evaluate only the glass frequency.

Numerical FEM analysis developed by Ansys software [22]. In accordance with the mechanical model, both glass and gas are modelled by 3D 8-node solid continuum elements. For this, relatively small meshes of 20.0 × 20.0 cm have been used, for a total of 900 mesh with 6834 nodes, and 450 mesh with 3534 nodes for TGU and DGU, respectively (Fig. 3a)). The glass is a simple supported plate (at all sides), with linear elastic material model, whereas the gas is a considered as adherent "added mass"[1] to the glass.

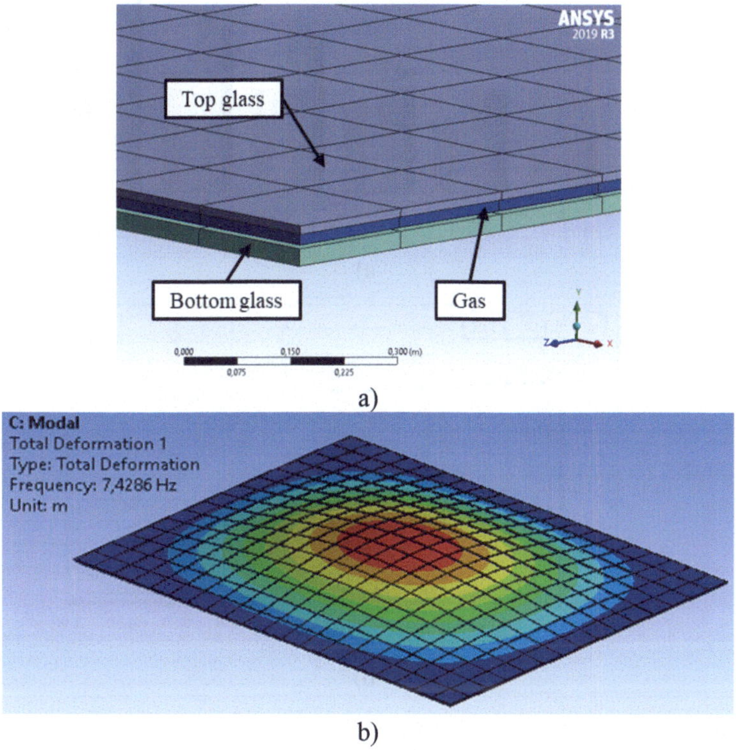

Fig. 3. FEM model of the TGU: a) mesh and elements; b) modal analysis of the gas-glass system (fundamental mode).

Figure 3b) shows the modal analyses (for brevity only one mode is shown), obtained with FEM, of the gas-glass system indicating values of the six frequencies, f_s, of 7.428–33.653 Hz (the gas is air). Note that f_s refers only to one plate where the gas act (already mentioned for Eq. (3)). This modal analysis by FEM model has been carried to calibrate it with respect the analytical results provided by [18].

Figure 4 shows the results in terms of vibration frequencies (Fig. 4a)), and their influences with respect ρ' and s' (Fig. 4b)).

Figure 4a) shows the vibration frequency for all considered gasses, and the influence of the non-dimensional mass ratio, ρ' $(= (\rho_0 \times a)/m_1)$, and gas thickness, s' (= s/a), on fundamental frequency of the glass plate only (without gas), f_1. The ratio of s' ranges between s' = 0.007–0.10, as considered in [18].

[1] Here the gas is considered fully incompressible, i.e., $\upsilon_{eq} \approx 0.50$ and $E_{eq} \approx 0$, where υ_{eq} is the Poisson's ratio of the equivalent mass, and E_{eq} is the elastic modulus equivalent to the bulk modulus. In this way, the pressure variations are null, thus the "added mass" approach is valid, and the gas is assumed to be rigid [23].

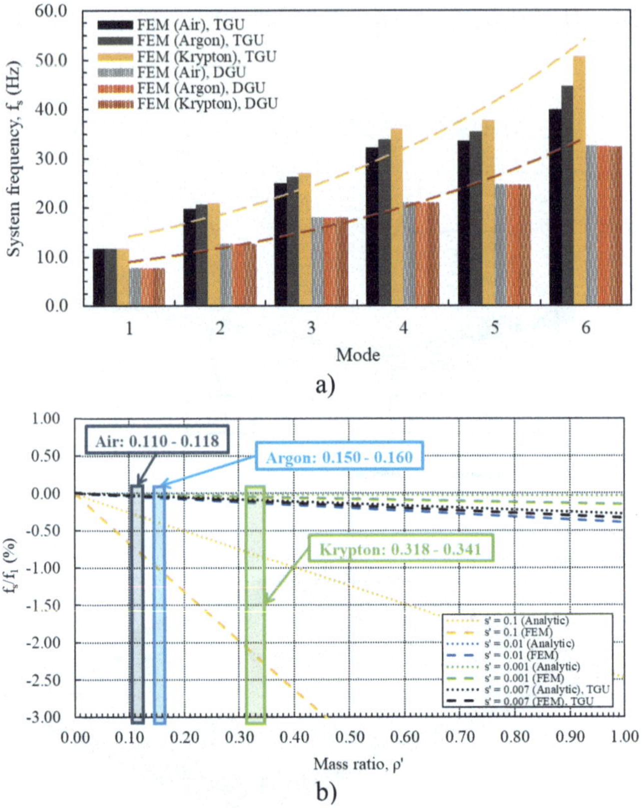

Fig. 4. Results in terms of a) vibration frequencies; b) their influences with respect ρ' and s' (the ρ' variations correlated to $\Delta\rho/\rho_0$ values, as estimated in Table 1, are highlighted).

In Fig. 4a) it is possible to note that a difference of ρ_0 does not affect in a significant manner the modal behaviour of the system. However, there is a difference with respect to the vibration frequency of only glass. In particular, the analytical solution with krypton can reach a value greater of ~5.0 than the vibration frequency f_1, indicating the importance in considering the dynamic analysis for insulating glasses (see the differences between two dashed curves). In general, a high value of ρ_0 tends to increase the frequency vibrations, therefore its increasing due to a temperature variation further raises its frequency; highlighted also in Fig. 4b) by histograms.

The analytical solutions (Eq. (7)) have a good agreement with the approximated solutions provided in [18], thus the small differences are mainly correlate to the fact that the latter solution considers only the first vibration mode. Note that the vertical axis indicates a very small range of f_s/f_1 ratio (up to 3.0%); in this sense these analyses should be considered accurate.

It is possible to see that the influence of ρ', thus of ρ_0 (the parameters, m_1 and a, are maintained fixed), is low for FEM analyses with high slopes.

Another consideration regards the s' ratio effects; in fact, the thickness of the cavity, s, alters the modal behaviour of the system. Considering the results in Fig. 4b) it is evident that for s' $\geq$ 0.01 the influence of the gas on the vibration system should be considered.

4 Conclusions

In this paper, the dynamic vibrations of insulating glasses by considering different gas temperature and densities have been studied. From analytical and numerical analyses, the real influence of the gas-glass interactions can be detected. The analytical solution with krypton can reach a value greater of ~5.0 than the vibration frequency of the glass plate only, indicating the importance in considering the dynamic analysis for insulating glasses. In general, a high gas density tends to increase the frequency vibrations. For DGUs with different thickness, a further increasing of the gas-glass frequency is possible, due to glass-gas-glass double interactions. The $\Delta\rho/\rho_0$ ratio varies from -6.50% to $+7.50\%$ altering its real contribution on the system, thus it should not be neglected.

The quantification of a correct gas-glass system frequency can be relevant to estimate not only the mechanical behaviour but also the IGU performance for sound isolations and blast resistances, as well as the influence of the interlayer, when the insulating glass is made by laminated glass. The main limitations of this study could regard the influence of the spacers, the more refined and advanced gas modelling (e.g., by using computational fluid dynamics, etc.), and the double interaction between superior and inferior glass by analytical equations. We hope in the future to design and carry out an experimental campaign allowing to compare and validate these models.

Acknowledgments. The first author thanks the University of Salamanca (USAL), Spain, for the Wolfram Mathematica license and to pay the rights (when applicable) to completely download all papers in the references. The second author received support from CNPq and FAPESP, both Brazilian research funding agencies.

References

1. Banionis, K., Kumziene, J., Burlingis, A., Ramanauskas, J., Paukstys, V.: The changes in thermal transmittance of window insulating glass unites depending on outdoor temperatures in cold climate countries. Energies. **14**, 1–22 (2021)
2. Feldmeier, F.: How to handle climatic loads in the design of insulating glass units. Stahlbau. **65**(8), 285–290 (1996)
3. Intergovernmental Panel on Climate Change (IPCC). 2021. Report (accessed on 10/2021). https://www.ipcc.ch/2021/08/09/ar6-wg1-20210809-pr/
4. Zhang, J., Wang, X., Wang, L., Su, P.: Experimental study and theoretical verification of explosion-proof performance of insulated glass. Shock. Vib. **2020**, 1–6 (2020)
5. Badalassi, M., Biolzi, L., Royer-Carfagni, G., Salvatore, W.: Safety factors for the structural design of glass. Constr. Build. Mater. **55**, 114–127 (2014)
6. Heiskari, J., Romanoff, J., Laakso, A., Ringsberg, J.W.: On the thickness determination of rectangular glass panes in insulating glass units considering the load sharing and geometrically nonlinear bending. Thin-Walled Struct. **171**, 1–13 (2022)

7. Galuppi, L., Zacchei, E.: Analytical solutions for plates connected by edge beams, under various loading conditions: an application for insulating glass units. Thin-Walled Struct. **188**, 1–16 (2023)
8. Zhang, X., Hao, H., Ma, G.: Laboratory test and numerical solution of laminated glass window vulnerability to debris impact. Int. J. Impact Eng. **55**, 49–62 (2013)
9. Zhang, X., Liang, J., Huang, D.: Study on the mechanical response of anticlastic cold bending insulating glass and its coupling effect with uniform load. PLoS One. **16**, 1–22 (2021)
10. Morse, S.M., Norville, H.S.: Comparison of methods to determine load sharing of insulating glass units for environmental loads. Glass Struct. Eng. **1**, 315–329 (2016)
11. Galuppi, L.: Practical expressions for the design of DGUs. The BAM approach. Eng. Struct. **221**, 1–21 (2020)
12. Galuppi, L., Zacchei, E., Esteves, M., Ferrão, J., Simões, N.: Experimental validation of the Betti's analytical method for double glass units. Eng. Struct. **315**, 1–14 (2024)
13. Bedon, C., Amadio, C.: Mechanical analysis and characterization of IGUs with different silicone sealed spacer connections - part 1: experiments. Glass Struct. Eng. **5**, 301–325 (2020)
14. Louter, C., Belis, J., Veer, F., Lebet, J.P.: Structural response of SG-laminated reinforced glass beams; experimental investigations on the effects of glass type, reinforcement percentage and beam size. Eng. Struct. **36**, 292–301 (2012)
15. Pelayo, F., Lamela-Rey, M.J., Muniz-Calvente, M., López-Aenlle, M., Alvarez-Vazquez, A., Fernandez-Canteli, A.: Study of the time-temperature-dependent behaviour of PVB: application to laminated glass elements. Thin-Walled. Struct. **119**, 324–331 (2017)
16. Bedon, C., Fasan, M., Amadio, C.: Vibration analysis and dynamic characterization of structural glass elements with different restraints based on operational modal analysis. Buildings. **9**, 1–23 (2019)
17. Zacchei, E., Simões, N., Vieira, A., Esteves, M., Silva, H.: Modelling of layers interactions on the structural behaviour of insulating glasses. Vertical deflection analyses. Case Stud. Construct. Mater. **18**, 1–17 (2023)
18. Vann, W.P., Das, Y.C.: Free vibration analysis of insulating glass units. J. Wind Eng. Ind. Aerodyn. **36**, 1145–1153 (1990)
19. Clough, R.W., Penzien, J.: Dynamics of Structures, p. 752. McGraw-Hill, New York (2003)
20. Galuppi, L., Royer-Carfagni, G.: Betti's analytical method for the load sharing in double glazed units. Compos. Struct. **235**, 1–14 (2020)
21. Galuppi, L., Manara, G., Royer-Carfagni, G.: Practical expression for the design of laminated glass. Compos. Part B. **45**, 1677–1688 (2013)
22. Ansys R3. 2019. Academic, software
23. Zacchei, E., Molina, J.L.: Shape optimization of double-arch dams by using parameters obtained through Bayesian estimators. Iran. J. Sci. Technol. Trans. Civ. Eng. **43**(4), 649–662 (2018)
24. Wright, J.L., Kotey, N.A.: Solar absorption by each element in a glazing/shading layer array. ASHRAE Trans. **112** (2006)
25. Galuppi, L., Maffeis, M., Royer-Carfagni, G.: Engineered calculation of the uneven in-plane temperatures in insulating glass units for structural design. Glass Struct. Eng. **7**(1), 71–99 (2022)
26. Image creator in Bing. https://www.bing.com/images/create?FORM=GENILP
27. Zacchei, E., Brasil, R.: Analytical and numerical analyses for modal responses of insulating glass units. J. Archit. Eng. **31**, 1–12 (2025)

Solutions for Mitigating Adverse Effects of Climate Change on Crop Production in Water Shortage Environment: An Example of Semi-Arid Konya Closed Basin of Türkiye

Bilal Acar(✉)

Department of Farm Buildings and Irrigation, Faculty of Agriculture, University of Selçuk, Konya, Turkey
biacar@selcuk.edu.tr

Abstract. Agro-production relies on mainly environmental conditions and climate change has huge impacts on agriculture particularly on water resources. Türkiye, considered as among high risk countries affected from climate change, is situated at eastern part of Mediterranean region of world. In this work, impacts of agricultural drought on agro-production of Konya Closed Basin (KCB) were assessed comprehensively and some practical solutions for productive use water resources were suggested. In result, water scarcity, becoming an increasingly serious problem in this basin day by day, has huge effects on agricultural activities. Unconscious irrigation practices result reductions in water levels. Overwater use, due to the expansion of cropped lands with high water consumption, from groundwater reservoir causes sinkholes in some part of KCB such as in Karapınar surroundings. Water use in irrigation processes is almost 75% in Konya plain and due to the severe agricultural drought in basin at recent years, it is impossible for famers to obtain economical income from agricultural activities without irrigation even from winter cereals. Development of drought-resistant plant species and widening of their cultivation is one of the important solutions for mitigating adverse effects of drought on water resources. In addition, reducing planting areas with high water consuming crops such as corn or in the other word increasing planting areas of less water consuming crops is another viable solution for sustainable agricultural water management in semi-arid KCB. Deficit irrigation is another important strategy for achievement better water productivity with application of unit water especially in water shortage ecologies. As a result, climate change is reality and agro-production program must be done by considering current water resources in water-starved areas.

Keywords: Agricultural drought · Climate change · Sustainable solutions · Water management

1 Introduction

Perhaps the most important adverse effect of climate change is on water cycle and consequently on water resources. Climate change may cause number of events including drought, flood, immigration and so on. As we all know that agro-production needs many

J. C. Casila et al. (Eds.): CCORE 2024, SPEES, pp. 103–110, 2025.
https://doi.org/10.1007/978-3-031-93887-0_12

inputs such as water, soil, sunshine and temperature. In that context, climate change sometimes can have huge risks on agro-production due to unpredictable uncertainty.

Approximately 2.5% of the water on earth is fresh and it's about 69% of this is ice caps, 30% is as underground and 1% is as surface water bodies [1]. In that regard, it is obvious that sustainable use of limited fresh water is an indispensable necessity in all sectors particularly in irrigation at arid and semi-arid regions (Fig. 1).

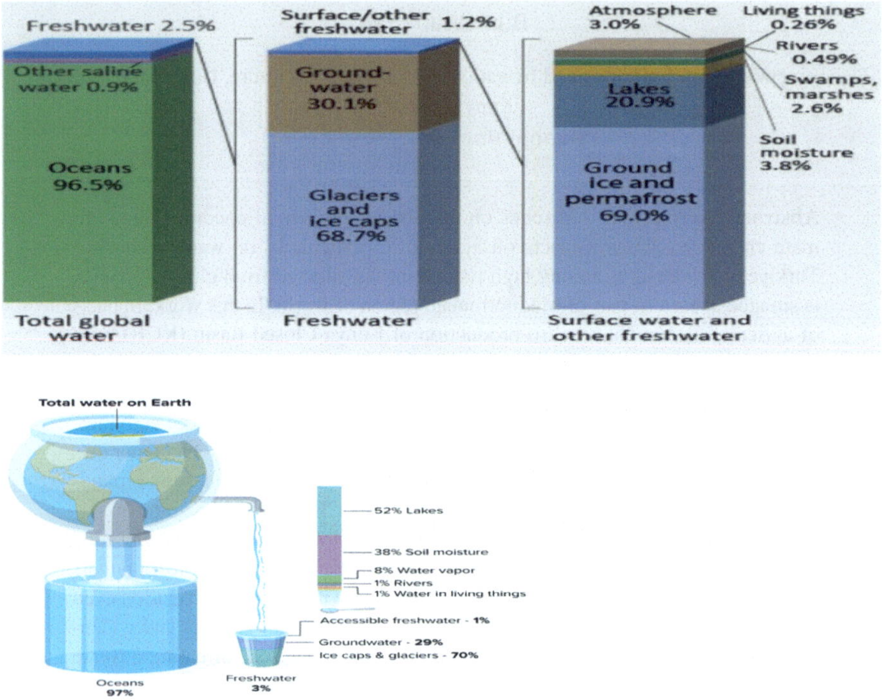

Fig. 1. Water distribution on earth [1].

Available water resources are direct related to rainfall regimes. Türkiye is a semi-arid climate characteristic with long-term average rainfall of 574 mm producing annual 450 billion m^3 water [2]. There are 25 river basins, 320 natural lakes and plenty water storage structures in Türkiye [3] (Fig. 2).

In Türkiye, annual amount of usable water per capita is around 1346 m^3 regarding to 2020-record and classified as water shortage country among countries [3]. By experiencing the last 20-year meteorological data, it is estimated that such country is highly exposing from climate change, and will face serious water scarcity problems. In that regard, economical, rational and effective use water resources seem imperative.

It is located in the eastern part of Mediterranean region so effects of climate change will be felt more intensely. After 2020, Türkiye has exposed more severe droughts, some floods and plenty forest fires. Therefore, it is an absolute necessity to always have dynamic solutions against negative natural events caused by climate change.

Fig. 2. Main river basins in Türkiye [4].

2 Analysis of Türkiye by Considering Climate Change

As mentioned above, negative results of climate change are forest fires, floods, ecosystem losses and so on in Türkiye. Drought with a very short definition is "a lack of rainfall". If there is 50–60% less rainfall than long-term average, it is said drought is existing. The effects of drought depend on its intensity, duration and geographical extent. In general, insufficient rainfall is called "meteorological drought", decrease in soil moisture status is called "agricultural drought", and decreases in water level in rivers, lakes, dams and groundwater is called "hydrological drought" [5].

While Türkiye has dry years in the last years, the severity and duration of droughts are getting increase dramatically. Increase in temperatures, decreases in precipitation and snow cover is expected over next 26 years. In Türkiye, precipitation reduced by 6% by comparison to long-term average in 2023. In recent years, the news relevant to the lakes drying up, decreasing river water levels and decreasing volume of water in dams have been encountered more frequently. If these trends continue, in just after 6 year, in 2030, the annual water amount per capita may reach up to below 1000 m^3 and Türkiye will become more noticeably among the water poor countries. Similarly irrigated areas will be in danger of water shortage [6]. In addition, as climate change will occur more severely, especially southern neighbors of Türkiye, remarkable depletion in surface waters of Türkiye will lead to more critical international water conflicts.

Similarly, in accordance of European Environment Agency Report of 2016 [7], due to climate change, Türkiye could have increase in extreme temperatures, decreases in precipitation and water flow, increases in drought events as well as agricultural water demand, biodiversity loss, forest fires, and even conflict between water users.

3 Assessment of Konya Closed Basin (KCB) in Regard to Agricultural Drought

Water is one of the most important inputs for crop production and irrigation is necessary when precipitation is insufficient for meeting the whole crop water needs. The stress on limited water resources is increasing day by day due to widening areas of irrigated agriculture in KCB for recent years. It has been witnessed that this pressure raises gradually with climate change consequently drought events becoming more noticeable. For this reason, water scarcity is becoming an increasingly serious problem in this basin

[8]. Global warming and unconscious irrigation process are leading decrements in water levels in some rivers, lakes and dams. Similarly, a notable part of the lakes and wetlands in granary Konya plain have been disappeared (Fig. 3).

Fig. 3. Three-dimensional view of KCB [9].

Long-term average annual rainfall of Konya plain is around 300–350 mm. Total annual available surface and groundwater amounts are 1.77×10^9 m^3 and 1.68×10^9 m^3 (total of 3.46×10^9 m^3), respectively. However, current annual water use is around 4.07×10^9 m^3 and about 0.99×10^9 m^3 over water extraction is present from the groundwater reservoir (Table 1). In the outcome of this analysis, there is 1 m or 2 m water level reduction in irrigation wells at KCB each year. Reductions in groundwater levels cause sinkholes in some part of Konya plain such as in Karapınar around. The number and size of the sinkholes are getting increase recently and people are very stressful about that environmental problem (Table 2).

Table 1. Water potential of Konya [9]

Long-term average precipitation (mm)	360.20
Total volume of rainfall ($\times 10^9$ m^3)	14.77
Evaporation ($\times 10^9$ m^3)	7.01
Deep percolation ($\times 10^9$ m^3)	2.22
Surface runoff ($\times 10^9$ m^3)	5.47
Available surface water resources ($\times 10^9$ m^3)	1.77
Annual extractable groundwater ($\times 10^9$ m^3)	1.68
Total available water ($\times 10^9$ m^3)	3.46
Current water use ($\times 10^9$ m^3)	4.07
Over water consumption from groundwater ($\times 10^9$ m^3)	**0.99**

In Konya plain, April of 2024 was recorded as the hottest April of the last 53 years and there was a decrease of up to 50% in precipitation. In Central Anatolia where precipitation decreased by 65% by comparison to normal, the grain warehouse Konya

Table 2. Surface and groundwater potential of Konya plain [9]

Surface water resources ($\times 10^9$ m^3)	2.94
Groundwater resources ($\times 10^9$ m^3)	1.51
Total	**4.45**

plain alarm bells are ringing. Konya covers 7.84% of the grain production areas of Türkiye and ranks first in Türkiye in terms of production amount of 13 different agro-products. In drought affected parts of Konya plain, there were around 20–30% yield loss in barley and 10–20% yield loss in wheat. The rainfall of May 2024 was recorded as 54.2 mm and it prevented further loss of yield in grain in such plain [10].

Water consumption for irrigation is around 75% in Konya plain [11] and it is impossible to produce economical yield without irrigation in such environment. There are two main reasons affecting sustainable utilization of water resources particularly groundwater namely widening of irrigation areas not caring current water supplies, and increasing the planting areas in favor of high water consumption crops such as both the grain and silage corn. In that case practical solution for sustainable water resources is re-planning crop patterns in the advantage of less water consuming crops such as all cereals and legumes.

4 Precautions Against to the Agricultural Drought in Arid or Semi-Arid Regions

In world general, irrigation or agriculture is the highest water user activity; 71% is used in irrigation, 18% in industry, and 11% in residential purposes. In Türkiye, 73% is in irrigation, 11% in industry and 16% in domestic needs [12]. Agricultural drought is inevitable when soil moisture is not sufficient to meet all crop water consumption during growth cycles. Farmers performing rain-fed farming are more affected by drought and have limited solutions to do. As long as there is sufficient water resources, there are more precautions for farmers doing irrigated agriculture.

It can be noted that maximum water productivity can be obtained from correct design and well management of the irrigation systems [13, 14] and pressurized irrigation systems are very important role to play for application of irrigation water with high uniformity in crop rooting environment [15]. Water application performance of sprinkler irrigation systems should be greater than 75% especially in arid and semi-arid regions for achievement of water economy [16]. Similarly drip irrigation system is superior over other irrigation systems in regard to water saving under high water management and can be highly recommended for row crops or fruit trees particularly for areas with water scarcity [17, 18].

Developing drought-resistant plant species and expanding their cultivation areas is another important solution for reducing the adverse effects of drought. Reducing planting areas of corn and increasing the farming areas of cereals and legumes are also strongly recommended for maximal water efficiency in water scant regions (Figs. 4 and 5).

Fig. 4. Wheat production in KCB [19].

Fig. 5. A view from maize farming in KCB [20].

Collecting and storing rain and snow water and presenting it to different uses including irrigation is named as "water harvesting". The amount of water obtained from rainwater harvesting should not be underestimated or ignored [5].

The other practical solution is deficit irrigation program. In this way, crops are forced to use water from rooting zone with maximal efficiency. In result, irrigation water use efficiency increases. In the other explanation, unit applied water leads to highest crop yield. There are number of field trials performed by our team relevant to the deficit irrigation practices for potato, sugar beet, grain corn, melon, water melon, dry bean, and sunflower in Konya plain of Türkiye. In accordance of our field experiments, we have found that water deficiency should be done at crop growth cycles that are not sensitive to the water stress cycle. Instead of fixed water deficiency in all irrigation events, deficit irrigation should be done by considering crop growth cycles. Acar et al. [21] suggested that in fixed deficit irrigation, water limitation should be done maximum 25% not resulting significant yield reductions.

Irrigation program is necessary for optimal use of water resources. Considering the average findings of two-year field studies, 25.5% water savings and 23.2% electrical energy savings were found in sugar beet with scheduled drip irrigation compared to conventional irrigation; 23.5% water savings and 23.9% electrical energy savings were achieved in grain corn and 20.1% water savings and 16.4% electrical energy savings were achieved in silage corn. In addition to all these savings, 9.0% yield increase in sugar beet, 29.2% yield increase in grain corn, and 32.9% yield increase in silage corn was achieved via application of programmed irrigation practices [8] (Table 3).

Table 3. Advantages of scheduled irrigation over conventional irrigation [8].

Criteria	Sugar beet	Grain corn	Silage corn
Water saving (%)	**25.5**	**23.5**	**20.1**
Electrical saving (%)	23.2	23.9	16.4
Yield increase (%)	9.0	29.2	32.9
Applied water with scheduled irrigation (mm)	**918**	**839**	**754**
Applied water with conventional irrigation (mm)	**1234**	**1098**	**944**

5 Conclusion

Limited water resources in the Middle Anatolia Region have forced producers to use pressurized irrigation systems due to the high water productivity. Konya plain is one of the well-known agro-production centers in Türkiye in accordance of using pressurized irrigation systems and all innovative farming technologies. There is a gradual decrease in water level of irrigation wells due to over water extraction for irrigation of current crop patterns. It is urgent for water managers to make the new crop patterns based on the current available water resources of KCB. Otherwise it is inevitable to be witness a lot of sinkholes formations in somewhere of such basin. In that context, farming land sizes with low water consumption crops such wheat, barley and all legumes must be widened. Finally, deficit irrigation also could be good solution for better water economy in water poor ecologies. Water resources are backbone of the all nations and we must use them with great care in all sectors particularly at irrigation.

References

1. Desoine, D.: Distribution of water on earth (2022). https://Flexbooks.Ck12.Org/Cbook/Ck-12-Middle-School-Earth-Science-Flexbook-2.0/Section/8.1/Primary/Lesson/Distribution-Of Water-On-Earth-Ms-Es/-access. Accessed 27 Aug 2024
2. DSI: Faaliyet Raporu, Toprak ve Su Kaynakları (2021). https://cdniys.tarimorman.gov.tr/api/File/GetFile/425/Sayfa/759/1107/DosyaGaleri/2021_yili_faaliyet_raporu.pdf#page=42-erişim-tarihi-access. Accessed 19 Sept 2022. (In Turkish)
3. DSI: Toprak Su Kaynakları (2022). https://www.dsi.gov.tr/Sayfa/Detay/754-erişim-tarihi-access. Accessed 19 Sept 2022. (In Turkish)
4. https://www.tarimorman.gov.tr/SYGM/Sayfalar/AlbumDetay.aspx?OgeId=974-access. Accessed 4 Sept 2024
5. https://kriterdergi.com/ekonomi/tarimsal-kuraklik-ongoruler-onlemler-oneriler-access. Accessed 28 Aug 2024. (In Turkish)

6. https://etelgraf.com/tarim-bakani-alti-yil-sonra-turkiyenin-yarisi-su-kitligi-cekecek/-access. Accessed 27 Aug 2024. (In Turkish)
7. Climate Change: Impacts and Vulnerability in Europe 2016—An Indicator-Based Report, p. 334 (2017)
8. Süheri, S., Yavuz, D.: Konya Kapalı Havzasında Programlı ve Etkin Damla Sulama İçin Rehber. Sürdürülebilir Arazi Yönetimi ve İklim Dostu Tarım Projesi. Food and Agriculture Organization of the United Nations, T.C. Tarım ve Orman Bakanlığı, Ankara, 25 s (2021). (In Turkish)
9. https://konya.tarimorman.gov.tr/Belgeler/liflet/kitaptakepsons1.pdf-access. Accessed 28 Aug 2024. (In Turkish)
10. https://www.yenimeram.com.tr/konya-ovasi-nda-kuraklik-alarmi-arpada-yuzde-30-bugdayda-yuzde-20-verim-kaybi-yasanabilir-541995.htm-access. Accessed 28 Aug 2024. (In Turkish)
11. Yavuz, D., Yavuz, N., Seymen, M., Türkmen, Ö.: Evapotranspiration, crop coefficient and seed yield of drip irrigated pumpkin under semi-arid conditions. Sci. Hortic. **197**, 33–40 (2015)
12. Nayır, M., Somuncu, M.: Ankara İlinde Tarımsal Su Yönetimi. In: International Geography Symposium on the 30th Anniversary of TUCAUM, 36 October, Ankara, pp. 772–791 (2018) (In Turkish)
13. Acar, B., Direk, M., Yurteri, Y.D., Sayman, O., Muradi, S.M.: Field water management for saving water in water-starved environments such as Konya plain, Türkiye. Green Rep. **3**(8), 48–50 (2022)
14. Acar, B., Yurteri, Y.D.: Efficacy of single drip lateral design for two plant rows of grain maize in Konya-Kadınhanı-Kolukısa province, Turkey. Ann. Univ. Craiova Biol. Horticult. Food Prod. Process. Technol. Environ. Eng. **28**(64), 5–13 (2023)
15. Acar, B., Riaz, N., Gökçe, A., Dinçer, H., Ahmad, A.H.: Water application performance of sprinkler irrigation systems under different layout and operating conditions. In: 3rd International Conference on Sustainable Ecological Agriculture (ICSEA), March 6-8, Konya, Türkiye, pp. 27–35 (2024)
16. Montazar, A., Moridnejad, A.M.: Influence of wind and bed slope on water and soil moisture distribution in solid-set sprinkler systems. Irrig. Drain. **57**, 175–185 (2008)
17. Acar, B.: Efficacy of drip irrigation system for some vegetables: assessments of yield and water productivity. Int. J. Agric. Econ. Dev. **8**(2), 1–7 (2020)
18. Asres, L.A.: Alternative techniques of irrigation water management for improving crop water productivity. Rev. Agric. Sci. **11**, 36–53 (2023). https://doi.org/10.7831/ras.11.0_36
19. https://www.turkishagrinews.com/wheat-yield-falls-in-konya-plain/-access. Accessed 9 Sept 2024. (In Turkish)
20. https://selcuk.edu.tr/Birim/arastirma-ve-uygulama_merkezleri/tarimsal_uygulama_ve_arastirma_merkezi/2099/arazi-varligi/52154-access. Accessed 9 Sept 2024. (In Turkish)
21. Acar, B., Topak, R., Yavuz, D., Kalender, M.A.: Is drip irrigation technique sustainable solution in agriculture for semi-arid regions? A case study of Middle Anatolian Region, Turkey. Int. J. Agric. Econ. Dev. **2**(2), 1–8 (2014)

Climate Change and Cryosphere: Risks of Frozen Pathogens

Ahmed K. Shousha, Samah M. Ali, Duha A. Ghanem, and Doaa A. El-Emam(✉)

Environmental Sciences Department, Faculty of Science, Damietta University, New Damietta, Egypt
doaa_elemam17@yahoo.com

Abstract. Climate change is significantly impacting the cryosphere, which includes glaciers, ice caps, and permafrost, leading to the potential release of ancient pathogens. As global temperatures rise, these frozen environments are melting, releasing microorganisms that have been preserved for millennia. This phenomenon poses serious risks to human and animal health, as these pathogens may be unfamiliar to contemporary immune systems, potentially leading to new disease outbreaks. The melting of glaciers not only contributes to rising sea levels and alters ecosystems but also mobilizes pollutants and pathogens previously trapped in ice. The chapter discusses the mechanisms through which climate change facilitates the release of pathogens, including the accelerated melting of glaciers and thawing of permafrost. Historical cases, such as the anthrax outbreak in Siberia and the revival of ancient viruses, underscore the urgency of this issue. The environmental and economic impacts of these pathogens are profound, affecting biodiversity, ecosystem stability, and healthcare costs. Effective management strategies are essential to mitigate the risks posed by thawing pathogens. This includes robust monitoring systems, public health education, and international cooperation to address the transboundary nature of emerging diseases. The chapter emphasizes the need for sustainable practices to combat climate change, highlighting the importance of transitioning to renewable energy, preserving natural ecosystems, and fostering community awareness. Ultimately, addressing the challenges of climate change and its impact on the cryosphere is crucial for safeguarding public health and ensuring ecological resilience.

Keywords: Frozen Pathogens · Melting Ice · Greenhouse Gas Emissions · Public Health Risks · Ancient Pathogens · Climate Mitigation Strategies

1 Introduction

Climate change is an accelerating global phenomenon that presents significant challenges across various regions. A notable consequence of climate change is the melting of the cryosphere, which is intricately associated with the resurgence of ancient frozen pathogens. The cryosphere encompasses the frozen components of the Earth's surface, including polar ice caps, glaciers, permanent snow cover, and sea ice [1]. Throughout geological history, the Earth has undergone natural climate fluctuations, resulting in the

J. C. Casila et al. (Eds.): CCORE 2024, SPEES, pp. 111–135, 2025.
https://doi.org/10.1007/978-3-031-93887-0_13

formation and disappearance of extensive ice and snow coverage. During glacial periods, vast ice sheets extended across large portions of the Earth, impacting both polar and temperate zones by lowering temperatures. In contrast, interglacial periods have been characterized by significant ice melt and rising global temperatures [2]. The combustion of fossil fuels for electricity and heat generation is a primary driver of global greenhouse gas emissions. A substantial proportion of the world's electricity continues to be produced through the burning of coal, oil, and natural gas, releasing considerable amounts of carbon dioxide and nitrous oxide potent greenhouse gases that contribute to the greenhouse effect by trapping solar radiation. Currently, renewable energy sources, including wind and solar power, account for just over a quarter of global electricity generation. Unlike fossil fuels, these renewable sources emit minimal to no greenhouse gases or pollutants, underscoring their critical role in mitigating climate change ([3, 4]). This understanding of climate dynamics, energy production, and their interdependent effects is essential for formulating effective strategies to combat climate change and protect global health [5].

Manufacturing and industry produce emissions, mostly from burning fossil fuels to produce energy for making things like cement, iron, steel, electronics, plastics, clothes, and other goods. Mining and other industrial processes also release gases, as does the construction industry. Machines used in the manufacturing process often run on coal, oil, or gas; and some materials, like plastics, are made from chemicals sourced from fossil fuels. The manufacturing industry is one of the largest contributors to greenhouse gas emissions worldwide [6].

Cutting down forests to create farms or pastures, or for other reasons, causes emissions, since trees, when they are cut, release the carbon they have been storing. Each year approximately 12 million hectares of forest are destroyed. Since forests absorb carbon dioxide, destroying them also limits nature's ability to keep emissions out of the atmosphere. Deforestation, together with agriculture and other land use changes, is responsible for roughly a quarter of global greenhouse gas emissions [7].

Studying climate change is important because it has significant implications for the world we live in and for future generations. Climate change can have significant impacts on ecosystems, including changes in the timing of seasonal events, shifts in species distributions, and increases in the frequency and intensity of extreme weather events such as hurricanes, floods, and wildfires. These impacts can disrupt the delicate balance of ecosystems, leading to the loss of biodiversity and the potential collapse of entire ecosystems. Additionally, climate change can also have severe consequences for human societies, including threats to food security, water resources, and public health. Therefore, understanding and addressing climate change is crucial for ensuring a sustainable and resilient future for both nature and humanity [8, 9].

The cryosphere, which includes ice caps, glaciers, and permafrost, is particularly vulnerable to the effects of climate change. Rising temperatures can cause the melting of ice and thawing of permafrost, releasing long-frozen pathogens into the environment. This poses a significant risk to human and animal health as these pathogens may be unfamiliar to our immune systems and could potentially lead to new disease outbreaks. Therefore, studying the impacts of climate change on the cryosphere and developing strategies to mitigate these risks is essential for safeguarding public health [10].

The aim of this chapter is to examine the impacts of climate change on the cryosphere, which includes frozen regions such as glaciers, ice caps, and permafrost. These frozen areas act as natural storage for various pathogens and microorganisms that have been trapped in the ice for centuries. As global temperatures rise, the melting of these frozen regions could potentially release these pathogens into the environment, posing a significant threat to human and animal health. By understanding this potential risk and developing strategies to mitigate it, we can better prepare for the challenges posed by climate change on both ecological and public health fronts.

2 Cryosphere

The cryosphere refers to the portions of the Earth where water is permanently frozen, predominantly in the form of glaciers and ice sheets. This term is closely associated with the hydrosphere but specifically signifies poly-extreme environments characterized by extreme low temperatures, elevated ultraviolet (UV) radiation levels, oligotrophic conditions, and reduced water activity [11]. The cryosphere includes a diverse range of frozen water forms, such as sea ice, lake ice, river ice, ice caps, ice sheets, glaciers, permafrost, and seasonal snow. The latest report from the Intergovernmental Panel on Climate Change (IPCC) indicates that the cryosphere is undergoing significant, widespread, and pervasive transformations. These changes can be observed across various landscapes, from high mountain ranges to polar regions, coastal areas, and even deep oceanic environments. Recent research by Slater et al. (2021) has combined satellite observations with numerical modeling to quantify global ice loss, revealing approximately 28 trillion tons of ice loss between 1994 and 2017. This trend is further underscored by recent significant climatic events observed worldwide, which corroborate the ongoing changes in the cryosphere [12]. The implications of such extensive ice loss for human societies are profound, necessitating adaptations to emerging and unforeseen challenges. Accelerated cryosphere melting might have far-reaching effects on ecosystems, sea levels, and weather patterns around the world, in addition to the hydrological impacts already mentioned [13]. To create well-informed policies and plans to lessen the impact of climate change on ecosystems and humans alike, it is essential to have a firm grasp of these dynamics.

3 Impact of Climate Change on Cryosphere

Please note that an indent, either. The cryosphere, which includes ice caps, glaciers, and permafrost, is particularly vulnerable to the impacts of climate change (see Fig. 1). Rising temperatures are causing widespread melting of glaciers and ice caps, leading to rising sea levels and increased risk of coastal flooding. Additionally, the thawing of permafrost releases large amounts of greenhouse gases into the atmosphere, further exacerbating climate change. These changes in the cryosphere not only have significant implications for global climate patterns but also pose threats to communities and ecosystems that rely on these frozen landscapes for their livelihoods and survival [14].

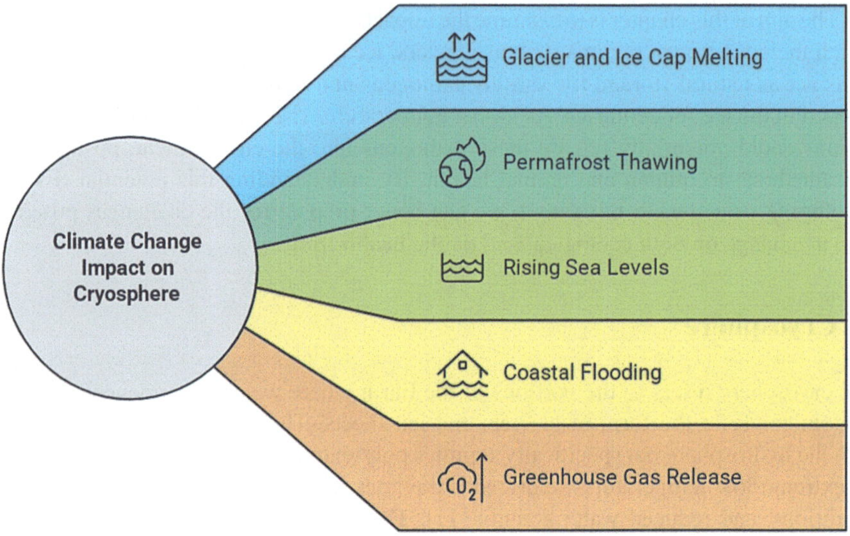

Fig. 1. Impact of climate change on cryosphere.

3.1 Melting Glacier and Ice Sheets

Antarctica and Greenland are undergoing significant ice mass loss, with Antarctica melting at approximately 150 billion tons per year and Greenland at about 270 billion tons annually. This melting is a major contributor to global sea level rise and has been monitored since 2002 by NASA's GRACE and GRACE Follow-On satellites (see Fig. 2). Collectively, the ice sheets of these regions store around two-thirds of the Earth's fresh water, and their loss is largely driven by rising temperatures of the Earth's surface and oceans, accounting for approximately one-third of the global average sea level rise since 1993. The current rapid retreat of glaciers and ice sheets represents one of the most evident signs of climate change. Many glaciers are expected to significantly shrink over the coming decades, with low-elevation glaciers potentially disappearing altogether within this century. This phenomenon is prevalent across the globe, notably in areas such as the Alps, Greenland, and the Central and Southern Andes. In the Alps, for example, glaciers below 3,000 m are experiencing the most significant reductions in area. Historically, glacier retreat and expansion have varied, with several periods documented throughout Earth's history. However, present-day glacier retreat is markedly accelerated due to anthropogenic factors, with rates higher than those seen in the past [15]. The notable glacier retreats in the Alps since the mid-nineteenth century is attributed to increased black carbon emissions due to industrialization in Western Europe. Consequently, the melting glaciers and ice sheets lead to long-term loss of natural freshwater storage, which constitutes approximately 70% of the Earth's freshwater. This alteration of the hydrological cycle poses strong societal impacts globally. Moreover, the current melting has facilitated the mobilization and transport of various inorganic and organic molecules, as well as pollutants. For instance, melting rock glaciers in the Alps have notably raised nickel and other metal concentrations in lakes, surpassing drinking water standards, with unknown biological effects [16]. Similar impacts have been observed in the Cordillera Blanca in Peru, where high levels of lead and nickel have impaired water quality. Additionally, glacier retreat exposes soils, plant detritus, and nutrient resources previously locked under ice, which are essential for the succession of biota in both terrestrial and aquatic ecosystems. In glacier forelands, soil microbial communities heavily rely on ancient carbon sources until primary succession occurs. The decline of glaciers is also impacting river and stream biodiversity, with more studies available compared to lakes [17]. An important consequence of glacier melting is the formation of new proglacial and ice-contact lakes, which can be unstable and prone to dangerous outburst floods. However, the increase in the number and area of these glacial lakes is a significant development [18].

3.2 Consequence of Melting

In the recent years, sea level rise induced by global warming and its impacts on coastal zones has become a question of growing interest in the scientific community, as well as the media and public. It is now well established that the Earth's climate is warming and that the main cause is the accumulation of greenhouse gases (GHGs) inside the atmosphere, produced by anthropogenic fossil fuel combustion and change in land use (mostly deforestation) Global warming has already given rise to several visible consequences, in particular increase of the Earth's mean surface temperature and of ocean heat content Hobbs and Willis, melting of sea ice and glaciers, and loss of ice mass from the Greenland and Antarctica ice sheets [20].

Global sea levels are rising at unprecedented rates due to human-induced climate change, a phenomenon not observed in over 2,500 years. This rise is primarily driven by two interconnected factors: the melting of ice sheets and glaciers, which contributes additional water to the oceans, and the thermal expansion of seawater as it warms. Since the mid-to-late nineteenth century, direct sea level observations from in situ tide gauges have provided valuable data, while satellite observations since 1993 have detailed changes in global sea level. Additionally, coastal tide gauge and satellite data from 1900 to 2018 reveal overall trends, showcasing strong regional variations in sea level changes (see Fig. 3). The factors influencing sea level change are classified into those that lead to increases (plus signs) and those that result in decreases (minus signs), with the impact of each factor plotted over time [21]. Modeling future climate scenarios suggests that sea levels will continue to rise over the coming decades and centuries [20]. This rise poses significant threats to coastal areas, where around 10% of the global population resides at elevations less than 10 m above sea level. Observations from the twentieth century indicate shoreline erosion in various regions; however, the extent to which this erosion is attributable to climate-related sea level rise versus local factors such as ground subsidence, coastal management practices, land use changes, and sediment supply deficiencies remains uncertain. Despite these uncertainties, it is virtually certain that the anticipated acceleration of sea level rise due to ongoing global warming will heighten the vulnerability of many low-lying, densely populated coastal regions and pose substantial threats to a significant portion of the human population in the near future [22, 23].

(a)

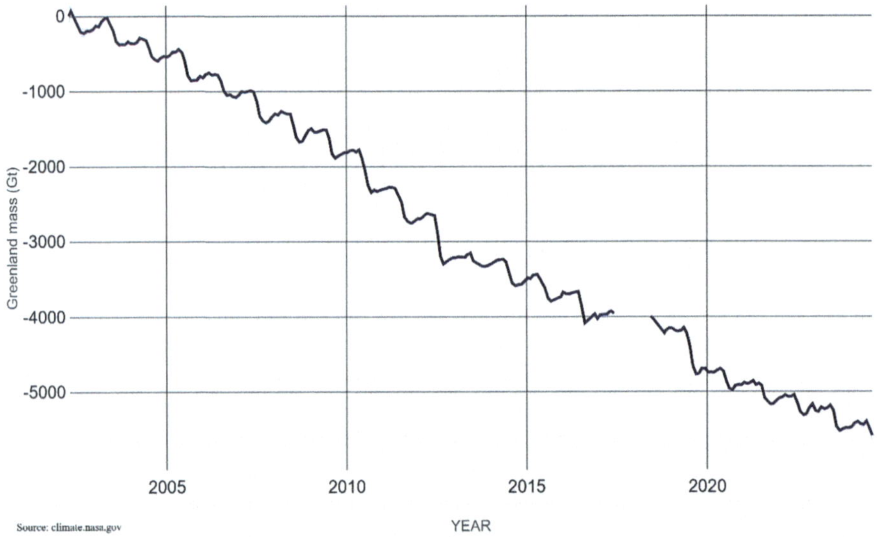

(b)

Fig. 2. Ice mass measurement by NASA's GRACE satellites since 2002 till 2024, (A: Antarctica mass variation; B: Greenland mass variation) [19].

3.3 Impact on Coastal Communities and Ecosystem

Coastal marine systems are among the most ecologically and socio-economically vital on the planet. Marine habitats from the intertidal zone out to the continenta shelf break

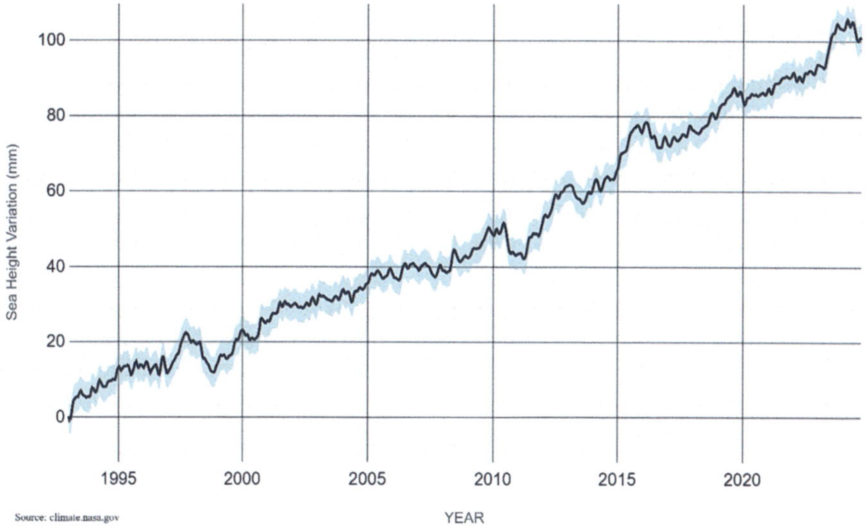

(a)

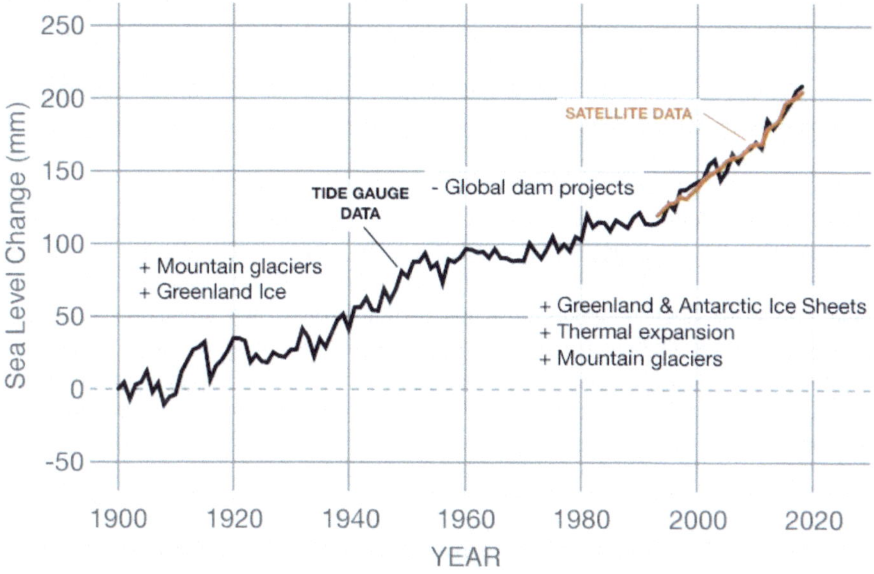

(b)

Fig. 3. Global sea levels satellite observations (1993–2024), (a) The first graph tracks the change in global sea level since 1993, as observed by satellites. (b) The second graph shows how much sea level changed from approximately 1900 to 2018, utilizing data from coastal tide gauges and satellite measurements [21, 24].

are estimated to provide over US$14 trillion worth of ecosystem goods (e.g. food and raw materials) and services (e.g. disturbance regulation and nutrient cycling) per year, or c. 43% of the global total. However, there is a strong scientific consensus that coastal marine ecosystems, along with the goods and services they provide, are threatened by anthropogenic global climate change. Recent climatic trends, which are only a fraction of the magnitude of predicted changes in the coming centuries, have already triggered significant responses in the Earth's biota. As these changes continue, we risk serious degradation of marine ecosystems, with far-reaching consequences for human health and welfare [22, 25].

Given their global importance, coastal marine environments are a major focus of concern regarding the potential impacts of anthropogenic climate change. More importantly, temperature is only one of a suite of potentially interacting climatic variables that will drive future ecological change in marine systems. Finally, studies conducted on population and community level processes suggest that climatic impacts on individual organisms do not necessarily translate directly into changes in distribution and abundance [5, 23].

4 Frozen Pathogens

Prehistoric organisms are living and growing today. We can revive them or they can spontaneously revive by melting out of their icy tombs. Mix 99 parts water with 1part microbes, stir and freeze. What you get is frozen microbes that will be alive and well long after you, your children, your grandchildren, and most of your descendants are gone. Some of the microbes may live on (if kept frozen) for millions of years. Some of the life on Earth has lived through a multitude of catastrophic events on Earth. Events that would have killed most, if not all, of the humans on Earth [26].

While ice contains a valuable record, it also encases microbes that are beneficial (e.g., those that recycle nutrients, those that fix carbon from carbon dioxide, those that help plants to soak up water and nutrients, those that produce antibiotics, and a multitude of other vital processes), as well as those that are dangerous. There are many pathogenic microbes encased in environmental ice around the world. All are released as glaciers melt. With the increases in melting of ice worldwide, more of these organisms are being released. The potential for disease outbreaks, epidemics, and pandemics is real, as is the loss of beneficial microbes. Disease outbreaks have likely occurred before from glacial melting, but it is unknown when and how frequently [1].

5 Pathogens Emergence Mechanism from the Cryosphere

Please Humans have used ice for centuries to preserve foods, but it has been a natural preservative of life for millions of years. Much evidence has accumulated over the past 200 years, primarily in the past 30 years, indicating that many microbes, as well as some multicellular organ-isms, can be frozen and then revived when the ice melts decades, centuries, or millennia later. Yearly, millions of tons of organisms are entrapped in ice, and currently many more are melting out of the ice due to global climate change, and the warming that it has caused. In the polar regions, much of the ice remains frozen, sometimes for millions of years. It continually builds up millennium after millennium, eventually melting or subliming, thus releasing its cargo of viable microorganisms [27].

Climate change can have significant impacts on the cryosphere, which includes frozen areas such as glaciers, permafrost, and ice caps. These frozen environments serve as reservoirs for various pathogens, including bacteria and viruses, that have been trapped and preserved for long periods. However, as the Earth's temperature rises due to global warming, the cryosphere undergoes melting and thawing processes, leading to the release of these ancient pathogens. There are several mechanisms through which climate change can facilitate the release of pathogens from the cryosphere (see Fig. 4):

- **Melting glaciers and ice caps:** As global temperatures increase, glaciers and ice caps melt at an accelerated rate. These frozen bodies of water can contain ancient microbial communities and viruses that have been dormant for centuries. When the ice melts, these pathogens are released into the surrounding environment, including water bodies and soil [28].
- **Thawing permafrost:** Permafrost is a permanently frozen layer of soil found in polar regions and high-altitude areas. It contains a vast amount of organic matter and microbial life that has been frozen for thousands of years. With climate change, permafrost thaws, causing the release of organic material and the activation of microbial communities. This includes potentially harmful bacteria and viruses that can pose a threat to human and animal health [29].
- **Disruption of ecosystems:** Climate change disrupts ecosystems and ecological balances. As temperatures rise, the distribution and abundance of plant and animal species change. This can lead to altered interactions between hosts, vectors, and pathogens, potentially favoring the spread and emergence of new diseases. Additionally, changes in precipitation patterns and water availability can create new breeding grounds for disease-carrying organisms [30].
- **Increased human exposure:** As the cryosphere thaws, previously isolated areas become more accessible to human populations. People living in or visiting these regions may come into contact with thawed pathogens, increasing the risk of disease transmission. This is particularly relevant for indigenous communities and Arctic populations who have traditional lifestyles closely connected to the cryosphere.

The release of pathogens from the cryosphere due to climate change raises concerns about the potential emergence and reemergence of infectious diseases. These diseases may affect both human and animal populations, and their spread can be facilitated by global travel and trade networks. Understanding and monitoring these processes are crucial for early detection, surveillance, and effective public health responses to mitigate the risks associated with the release of pathogens from the cryosphere [31]. When the permafrost thaws, it can release microorganisms that have been dormant for thousands or even millions of years. These microorganisms may include bacteria, viruses, and other infectious agents that were present in the environment at the time of freezing. As the permafrost melts, these pathogens can become reactivated and potentially pose a threat to human and animal health.

If we consider that the average density of microorganisms in glacier ice varies between 10^2 and 10^7 cells/mL, approximately 4×10^{21} microbes are released each year, due to the melting of non-polar ice. From an epidemiological point of view this is

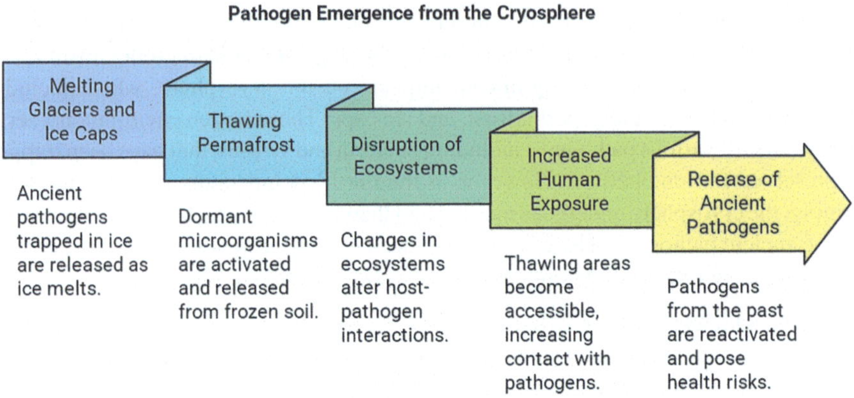

Fig. 4. Dynamics of Pathogen Release from the Cryosphere.

disturbing, since many of these microbes closely related to human pathogens are considered "ancient" forms of life that became extinct from the surface of the Earth thousands and even millions of years ago [32].

The survival and viability of ancient pathogens released from permafrost represent a complex interplay of molecular, physiological, and environmental factors that remain poorly understood. The preservation of microorganisms in a frozen state involves sophisticated cellular mechanisms that protect against the formation of damaging ice crystals, including the production of cryoprotectants and the regulation of membrane fluidity. However, the long-term effectiveness of these protective mechanisms over geological time scales remains uncertain. DNA damage accumulates even in frozen states due to background radiation and chemical degradation, raising questions about the capacity of ancient organisms to repair this damage upon thawing. Research has shown that some microorganisms possess remarkable DNA repair capabilities that activate upon reanimation, but the prevalence and efficiency of these mechanisms across different taxonomic groups are not well characterized [33].

The role of environmental conditions in determining post-thaw viability presents another layer of complexity. Temperature fluctuations during the thawing process may critically influence survival rates, yet the precise thermal parameters that optimize viability remain unclear. The availability of nutrients in the modern environment poses another significant unknown, as ancient microorganisms must adapt to potentially different nutrient profiles than those present during their initial freezing. Furthermore, the pH tolerance ranges of ancient pathogens may have evolved differently from their modern counterparts, affecting their ability to survive in contemporary ecosystems [34].

Cellular stress responses during the transition from frozen to thawed states represent a particularly challenging area of study. While some microorganisms enter specialized dormant states that enhance their survival prospects, the molecular mechanisms underlying successful reanimation remain poorly understood. The potential for viable but non-culturable states further complicates assessment of true survival rates, as traditional cultivation methods may underestimate the number of potentially viable organisms. Additionally, the role of sporulation and other resistant forms in long-term survival, while documented in some species, requires further investigation across a broader range of organisms [35].

The interaction between physical and chemical factors in the permafrost environment adds another dimension to viability uncertainties. The presence of minerals, organic compounds, and salt concentrations may provide protective effects during long-term freezing, but the extent of this protection and its mechanisms remain largely theoretical. Furthermore, the impact of freeze-thaw cycles, which may become more frequent with climate change, on the viability of these organisms represents a significant knowledge gap in our understanding of their survival potential [36].

6 Examples of Frozen Pathogens That Have Been Released

There have been a few documented cases of frozen pathogens being released due to thawing ice and permafrost. Here are a few notable examples:

Anthrax in Siberia: In 2016, an outbreak of anthrax occurred in Yamal Peninsula, Siberia. It was traced back to the thawing permafrost, which exposed the carcass of an infected reindeer that had died from anthrax approximately 75 years ago. The thawing released dormant spores of Bacillus anthracis, the bacteria responsible for anthrax, leading to the infection of both animals and humans [34].

Spanish Flu Virus: In 2005, scientists successfully extracted and sequenced the genetic material of the Spanish flu virus (H1N1 strain) from the lung tissue of a victim buried in permafrost in Alaska. The virus, which caused a devastating pandemic in 1918–1919, was preserved in the frozen environment and provided valuable insights into the virus's genetic makeup and evolution [37].

Ancient Viruses in Siberian Permafrost: Researchers have discovered ancient viruses preserved in Siberian permafrost. In 2014, scientists isolated and revived a 30,000-year-old giant virus called Pithovirus sibericum from the permafrost. This discovery raised concerns about the potential reactivation of other ancient viruses as the permafrost melts [38].

Microbes in Antarctic Ice: Antarctic ice cores have revealed the presence of various microbial species. In 2019, researchers reported the identification of unknown and potentially novel species of bacteria and fungi from a 15,000-year-old Antarctic ice core. These microbes have remained dormant in the ice and provide insights into microbial life in extreme environments.

It's important to note that the release of frozen pathogens is still a relatively rare occurrence, and the actual risk of their reactivation and transmission to humans is still being studied. These examples highlight the potential consequences of thawing ice and permafrost in the context of climate change and the need for ongoing research and monitoring [39, 40].

7 Environmental and Economic Impacts of Frozen Pathogens

The release of frozen pathogens, particularly those preserved in permafrost regions, presents complex environmental and economic challenges as global temperatures continue to rise. From an environmental perspective, thawing permafrost may release previously dormant microorganisms, including bacteria and viruses that have been trapped in ice for thousands or millions of years, potentially disrupting existing ecosystems and biodiversity. Contemporary organisms lack evolved defenses against these ancient pathogens, making them particularly vulnerable to new threats. Additionally, the release of these microorganisms could affect soil microbiota and biogeochemical cycles, potentially altering nutrient cycling and ecosystem functioning. Economically, the impacts span multiple sectors, requiring substantial investments in containment facilities, safety protocols, and advanced detection technologies. While studying these organisms could yield valuable insights for developing new antimicrobial compounds or understanding pathogen evolution, their potential reemergence poses significant risks to public health systems and economic productivity in affected regions. Arctic communities face particular vulnerability, as permafrost thaw may expose populations to novel health risks while simultaneously damaging infrastructure through ground destabilization. Addressing these challenges necessitates a comprehensive approach involving stringent regulations, proper waste management systems, heightened public awareness, and advanced monitoring technologies to detect and prevent pathogen spread. The implementation of these measures requires significant resource allocation but is essential for protecting both environmental and economic interests in affected regions [41].

7.1 Environmental Impacts

The environmental ramifications of frozen pathogens extend far beyond initial contamination events, creating complex cascading effects throughout ecosystems. When these pathogens are introduced into new environments, either through natural processes or human activities, they can trigger a series of interconnected environmental impacts that affect multiple ecological levels [42].

7.1.1 Ecosystem Disruption

The introduction of frozen pathogens into ecosystems represents a significant biological disturbance that can destabilize established ecological relationships [43]. These disruptions manifest through:

- Primary Effects: Direct infection of susceptible species, leading to immediate health impacts and population stress
- Secondary Effects: Alterations in species interactions and community dynamics
- Tertiary Effects: Long-term changes in ecosystem structure and function, potentially leading to regime shifts in affected areas

7.1.2 Resource Contamination

The contamination of environmental resources by frozen pathogens presents multiple challenges:

- Soil Contamination: Pathogens can persist in soil matrices, affecting soil microbiota and plant health
- Water System Impacts: Contamination of aquatic systems can affect both freshwater and marine ecosystems

- Agricultural Impact: Contaminated resources can compromise agricultural productivity and food safety
- Groundwater Concerns: Long-term persistence in groundwater systems can create ongoing contamination issues

7.1.3 Economic Costs

The economic implications of frozen pathogen outbreaks are extensive and multifaceted [44]:

- Direct Costs: Immediate expenses related to containment and treatment
- Indirect Costs: Loss of productivity in affected industries
- Long-term Economic Impact: Reduced property values in affected areas
- Industry-specific Impacts: Particularly severe effects on food production and processing sectors
- Healthcare System Burden: Increased strain on medical facilities and resources
- Research and Development Costs: Investment needed for prevention and treatment strategies

7.1.4 Waste Management Challenges

Disposal of contaminated goods poses challenges in waste management, requiring special protocols to prevent further contamination [45, 46]:

- Enhanced Safety Measures: Implementation of strict handling procedures
- Specialized Equipment: Need for dedicated containment and disposal facilities
- Training Requirements: Specialized staff training for handling contaminated materials
- Monitoring Systems: Continuous surveillance of disposal sites
- Emergency Response Protocols: Development of rapid response procedures for containment breaches

7.2 Disruption of Ecosystem

An ecosystem is disrupted by significant changes, such as invasive species, pollution, habitat destruction, climate change, or diseases, which disrupt the natural balance, structure, or functioning of an environment. This can lead to imbalances in species populations, food chain changes, biodiversity alterations, and even ecological collapse, affecting the health and stability of the ecosystem and potentially causing long-term negative environmental impacts. Especially The disruption of ecosystems due to frozen pathogens primarily occurs when these pathogens are introduced into new environments by different approaches [47].

7.2.1 Species Vulnerability

Certain pathogens possess the capacity to infect local species that lack natural defenses, resulting in illness and mortality within those populations. This, in turn, disrupts the ecological balance of the habitats they inhabit. A prominent example is the Chytrid fungus (Batrachochytrium dendrobatidis), which specifically targets amphibians. When this pathogen is introduced to new environments, it can lead to severe diseases and significant population declines among amphibian species that have not co-evolved with it, thereby destabilizing ecosystems where these species serve as keystone members [48].

The effects on vulnerable species warrant particular attention:

- **Amphibian Example:** The case of the Chytrid fungus illustrates how pathogens can inflict devastating impacts on populations that lack evolutionary defenses.
- **Transmission Dynamics:** Understanding the patterns of pathogen spread in naive populations is crucial for assessing potential risks.
- **Population Recovery Challenges**: Restoring affected species can be exceptionally difficult, often requiring extensive conservation efforts.
- **Genetic Diversity Impact:** The introduction of pathogens may lead to a loss of genetic variation within affected populations, further compromising their resilience.

7.2.2 Altered Food Chains

Diseases caused by frozen pathogens can affect species crucial for the food chain. For instance, if a pathogen impacts a species that is a primary food source for others, it can trigger imbalances and potentially harm other species dependent on it. Like Chronic Wasting Disease (CWD) in deer, caused by prions, can disrupt food chains. If CWD spreads to new areas, it could potentially affect populations of deer and related species, altering predator-prey dynamics and impacting the overall balance of the ecosystem [49]. The modification of food chain dynamics can occur through:

- Primary Producer Effects: Impacts on foundational species
- Consumer Level Changes: Alterations in predator-prey relationships
- Chronic Wasting Disease Example: Demonstrates how pathogens can affect multiple trophic levels
- Energy Flow Disruption: Changes in nutrient cycling and energy transfer patterns

7.2.3 Biodiversity Impact

In severe cases, widespread disease outbreaks caused by frozen pathogens can lead to significant declines in biodiversity within affected ecosystems. This decline may occur when a disease disproportionately impacts certain species, resulting in reduced population numbers. A pertinent example is White-Nose Syndrome, which is caused by the fungus Pseudogymnoascus destructans and has led to alarming declines in several bat species across North America. The reduction of insectivorous bat populations due to this disease can have profound effects on insect populations and disrupt ecosystem balance [50]. The implications for biodiversity are extensive and can be categorized as follows:

- Population-Level Effects: Direct impacts on species abundance and distribution.
- Community-Level Changes: Alterations in species interactions and overall community composition.
- Cascading Effects of White-Nose Syndrome: This example highlights the interconnectedness of species and how declines can ripple through ecosystems.
- Recovery Potential: An assessment of ecosystem resilience and the possibilities for restoration following species declines.

7.2.4 Ecosystem Functioning

Pathogen targets a species that plays a critical role in ecosystem functioning such as a key pollinator or predator it can compromise the overall health and stability of the ecosystem. This disruption often results in cascading effects on other species and ecological processes. A notable example is Sudden Oak Death, caused by the oomycete Phytophthora ramorum, which affects several tree species, particularly oaks. Upon introduction into new environments, this pathogen can significantly alter tree populations, modify forest structure, and impact the myriad species that rely on these trees for habitat and food [47].

The implications for ecosystem functioning include:

- **Structural Changes:** Alterations in physical habitat structure that influence species composition and interactions.
- **Functional Modifications:** Changes in key ecosystem processes and services, such as nutrient cycling and carbon storage.
- **Ecological Cascade:** The loss of keystone species, as illustrated by Sudden Oak Death, can lead to broader shifts across entire ecosystems.
- **Long-term Consequences:** Persistent alterations in ecosystem dynamics, potentially affecting stability and resilience.
- **Restoration Challenges:** Complexities involved in restoring affected ecosystems to their original state, given the intricacies of ecological interdependence.

7.3 Increased Healthcare Costs

Addressing the impact of increased healthcare costs due to frozen pathogens necessitates a balanced approach involving early detection, efficient management of outbreaks, investment in research, and public health interventions to minimize the burden on healthcare systems [51]. The increased healthcare costs due to frozen pathogens stem from various factors:

7.3.1 Outbreak Management

When frozen pathogens cause outbreaks of diseases, healthcare systems incur higher costs due to increased hospitalizations, treatments, and the need for medical supplies to manage the affected individuals. Influenza viruses, especially novel strains that can arise from frozen reservoirs, often lead to seasonal outbreaks. These outbreaks increase hospital visits, the need for antiviral medications, and sometimes ICU admissions, thereby escalating healthcare costs [52].

7.3.2 Diagnostic Procedures

Identifying and diagnosing illnesses caused by these pathogens often require specialized tests and diagnostic procedures, which can be expensive and resource-intensive for healthcare providers. For example, Hepatitis A virus (HAV) can be transmitted through contaminated frozen foods. Diagnosis often involves specific blood tests to detect antibodies or viral RNA, which can be costly, especially when outbreaks require widespread testing of potentially exposed individuals [52].

7.3.3 Research and Development

Healthcare institutions may invest significantly in research and development to understand these pathogens better, develop effective treatments, and prevent further spread. These costs contribute to the overall healthcare expenditure. Human Immunodeficiency Virus (HIV) has been known to persist in frozen tissues for some time. Research into treatments, vaccines, and preventive measures against HIV remains a substantial ongoing investment in healthcare due to its persistence and impact.

7.3.4 Public Health Programs

Public health initiatives aimed at mitigating the risks associated with frozen pathogens are critical for effective disease prevention and control. These programs require significant infrastructure and sustained financial investment to manage the complexities of pathogen surveillance and response [53].

Key components of these initiatives include comprehensive surveillance systems that facilitate early detection of pathogens through active monitoring, laboratory networks for rapid diagnostics, and integrated data management systems to track outbreak trends. Additionally, cross-border coordination and environmental monitoring are essential for proactive risk management. Preventive measures are vital, encompassing vaccination programs, standardized risk assessments, rigorous quality control for frozen products, and cold chain monitoring to ensure product integrity. Training programs for personnel handling frozen materials further enhance safety compliance. Public education initiatives are crucial for raising awareness about frozen pathogen risks, targeting community outreach, professional training, and effective risk communication strategies. Economic implications of these programs include direct costs, infrastructure investments, personnel training, emergency response capabilities, and ongoing research funding.

A case study on norovirus illustrates the importance of rapid detection systems, containment measures, and healthcare preparedness in managing outbreaks. Long-term sustainability of these public health programs hinges on resource allocation, program evaluation, adaptive management, international cooperation, and capacity building. Overall, comprehensive public health strategies addressing frozen pathogens represent a necessary investment in public health infrastructure, yielding significant societal benefits through disease prevention and outbreak control. Continuous evaluation and adaptation are essential to meet emerging challenges in pathogen management.

8 Potential Risks and Challenges Posed by Released Pathogens

Preservation and Reactivation: Pathogens preserved in frozen environments might remain viable for extended periods. This poses a risk if these pathogens can be reactivated or become infectious again when conditions are favorable.
Climate Change Impact: As global temperatures rise due to climate change, previously frozen regions may thaw, potentially releasing dormant pathogens. This could lead to the emergence of infectious diseases in new geographic areas.

Transportation and Trade: The global movement of frozen goods, including food and biological materials, increases the risk of spreading pathogens across borders. This is particularly relevant in today's interconnected world where goods are transported internationally.

Unknown Pathogens: There may be undiscovered or poorly understood pathogens preserved in frozen environments. As these environments change, there is a risk of uncovering new pathogens that may pose threats to human, animal, or plant health.

Ecological Impact: Changes in frozen environments can disrupt ecosystems and impact wildlife. This disruption may influence the transmission of diseases among animals, and potentially to humans, through zoonotic spillover events.

Antibiotic Resistance: Pathogens preserved in frozen environments may carry resistance genes. If these genes are activated, it could contribute to the spread of antibiotic resistance, making infections more challenging to treat.

Infrastructure and Preparedness: Many regions may not be adequately prepared to deal with the consequences of thawing frozen environments, including the potential release of pathogens. Lack of infrastructure, surveillance systems, and public health measures could exacerbate the risks.

Potential for Bioterrorism: The intentional release of frozen pathogens as a bioweapon is a major concern. The ease of storage and transportation of these pathogens makes them potentially attractive for malicious use, emphasizing the need for robust security measures and international cooperation to prevent bioterrorism [1].

Addressing the risks and challenges posed by the spread of frozen pathogens requires a multidisciplinary approach involving scientists, policymakers, public health officials, and international organizations. Implementing stringent biosecurity measures, enhancing surveillance and detection systems, and promoting global collaboration are crucial steps in mitigating the potential threats associated with frozen pathogens [54, 55].

9 Challenges in Managing and Control

9.1 Managing and Controlling Diseases Caused by Frozen Pathogens

The management and control of diseases caused by frozen pathogens present distinct challenges owing to their capacity to remain viable in cold conditions. Ensuring adherence to proper storage, transportation, and handling protocols for frozen goods is essential to prevent the proliferation of these pathogens. Additionally, fostering awareness regarding the risks associated with frozen pathogens and implementing stringent hygiene measures can play a critical role in mitigating their transmission [56].

9.1.1 Storage and Transportation

Maintaining the integrity of cold chains for frozen goods is essential. Temperature fluctuations during storage or transportation can compromise the viability of pathogens, thereby necessitating strict temperature control measures throughout the supply chain to ensure safety and quality [49].

9.1.2 Pathogen Viability

Frozen pathogens exhibit remarkable resilience, allowing them to survive for prolonged periods, which complicates traditional sanitation methods. This inherent longevity heightens the risk of contamination if proper handling procedures are not meticulously followed, underscoring the necessity for rigorous adherence to established protocols [57].

9.1.3 Awareness and Hygiene

Educating individuals about the risks associated with handling and consuming frozen foods contaminated with pathogens is vital. Emphasizing proper hygiene practices during food preparation and handling helps reduce the transmission of these pathogens.

Addressing each of previous challenges comprehensively is essential to effectively manage and control the spread of diseases caused by frozen pathogens. Managing and controlling Also involves a combination of preventive measures, surveillance, research, and international collaboration [58].

9.2 Monitoring and Surveillance

Implement robust monitoring and surveillance systems to track changes in frozen environments and detect the emergence of potential pathogens.

Conduct regular health assessments in regions prone to permafrost thaw, especially in areas with historical evidence of infectious diseases. Regular monitoring of frozen food products for pathogen presence and implementing robust regulations are essential. This includes testing protocols, labeling requirements, and strict adherence to international food safety standards [56].

9.3 Research and Early Detection

Research into advanced detection methods specific to frozen pathogens can significantly improve our ability to identify and mitigate risks associated with these microorganisms. Invest in scientific research to understand the types of pathogens present in frozen environments and their potential for reemergence. Develop early detection methods for identifying pathogens as they are released, enabling timely response measures [59].

9.4 International Cooperation

Foster international collaboration and information-sharing to address the transboundary nature of diseases emerging from thawing permafrost and melting ice.

Establish global networks for monitoring and responding to potential outbreaks, involving collaboration between governments, researchers, and international health organizations [60].

9.5 Public Health Education

Educate communities in vulnerable regions about the potential health risks associated with the release of frozen pathogens.

Promote hygiene practices and provide information on disease prevention to reduce the risk of transmission.

9.6 Infrastructure and Healthcare Capacity Building

Strengthen healthcare infrastructure in regions at risk, ensuring that facilities are equipped to handle potential outbreaks.

Build healthcare capacity to respond effectively to emerging diseases, including training healthcare professionals and providing necessary resources [61].

9.7 Climate Change Mitigation

Address the root cause by implementing measures to mitigate climate change, as the thawing of permafrost is closely linked to global warming.

Support international efforts to reduce greenhouse gas emissions and promote sustainable practices to slow down permafrost thaw [4].

9.8 Vaccination and Treatment Development

Invest in research and development of vaccines and treatments for potential diseases that may emerge from thawed permafrost.

Establish stockpiles of vaccines and medical supplies to facilitate a rapid response to outbreaks [62].

9.8.1 One Health Approach

Adopt a One Health approach that recognizes the interconnectedness of human, animal, and environmental health.

Collaborate across disciplines, involving human and animal health experts, ecologists, and environmental scientists in disease prevention and control efforts.

9.8.2 Regulatory Measures

Implement and enforce regulations governing activities in vulnerable regions to minimize the risk of pathogen release.

Establish guidelines for responsible research in these areas to prevent accidental exposure and release of pathogens [63].

10 Successful Initiative Examples

The implementation of frozen pathogen mitigation strategies has seen several notable successes through established programs and regional initiatives, demonstrating the feasibility of comprehensive approaches when properly executed. The Arctic Council's initiatives stand as particularly compelling examples of successful international cooperation in addressing frozen pathogen risks. Through its Working Groups, the Council has established a robust framework for monitoring and responding to potential threats.

The Arctic Monitoring and Assessment Programme (AMAP) has particularly distinguished itself by developing and implementing standardized protocols for permafrost monitoring across multiple nations, creating a cohesive network of observation stations that provide real-time data on permafrost conditions. This network has proven invaluable in early detection of potential risk areas and has facilitated rapid response capabilities across international boundaries.

The Sustainable Development Working Group (SDWG) has further demonstrated the effectiveness of community-integrated approaches through its comprehensive health monitoring systems. By combining traditional indigenous knowledge with modern scientific methods, these systems have successfully engaged local communities in the monitoring process while ensuring scientific rigor. The program's success is evidenced by its high rate of community participation and the quality of data collected, which has directly contributed to improved understanding of permafrost-related health risks in Arctic regions. Similarly, the Arctic Contaminants Action Program (ACAP) has established effective waste management protocols that have been successfully implemented across remote Arctic regions, significantly reducing the risk of pathogen exposure through improved handling of potentially contaminated materials.

Regional initiatives have also demonstrated remarkable success in implementing practical solutions adapted to local conditions. Alaska's Permafrost Health Monitoring Network represents a pioneering approach to integrated surveillance, successfully combining traditional ecological knowledge with contemporary scientific monitoring methods. This program has created an effective early warning system that has already identified several potential risk areas and facilitated preventive measures before serious issues could develop. The network's success lies in its adaptive approach, which allows for modifications based on local conditions while maintaining consistent scientific standards across different regions.

Greenland's Early Warning System for Zoonotic Diseases has emerged as another noteworthy success story, demonstrating the effectiveness of integrated surveillance approaches. The system has successfully combined environmental monitoring with health surveillance, creating a comprehensive framework for detecting potential disease emergence from thawing permafrost. Its success is particularly evident in its ability to track and respond to unusual patterns of animal illness that might indicate the emergence of frozen pathogens, while simultaneously monitoring human health indicators in nearby communities. The program's integration of environmental and health data has proven crucial in identifying potential risks before they develop into serious threats.

Norway's Arctic Health Risk Assessment Program has made significant strides in developing adaptable protocols that can be implemented effectively in resource-limited settings. The program's success lies in its tiered approach to risk assessment and response, which allows for meaningful action even in areas with limited infrastructure. By creating scalable protocols that can be adjusted based on available resources while maintaining essential monitoring and response capabilities, the program has demonstrated how effective surveillance can be achieved even in challenging circumstances. The program's protocols have been successfully adapted and implemented in several other Arctic regions, demonstrating their versatility and effectiveness across different contexts.

These successful initiatives share several key characteristics that have contributed to their effectiveness: strong community engagement, flexible implementation strategies, integration of traditional and modern knowledge systems, and robust international cooperation frameworks. Their achievements provide valuable lessons for future implementation efforts and demonstrate the feasibility of comprehensive frozen pathogen mitigation strategies when properly designed and executed [34, 62, 64].

11 Conclusion

The unprecedented rate of cryosphere melting represents one of the most significant emerging threats to global public health and security. Our analysis reveals that the reactivation of ancient pathogens from thawing permafrost with an estimated 4×10^{21} microorganisms released annually from non-polar ice alone constitutes a potentially catastrophic risk that demands immediate attention from the international scientific community. The discovery that many of these preserved microorganisms are closely related to human pathogens, combined with evidence suggesting their role in historical extinction events such as the Neanderthal population decline, underscores the gravity of this emerging crisis.

The implications of this phenomenon extend far beyond immediate public health concerns. Our examination demonstrates the complex interconnections between climate change, ecosystem stability, and human security. The release of ancient pathogens serves as a stark reminder of the unforeseen consequences of global warming, where changes in one system can cascade through others with potentially devastating effects. The identification of viable microorganisms in millennia-old ice samples, coupled with their demonstrated ability to remain infectious, highlights the urgent need for enhanced surveillance and preparedness measures.

These findings have profound implications for global health security infrastructure. The potential reemergence of extinct pathogens presents unique challenges to modern medical systems, as contemporary populations lack evolved immunity to these ancient threats. This vulnerability is compounded by the rapid pace of permafrost thaw, which outstrips our current capacity to monitor and respond to emerging pathogens. The situation is further complicated by the global nature of the threat, where pathogens released in remote Arctic regions could potentially spread worldwide through modern transportation networks.

The research presented in this chapter emphasizes the critical importance of immediate action on multiple fronts. First, there is an urgent need to enhance global surveillance systems capable of detecting and monitoring pathogen release from thawing permafrost. Second, international cooperation must be strengthened to develop coordinated response mechanisms for potential outbreaks. Third, significant investment in research is required to better understand the viability and virulence of ancient pathogens in modern ecosystems.

Perhaps most crucially, this analysis demonstrates that addressing the root cause climate change remains paramount. The continued thawing of permafrost not only releases ancient pathogens but also accelerates global warming through the release of greenhouse gases, creating a dangerous feedback loop that threatens to amplify both environmental and health risks. This underscores the inseparable link between climate action and global health security. The path forward requires unprecedented levels of international cooperation, resource mobilization, and scientific innovation. Success in addressing these challenges demands a holistic approach that combines aggressive climate change mitigation strategies with enhanced global health security measures. Without immediate and sustained action, the release of ancient pathogens from thawing permafrost may become one of the most significant public health challenges of the twenty-first century.

The evidence presented throughout this chapter compels us to act with urgency and purpose. The time for incremental responses has passed; we must now embrace transformative action to address both the immediate threats posed by ancient pathogens and the underlying climate crisis that drives their release. Our future depends on our ability to rise to this challenge with the scale and speed that the situation demands.

References

1. Yarzábal, L.A., Salazar, L.M.B., Batista-García, R.A.: Climate change, melting cryosphere and frozen pathogens: should we worry…? Environ. Sustain. **4**(3), 489–501 (2021)
2. Sajjad, W., Rafiq, M., Din, G., Hasan, F., Iqbal, A., Zada, S., et al.: Resurrection of inactive microbes and resistome present in the natural frozen world: reality or myth? Sci. Total Environ. **735**, 139275 (2020)
3. Pirani, S.: Burning Up: A Global History of Fossil Fuel Consumption. Pluto Press (2018)
4. El-Batrawy, O.A., Abd-El-Shafy, A.M., El-Emam, D.A., Hassan, I.A.: Assessment of carbon footprints as a response to climatic changes. In: The First International Conference & Expo on Green Science, pp. 245–265. Springer Nature, Cham (2023)
5. Elemam, D.A., Eldeeb, A.R.: Climate change in the coastal areas: consequences, adaptations, and projections for the Northern Coastal Area, Egypt. Sci. J. Damietta Fac. Sci. **12**(2), 19–29 (2023)
6. Saxena, A., Srivastava, A.: Industry application of green manufacturing: a critical review. J. Sustain. Environ. Manag. **1**(1), 32–45 (2022)
7. Rusetuka, M.: An investigation of cutting down of trees as human rights abuse in the global world. Int. J. Environ. Sci. **6**(2), 50–66 (2023)
8. Malhi, Y., Franklin, J., Seddon, N., Solan, M., Turner, M.G., Field, C.B., Knowlton, N.: Climate change and ecosystems: threats, opportunities and solutions. Philos. Trans. R. Soc. B. **375**(1794), 20190104 (2020)
9. Elemam, D., Hamam, E.: Using the vulnerability index to assess the risk of climate change impacts on agricultural and food security. Alfarama J. Basic Appl. Sci. **5**(2), 293–308 (2024)
10. Liu, S., Wu, T., Wang, X., Wu, X., Yao, X., Liu, Q., et al.: Changes in the global cryosphere and their impacts: a review and new perspective. Sci. Cold Arid Reg. **12**(6), 343–354 (2021)
11. Dimri, A.P., Allen, S., Huggel, C., Mal, S., Ballesteros-Canovas, J.A., Rohrer, M., et al.: Climate change, cryosphere and impacts in the Indian Himalayan Region. Curr. Sci. **120**, 775–790 (2021)

12. Slater, T., Lawrence, I.R., Otosaka, I.N., Shepherd, A., Gourmelen, N., Jakob, L., et al.: Earth's ice imbalance. Cryosphere. **15**(1), 233–246 (2021)
13. Haeberli, W., Whiteman, C.: Snow and ice-related hazards, risks, and disasters: facing challenges of rapid change and long-term commitments. In: Snow and Ice-related Hazards, Risks, and Disasters, pp. 1–33. Elsevier, Amsterdam (2021)
14. Levy, R., Naish, T., Lowry, D., Priestley, R., Winefield, R., Alevropolous-Borrill, A., et al.: Melting ice and rising seas–connecting projected change in Antarctica's ice sheets to communities in Aotearoa New Zealand. J. R. Soc. N. Z., 1–24 (2023)
15. Dawson, T.: Heritage destruction, natural disasters, and the environment: atmospheric disasters. In: The Routledge Handbook of Heritage Destruction, pp. 138–146. Routledge, London (2023)
16. Kumar, A., Nagar, S., Anand, S.: Climate change and existential threats. In: Global Climate Change, pp. 1–31. Elsevier (2021)
17. Brighenti, S., Tolotti, M., Bruno, M.C., Wharton, G., Pusch, M.T., Bertoldi, W.: Ecosystem shifts in Alpine streams under glacier retreat and rock glacier thaw: a review. Sci. Total Environ. **675**, 542–559 (2019)
18. Zhang, M., Chen, F., Zhao, H., Wang, J., Wang, N.: Recent changes of glacial lakes in the high mountain Asia and its potential controlling factors analysis. Remote Sens. (Basel). **13**(18), 3757 (2021)
19. NASA: NASA's GRACE and GRACE Follow-On Satellites. http://grace.jpl.nasa.gov. (2025)
20. Rignot, E.: Sea level rise from melting glaciers and ice sheets caused by climate warming above pre-industrial levels. Phys. Uspekhi. **65**(1) (2022)
21. Frederikse, T., Landerer, F., Caron, L., Adhikari, S., Parkes, D., Humphrey, V.W., et al.: The causes of sea-level rise since 1900. Nature. **584**(7821), 393–397 (2020)
22. Abd-Elhamid, H.F., Abdelfattah, M., Zeleňáková, M., Kabeel, A.E., Barańczuk, J., Gharbia, S.S., Mahdy, M.: Monitoring coastal changes in Port Said, Egypt using multi-temporal satellite imagery and GIS-DSAS. Model. Earth Syst. Environ. **11**(1), 56 (2025)
23. Goelzer, H., Nowicki, S., Payne, A., Larour, E., Seroussi, H., Lipscomb, W.H., et al.: The future sea-level contribution of the Greenland ice sheet: a multi-model ensemble study of ISMIP6. Cryosphere. **14**(9), 3071–3096 (2020)
24. NASA: Satellite Sea Level Observations (2025). https://climate.nasa.gov/vital-signs/sea-level/?intent=111
25. Griggs, G., Reguero, B.G.: Coastal adaptation to climate change and sea-level rise. Water. **13**(16), 2151 (2021)
26. Galán, A.S., Pérez, L.T.: Global warming impacts on earth: damage to human health. In: Health and Climate Change, pp. 3–30. Academic (2025)
27. Harris, P.T.: Frozen ocean: ice ages and climate change. In: Mysterious Ocean: Physical Processes and Geological Evolution, pp. 89–110 (2020)
28. Shahgedanova, M.: Climate change and melting glaciers. In: The Impacts of Climate Change, pp. 53–84. Elsevier (2021)
29. Rounce, D.R., Hock, R., Maussion, F., Hugonnet, R., Kochtitzky, W., Huss, M., et al.: Global glacier change in the 21st century: every increase in temperature matters. Science. **379**(6627), 78–83 (2023)
30. Yadav, N., Upadhyay, R.K.: Global effect of climate change on seasonal cycles, vector population and rising challenges of communicable diseases: a review. J. Atmos. Sci. Res. **6**(1) (2023)
31. Mekonnen, G.A., Berhe, N.: Climate change and emerging and re-emerging tropical infections. InTechOpen (2023)
32. Kant, L., Immanuel, G., Narain, J.P.: Disease pandemics and the threat of microbial emergence. In: Climate Change and the Health Sector: Healing the World (2021)

33. Ernakovich, J.G., Barbato, R.A., Rich, V.I., Schädel, C., Hewitt, R.E., Doherty, S.J., et al.: Microbiome assembly in thawing permafrost and its feedbacks to climate. Glob. Chang. Biol. **28**(17), 5007–5026 (2022)
34. Andersen-Ranberg, E., Nymo, I.H., Jokelainen, P., Emelyanova, A., Jore, S., Laird, B., et al.: Environmental stressors and zoonoses in the Arctic: learning from the past to prepare for the future. Sci. Total Environ. **957**, 176869 (2024)
35. Kramer, A., Lexow, F., Bludau, A., Köster, A.M., Misailovski, M., Seifert, U., et al.: How long do bacteria, fungi, protozoa, and viruses retain their replication capacity on inanimate surfaces? A systematic review examining environmental resilience versus healthcare-associated infection risk by "fomite-borne risk assessment". Clin. Microbiol. Rev. **37**(4), e00186–e00123 (2024)
36. Zawierucha, K., Vecchi, M., Takeuchi, N., Ono, M., Calhim, S.: Negative impact of freeze–thaw cycles on the survival of tardigrades. Ecol. Indic. **154**, 110460 (2023)
37. Dias, R.A.: Towards a comprehensive definition of pandemics and strategies for prevention: a historical review and future perspectives. Microorganisms. **12**(9), 1802 (2024)
38. Manuel, R.D., Snyder, J.C.: The expanding diversity of viruses from extreme environments. Int. J. Mol. Sci. **25**(6), 3137 (2024)
39. Ezhova, E., Orlov, D., Suhonen, E., Kaverin, D., Mahura, A., Gennadinik, V., et al.: Climatic factors influencing the anthrax outbreak of 2016 in Siberia, Russia. Ecohealth. **18**(2), 217–228 (2021)
40. Tamang, S., Sharma, P., Kumar, S., Thakur, N.: Bacterial community structure, adaptations and prevalence of antimicrobial resistance in bacteria from Antarctica: a review. Pol. Sci. **40**, 101034 (2023)
41. Ghosh, M., Heo, Y., Pulicherla, K.K., Ha, M.W., Do, K., Son, Y.O.: Cold-active enzymes from deep marine psychrophiles: harnessing their potential in enhanced food production and sustainability. Crit. Rev. Biotechnol., 1–25 (2025)
42. Khan, I., Rehman, A., Zia, K., Naveed, U., Bibi, S., Sherazi, R., et al.: Microbes and environment: global warming reverting the frozen zombies. In: Environment, Climate, Plant and Vegetation Growth, pp. 607–633 (2020)
43. Kim, D., Kim, M., Woo, S., Nam, S., Myeong, N.R., Kim, E., Lee, Y.M.: Potential risks of bacterial plant pathogens from thawing permafrost in the Alaskan Tundra. Ecotoxicol. Environ. Saf. **289**, 117531 (2025)
44. Melo, J., Quintas, C.: Minimally processed fruits as vehicles for foodborne pathogens. AIMS Microbiol. **9**(1), 1 (2023)
45. El-Emam, D.A.: Valorization of solid waste products from a variety of sectors in wastewater treatment processes. In: Technical Landfills and Waste Management: Volume 2: Municipal Solid Waste Management, pp. 207–257. Springer Nature Switzerland, Cham (2024)
46. Aboghrbia, S.A., El-Emam, D.A.: E-waste valorization and management for precious metals mining: towards a circular economy. In: Haddout, S., Priya, K., Hoguane, A.M. (eds.) Proceedings of The 2nd International Conference on Climate Change and Ocean Renewable Energy. CCORE 2022. Springer Proceedings in Earth and Environmental Sciences. Springer, Cham (2024). https://doi.org/10.1007/978-3-031-71555-6_17
47. Revich, B.A., Eliseev, D.O., Shaposhnikov, D.A.: Risks for public health and social infrastructure in Russian Arctic under climate change and permafrost degradation. Atmos. **13**(4), 532 (2022)
48. Zhou, X., Wu, Y., Zhu, Z., Lu, C., Zhang, C., Zeng, L., et al.: Mucosal immune response in biology, disease prevention and treatment. Signal Transduct. Target. Ther. **10**(1), 7 (2025)
49. Zhang, Y., Liu, M.X., Yu, Y.L., Chen, S.: Frost-resistant hydrogel colorimetric sensing array for accurate detection of foodborne pathogens in cold chain systems. Biosens. Bioelectron. **271**, 116990 (2025)

50. Adla, K., Dejan, K., Neira, D., Dragana, Š.: Degradation of ecosystems and loss of ecosystem services. In: One Health, pp. 281–327. Academic (2022)
51. Choudhury, P., Bera, T., Pratihar, S., Das, S., Chatterjee, A.: mitigation of pathogenic microorganisms in infectious biomedical waste: biomedical waste management. In: Hospital Waste Management and Toxicity Evaluation, pp. 163–190. IGI Global Scientific Publishing (2025)
52. Gallo, M., Ferrara, L., Calogero, A., Montesano, D., Naviglio, D.: Relationships between food and diseases: what to know to ensure food safety. Food Res. Int. **137**, 109414 (2020)
53. Sarno, E., Pezzutto, D., Rossi, M., Liebana, E., Rizzi, V.: A review of significant European foodborne outbreaks in the last decade. J. Food Prot. **84**(12), 2059–2070 (2021)
54. Miner, K.R., D'Andrilli, J., Mackelprang, R., Edwards, A., Malaska, M.J., Waldrop, M.P., Miller, C.E.: Emergent biogeochemical risks from Arctic permafrost degradation. Nat. Clim. Chang. **11**(10), 809–819 (2021)
55. Kaur, S., Thakur, H., Singh, A., Ramasamy, V., Mudgal, G.: Poisoned seas: chemical threats to marine life and human health. In: Sustainable Development Goals Towards Environmental Toxicity and Green Chemistry, pp. 167–200. Springer, Cham (2025)
56. Wu, R., Trubl, G., Taş, N., Jansson, J.K.: Permafrost as a potential pathogen reservoir. One Earth. **5**(4), 351–360 (2022)
57. Mather, A.E., Gilmour, M.W., Reid, S.W., French, N.P.: Foodborne bacterial pathogens: genome-based approaches for enduring and emerging threats in a complex and changing world. Nat. Rev. Microbiol. **22**, 543 (2024)
58. Çelik, P.A., Manga, E.B., Çabuk, A., Banat, I.M.: Biosurfactants' potential role in combating COVID-19 and similar future microbial threats. Appl. Sci. **11**(1), 334 (2020)
59. Mora, C., McKenzie, T., Gaw, I.M., Dean, J.M., von Hammerstein, H., Knudson, T.A., et al.: Over half of known human pathogenic diseases can be aggravated by climate change. Nat. Clim. Chang. **12**(9), 869–875 (2022)
60. Bouffard, T.J., Uryupova, E., Dodds, K., Romanovsky, V.E., Bennett, A.P., Streletskiy, D.: Scientific cooperation: supporting circumpolar permafrost monitoring and data sharing. Landscape. **10**, 590 (2021)
61. Rieckert, A., Schuit, E., Bleijenberg, N., Ten Cate, D., de Lange, W., de Man-van Ginkel, J.M., et al.: How can we build and maintain the resilience of our health care professionals during COVID-19? Recommendations based on a scoping review. BMJ Open. **11**(1), e043718 (2021)
62. Lohmann, R., Beatty, B., Graybill, J., Grigorieva, E., Hansen, K.L., Soikkeli, A.: Perspective: dimensions of environment and health in arctic communities. Environ. Health. **2**(1), 3–10 (2023)
63. Gostin, L.O.: Global Health Security: A Blueprint for the Future. Harvard University Press (2021)
64. WHO: Global Research Agenda for Antimicrobial Resistance in Human Health. World Health Organization (2024)

Influence of Climate Change on Agriculture in Sirmaur District of Himachal Pradesh in India

Shanta Kumari(✉) and Kajal Choudhary

Department of Economics, Eternal University, District Sirmaur, Himachal Pradesh, India
shantakaushalchauhan@gmail.com

Abstract. The research investigated the influence of climate change on agriculture in Sirmaur district of Himachal Pradesh in India. Secondary data pertaining to rainfall and temperature trends in the area were analyzed for the period spanning 2010 to 2018. The primary focus was on the climatic parameters, specifically rainfall and temperature and their repercussions on the production and productivity of key crops such as wheat, paddy, maize, potato, and ginger. The findings underscored greater fluctuation in rainfall during the Kharif season compared to the Rabi season. Meanwhile, temperature exhibited relatively less variation across both seasons during the same period. These climatic shifts exerted notable impacts on the production of major crops, with paddy being the most affected, followed by wheat, ginger, maize, and potato. Notably, the decline in productivity was most pronounced in paddy, followed by wheat and potato. To mitigate the adverse effects of climate change the study suggested that concerned authorities should strengthen extension facilities and timely updates on meteorological information to farmers etc. are crucial steps. It will empower farmers to adopt appropriate strategies in a timely manner and thereby reduce climate-related risks to agriculture. Furthermore, collaborative multidisciplinary efforts hold significant promise in safeguarding agriculture from the detrimental impacts of climate change.

Keywords: Climate Change · Agriculture · Rainfall and Temperature Trends · Crop Productivity · Adaptation Strategies

1 Introduction

Agriculture is the backbone of India's economy, supporting the livelihoods of a significant portion of the population. In regions like Himachal Pradesh, particularly in the Sirmaur district, agriculture is not only an economic activity but also an integral part of the community's culture and sustenance. However, the growing challenges posed by climate change are disrupting traditional agricultural practices, with severe implications for production and productivity [1, 2]. Climate change involves long-term shifts in temperature, precipitation patterns and other atmospheric conditions. For agriculture, these changes manifest as unpredictable weather events, altered growing seasons and an increased frequency of extreme climatic conditions [3, 4]. These variations directly affect

J. C. Casila et al. (Eds.): CCORE 2024, SPEES, pp. 136–147, 2025.
https://doi.org/10.1007/978-3-031-93887-0_14

crop yields, soil health and water resources posing significant challenges to farmers. For instance, studies in the Shimla district have reported an annual increase in mean maximum temperature by 0.05°C during the Kharif season, adversely impacting crops such as rice and maize [4–7]. Sirmaur district, located in the southeastern region of Himachal Pradesh, features diverse agro-climatic zones that support the cultivation of crops like wheat, paddy, maize, potato and ginger [8, 9]. This agrarian economy is heavily reliant on monsoon rains, making it particularly vulnerable to climatic disruptions. Recent assessments have highlighted increased rainfall variability, especially during the Kharif season, along with rising temperatures and reduced snowfall. These climatic trends disrupt sowing and harvesting schedules, ultimately affecting crop productivity [10]. The major crops grown in Sirmaur are highly sensitive to climatic fluctuations. Paddy for instance, requires consistent water availability, making it particularly vulnerable to erratic rainfall patterns. Similarly, wheat and maize are adversely affected by temperature extremes. Previous studies have indicated that cereal productivity in Himachal Pradesh is below the national average, with climate variability being a significant contributing factor [11, 12].

1.1 Statement of the Problem

The issue of climate change in Sirmaur district is characterized by variations in rainfall and temperature that directly impact agricultural production and productivity. The crops affected include wheat, paddy, maize, potato and ginger. These challenges are contextualized within a specific time frame (2010–2018), providing a clear basis for analysis. Additionally, the differentiation between Kharif and Rabi seasons adds depth to the discussion, emphasizing the relevance and specificity of the problem. Given the critical importance of agriculture to the region and the evident threats posed by climate change, it is imperative to analyze these impacts comprehensively. Such an understanding is essential for developing targeted adaptation and mitigation strategies to safeguard agriculture in Sirmaur.

1.2 Objective of the Study

The present study aims to quantify the changes in key climatic parameters and their effects on the production and productivity of major agricultural crops in Sirmaur district, Himachal Pradesh.

2 Methodology

The study was conducted in Sirmaur district (Fig. 1) of Himachal Pradesh, as there was a lack of research addressing the impacts of climate change on agriculture at the farmer level in this region. The focus of the study was to examine changes in rainfall and temperature and their effects on the production and productivity of major agricultural crops in the district.

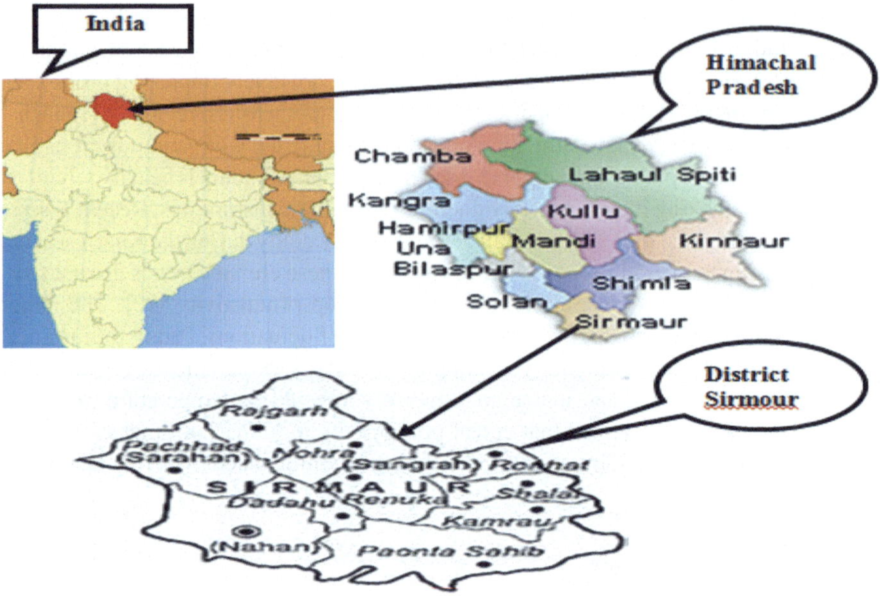

Fig. 1. Location of the study area in the map.

2.1 Data Collection

Secondary data on rainfall and temperature for the period 2010–2018 were obtained from the Nahan Meteorological Agency. The data were used to analyze climatic trends in the district. Additionally, data on the production and productivity of major crops, including wheat, paddy, maize, potato and ginger were sourced from the statistical abstracts of Himachal Pradesh [13–17]. These datasets provided insights into the relationship between climatic variations and agricultural performance in the study area.

2.2 Data Analysis

The collected data were analyzed using statistical tools such as averages, percentages and coefficients of variation. Deviations in actual rainfall from normal and previous years rainfall were calculated to identify patterns and anomalies. The results were visualized through graphs and charts effectively to present trends and correlations aligning with the study's objectives.

Following formulations were used to find the result of the study:

(a) Arithmetic Mean $= \frac{\sum Xi}{N}$

(b) CV $= \frac{\sigma}{Arithmetic\ mean} * 100$

CV is coefficient of variation and σ is standard deviation.

(c) Deviation in actual rainfall from normal rainfall (%)

$$= \frac{Current\ year\ rainfall - Normal\ year\ rainfall^{*}}{Normal\ Year\ rainfall} 100$$

(d) Deviation in actual rainfall from previous year rainfall (%)

$$= \frac{Current\ year\ rainfall - Previous\ year\ rainfall}{Previous\ year\ rainfall}{}^{*}100$$

3 Results and Discussion

Table 1 shows the trend analysis of rainfall and temperature during the rabi and kharif season from 2010 to 2018. The average rainfall in district Sirmaur during Kharif season was 340.53 mm in 2010 which declined to 336.50 mm in 2011. The coefficient of variation during the year 2010 in kharif season was 96.15% whereas in 2011 it was 75.58%. In 2012 the average rainfall declined to 191.02 mm and coefficient of variation during this year was 116.57%. The average rainfall has increased and shown an increasing trend during the year 2013 which stood at 300.33 mm with coefficient of variation 101.11%. In year 2014, the pattern of average rainfall again shown a declining trend which stood at 193.53 mm with coefficient of variation at 95.41%. The average rainfall increased again during the year 2015 which was 205.25 mm with coefficient of variation 123.78%. In 2016, average rainfall increased to 218.85 mm while coefficient of variation stood at 84.43%. Therefore, the average rainfall during the year 2017 stood at 294.48 mm which has shown an increasing trend with coefficient of variation 84.43% variation during the year. In year 2018, average rainfall was 273.32 mm with and 97.73% coefficient of variation.

Table 1. Changes in the pattern of average rainfall and temperature of District Sirmour, 2010–18.

Year	Rainfall (mm)				Temperature (°C)			
	Kharif Season		Rabi Season		Kharif Season		Rabi Season	
	Mean (mm)	CV (%)	Mean (mm)	CV (%)	Mean (°C)	CV (%)	Mean (°C)	CV (%)
2010	340.53	96.15	13.32	127.19	26.58	9.84	19.00	28.88
2011	336.50	75.58	8.88	98.60	26.42	7.89	17.83	25.18
2012	191.02	116.57	14.78	101.54	25.06	14.32	16.31	27.51
2013	300.33	101.11	53.20	130.06	25.67	11.13	17.75	24.08
2014	193.53	95.41	70.37	67.98	23.43	12.45	14.47	27.25
2015	205.25	123.78	60.37	128.26	24.71	8.37	17.06	22.47
2016	218.85	84.43	9.82	154.12	24.46	7.36	17.63	23.26
2017	294.48	80.40	38.28	98.09	24.27	7.60	16.91	23.25
2018	273.32	97.73	21.23	59.92	23.64	10.98	16.65	29.60

Source: Author's own calculation from the data of meteorological Agency, Nahan, 2021
Note: Average rainfall and temperature are calculated from the data of months from May to Oct. (kharif season) and Nov. to April (Rabi season) for each year (2010–18)
CV = Co-efficient of variation

In Rabi season, in 2010, the average rainfall pattern of district Sirmaur was 13.32 mm which declined to 8.88 mm in the year 2011. Therefore, coefficient of variation during the year rabi season was 127.19%. In 2011 it was 98.60%. In 2012 the average rainfall increased to 14.78 mm with coefficient of variation 101.54%. The average rainfall has increased and shown an increasing trend during the year 2013 which was 53.20 mm and its coefficient of variation was 130.06%. In year 2014, the pattern of average rainfall again has shown an increasing trend at 70.37 mm with coefficient of variation at 67.98%. The average rainfall pattern has shown a decreasing trend during the year 2015 which was 60.37 mm with 128.26% of coefficient of variation. In 2016, average rainfall again recorded a decreasing trend with 9.82 mm and coefficient of variation at 154.12%. Therefore, average rainfall and coefficient of variation during the year 2017 was 38.28 mm which has shown an increasing trend in average rainfall and 98.09% variation during the year. In the year 2018, average rainfall was at 21.23 mm with 59.92% coefficient of variation during the year.

The above table also shows the trend analysis of temperature during kharif and rabi season for the year 2010–2018 which depicts that the average temperature in year 2010 during kharif season was 26.58°C and that during rabi season it was 19°C while the coefficient of variation during the year was 9.84% for kharif and 28.88% for rabi season. In year 2011, average temperature during kharif season was at 26.42°C and that during rabi season it was 17.83°C which has shown a declining trend while the coefficient of variation for kharif and rabi season stood at 7.89% for kharif and 25.18% for rabi season. Therefore, the average temperature during kharif and Rabi season for the year 2012 remained at 25.06°C and 16.31°C and coefficient of variation for kharif and rabi seasons was 14.32% and 27.51%. In year 2013, during kharif season the average temperature was 25.67°C and that during rabi season it was 17.75°C while the coefficient of variation during the year was at 11.13% for kharif and 24.08% for Rabi season. The average temperature during kharif and rabi seasons for the year 2014 was 23.43°C and 14.47°C, respectively. While the coefficient of variation for kharif and rabi seasons was 12.45% and 27.25%, respectively.

In year 2015, average temperature during kharif season was 24.71°C and that during rabi season it was 17.06°C which has shown a declining trend while the coefficient of variation during the year stood at 8.37% for kharif season and 22.47% for rabi season. Therefore, the average temperature during kharif and rabi season for the year 2016 stood at 24.46°C and 17.63°C and coefficient of variation for kharif and rabi seasons was for 7.36% and 23.26%, respectively. In the year 2017, average temperature during kharif and rabi seasons was for 24.27°C and 16.91°C, respectively. Whereas the coefficient of variations for rabi and kharif seasons stood at 7.60% and 23.25%. Therefore, the average temperature during kharif and rabi season during the 2018 accounted was 23.64°C and 16.65°C with coefficient of variation for kharif and rabi seasons remaining at 10.98% and 29.60%, respectively.

The Fig. 2 reflected that there are more fluctuations in the rainfall in the kharif season in comparisons to rabi season. Whereas, in temperature (Fig. 3) less variations has been found in both the season during the year 2010 to 2018. Thus it is concluded that rise in temperature results warmer air and more evaporation of moisture from both land and water into the atmosphere. If moisture is more in the air there will be more rainfall and if it is less then there will be less rainfall. So the temperature and rainfall are related to each other. Rise in temperature is due to accumulation of more CO_2 in the environment which is mainly produced due to the pollution generated by the setting up of industries and deforestation.

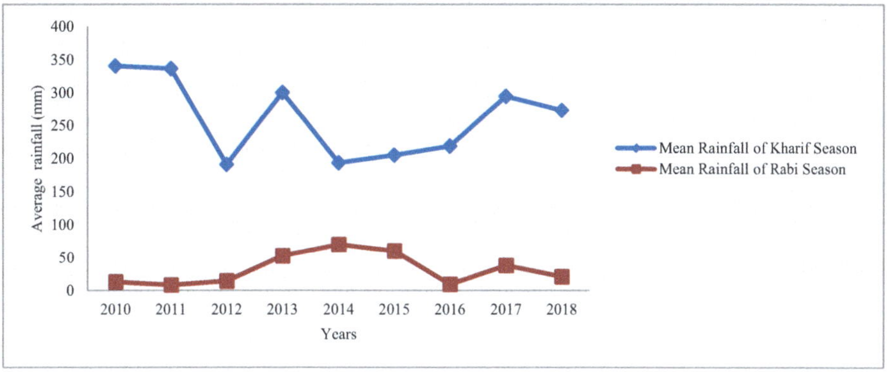

Fig. 2. Trend analysis of rainfall pattern, 2010–18. Source: Meteorological Agency Nahan, 2021.

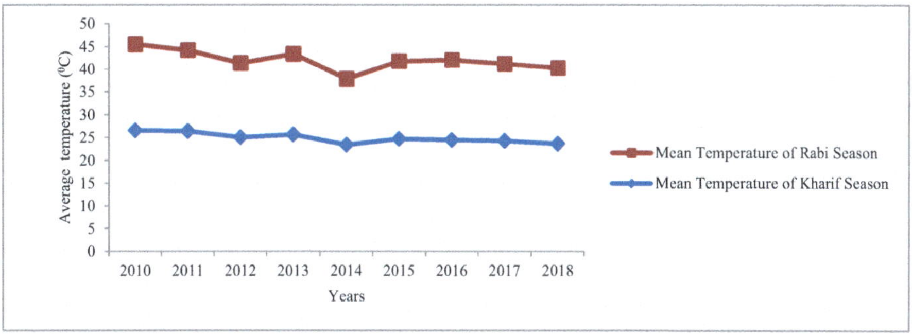

Fig. 3. Trend analysis of pattern of temperature, 2010–18. Source: Abstract of Himachal Pradesh, Department of Economics and Statistics of Himachal Pradesh, Shimla, https://en.climate-data.org, https://timeanddate.com > nahan Weather by Custom weather, ©2021.

The Table 2 given below has shown variations in average rainfall pattern from 2010–18. As the actual rainfall is compared with the normal rainfall, the deviation over normal rainfall is calculated in percentage terms while deviation over previous period is also given in percentage terms. It has been observed that, in 2010 actual rainfall was 176.93 mm whereas deviation over normal rainfall was 30.72%. It has shown a minor decline in actual rainfall i.e., 172.69 mm in 2011 whereas deviation from normal rainfall was 27.50% while deviation over the previous period was −2.39%. In year 2012, it is found that in comparison of different years there have been a severe decline in the in the rainfall which stood at 102.90 mm, while deviation from normal rainfall was −23.97% and deviation over the previous period was −40.41%. In year 2013, the amount of pattern of rainfall increased to 176.77 mm while the deviation has also up stretched over normal rainfall and over the previous year which stood at 30.60% and 71.79%. Therefore, it has again declined to 131.95 mm and deviation over normal rainfall and previous years stood at 2.51% and −25.35% in year 2014. In year 2015 a slight increase in rainfall amount took place which stood at 132.81 mm and deviations over normal rainfall and over the previous year was at 1.88% and 0.64%.

Table 2. Deviation of actual rainfall from the normal rainfall, 2010–18.

Year	Actual rainfall (mm)	Deviation from normal rainfall (%)	Deviation from previous period (%)
2010	176.93	30.72	–
2011	172.69	27.50	−2.39
2012	102.90	−23.97	−40.41
2013	176.77	30.60	71.79
2014	131.95	2.51	−25.35
2015	132.81	1.88	0.64
2016	114.33	15.53	−13.91
2017	166.38	22.93	45.52
2018	147.28	8.81	−11.47

Source: Author's own calculations from the data of Statistical Abstract of Himachal Pradesh, Department of Economics and Statistics of Himachal Pradesh, Shimla Meteorological Agency Nahan, 2021
Note: Average normal rainfall is 135.35 mm

Average rainfall for year 2016 declined to 114.33 mm and deviations over normal rainfall and over previous period stood at 15.53% and −13.91%, respectively. For the year 2017, it rose to 166.38 mm and deviations over normal rainfall and previous period stood at 22.93% and 45.52%. In 2018, actual average rainfall again declined to 147.28 and deviation over normal rainfall and previous period was 8.81% and −11.47%.

The Table 3 shows the variations in production pattern of major crops during year 2010–18. In 2010, the production of paddy, maize, wheat was at 13998, 59065 and 50904 MT (Metric Tonnes) while the production of wheat declined to 47177 MT and that of paddy and maize it has increased to 14322 and 63383 MT in year 2011. I terms of growth rate, it stood at 2.31% for paddy and 7.31% for maize while for wheat it has declined to −7.32%. In 2012, the production of paddy and maize declined to 13673 and 54746 MT but the production of wheat increased to 54631 MT. In terms of growth rate, it was −4.53% for paddy and −13.63% for maize and for wheat it was 15.80%.

In year 2013, production of paddy, maize and wheat was 14550, 53984 and 59434 MT, respectively. It has been observed that the growth rate for paddy and wheat increased and stood at 6.41% and 8.79% but for maize a slightly decline in growth rate is observed which was −1.39%. Production of paddy, maize and wheat for the year 2014 was 15426, 53222 and 64236, respectively whereas it has shown a negative growth rate for maize which was −1.41%. The grow rate was positive for paddy and wheat which was 6.02%

Table 3. Changes in the production of major crops of District Sirmour, 2010–18.

Year	Paddy		Maize		Wheat		Potato		Ginger	
	P (MT)	% Δ	P (MT)	% Δ	P (MT)	% Δ	P (MT)	% Δ	P (MT)	% Δ
2010	13998	–	59065	–	50904	–	17335	–	6134	–
2011	14322	2.31	63383	7.31	47177	−7.32	16980	−2.05	2374	−61.30
2012	13673	−4.53	54746	−13.63	54631	15.80	17690	4.18	9894	316.76
2013	14550	6.41	53984	−1.39	59434	8.79	18400	4.01	9636	−2.61
2014	15426	6.02	53222	−1.41	64236	8.08	19200	4.35	9377	−2.69
2015	21975	42.45	60949	14.52	63442	−1.24	20000	4.17	11031	17.64
2016	15906	−27.62	64872	6.44	48509	−23.54	20150	0.75	12134	10.00
2017	9836	−38.16	68794	6.05	33575	−30.79	20000	−0.74	11583	−4.54
2018	12871	30.86	66833	−2.85	41042	22.24	20075	0.38	11859	2.38

Source: Statistical Abstract of Himachal Pradesh, Department of Economics and Statistics of Himachal Pradesh, Shimla
Note: P*- Production in Metric Tonnes (MT)

and 8.08% respectively. In 2015, production of paddy and maize was high in comparison to other years from 2010–18 and was 21975 and 60949 MT, respectively, while that of wheat was slightly reduced to 63442 MT. The growth rate stood at 42.45% and 14.52% for paddy and maize while for wheat it was −1.24%. The production of paddy, maize, wheat and pulses for year 2016 was 15906, 64872 and 48509 MT, respectively. Therefore, from the given table, it is also observed that the growth rate of paddy and wheat had shown a decline by in −27.62 and −23.54% while the production of maize has recorded a growth rate of 6.44% (Fig. 4).

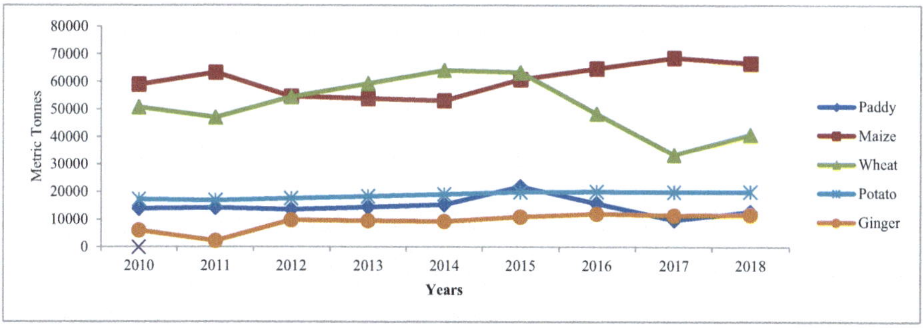

Fig. 4. Trend analysis of production of major crops, 2010–18.

In year 2017, production of paddy and wheat registered a significant decline which decreased to 9836 and 33575 MT, respectively while the production of maize increased to 68794 MT. Therefore, growth rates of paddy and wheat were negative trend −38.16% and

−30.79% while for maize it stood at 6.05%. But in 2018, production of paddy and wheat again increased to 12871 (30.86%) and 41042 (22.24%) MT while that of maize has declined to 66.833 (−2.855%). The production of potato and ginger in 2010 was 17335 and 6134 MT, respectively. Therefore, in year 2011 the growth rate of production of potato and ginger was −2.05% and −61.30%, respectively. In year 2012, the production of potato and ginger was 17690 MT (4.18%) and 9894 (316.76%). Production of potato and ginger in year 2013 stood at 18400 MT (4.01%) and 9636 MT (−2.61%) MT. But in 2014 production of ginger declined by −2.69% (9377 MT) whereas, potato recorded an increased trend by 4.35% (19200 MT).

In 2015 production of potato and ginger, increased by 4.17% (20000 MT) and 17.64% (11031 MT), respectively. In 2016 production of potato and ginger had increased by 0.75% (20150 MT) and 10.00% (12134 MT), respectively. In year 2017, the production of potato and ginger had declined by −0.74% and −4.54%. In year 2018, the production of potato and ginger had increased and was 20075 MT (0.38%) and 11859 (2.38%), respectively. Table 4 presents the variations in the yields of major crops during year 2010–18. In 2010 the yield of paddy, maize and wheat stood at 2.20, 2.82 and 1.98 MT (metric tonnes) while the yield of wheat has declined to 1.87 MT and that for paddy, maize has increased and stood at 2.20 MT and 2.99 MT in year 2011 and in terms of growth rate, it stood at 3.64% of paddy and 6.03% of maize while for wheat it has declined to −5.56%. In 2012, the yield of paddy and maize declined to 2.12 and 2.64 MT but the yield of wheat has increased to 2.09 MT. In terms of growth rate, it was −7.02% for paddy and −11.71% for maize and for wheat it stood at 11.76%.

Table 4. Changes in the yields of major crops of District Sirmour, 2010–18. (Metric Tonnes).

Year	Paddy		Maize		Wheat		Potato		Ginger	
	Yield	% Δ	Yield	% Δ	Yield	% Δ	Yield	% Δ	Yield	% Δ
2010	2.20	–	2.82	–	1.98	–	12.21	–	4.37	–
2011	2.28	3.64	2.99	6.03	1.87	−5.56	12.04	−1.39	1.84	−57.89
2012	2.12	−7.02	2.64	−11.71	2.09	11.76	12.37	2.74	6.54	255.43
2013	2.28	7.55	2.86	8.33	2.28	9.09	12.69	2.59	6.54	0.00
2014	2.43	6.58	3.12	9.09	2.47	8.33	11.13	−12.29	6.54	0.00
2015	3.12	28.40	3.24	3.85	2.47	0.00	10.00	−10.15	6.54	0.00
2016	2.56	−17.95	3.27	0.93	1.99	−19.43	13.00	30.00	6.54	0.00
2017	1.83	−28.52	3.30	0.92	1.45	−27.14	12.50	−3.85	6.54	0.00
2018	2.22	21.31	3.29	−0.30	1.73	19.31	12.75	2.00	6.54	0.00

Source: Statistical Abstract of Himachal Pradesh, Department of Economics and Statistics of Himachal Pradesh, Shimla
Note: Δ-Change

In year 2013, yield of paddy, maize and wheat was 2.28, 2.86 and 2.28 MT, respectively. It has been observed that the growth rate for paddy, maize and wheat has increased and stood at 7.55%, 8.33% and 9.09%, respectively. The yield of paddy, maize and wheat for year 2014 was 2.43, 3.12 and 2.47 MT whereas it shows a positive growth rate for all crops i.e., paddy, maize and wheat which are 6.58%, 9.09%, and 8.33%, respectively. The reason for increase in yield or productivity is decline in area and increase in the production due to use of HYV seeds. In 2015, yield of paddy and maize increased which stood at 3.12 and 3.24 MT, while that for wheat it had remained same at 2.47MT, recording a 0% growth rate. In case of growth rate for other crops, it was 28.40% in case of paddy which was highest among all the different years under consideration and 3.85% for maize. The yield of paddy, maize and wheat for year 2016 was 2.56, 3.27 and 1.99 MT, respectively. Therefore, from the given table it is also observed that the growth rate of paddy and wheat had shown a decline of −17.95% and −19.43% while maize yield has increased by 0.93%. In year 2017, yield of paddy and wheat, shows a severe decline which was 1.83 MT and 1.45 MT, respectively and maize production has increased to 3.30 MT. Therefore, growth rate of paddy and wheat shows a negative trend i.e., −28.52% and −27.14% while for maize it stood at 0.92%. But in 2018, yield of paddy and wheat again increased to 2.22 MT (21.31%) and 1.73 MT (19.31%), while that of maize has declined to 3.29 (−0.30%).

In 2010, yield of potato and ginger was 12.21 and 4.37 MT respectively. Therefore, in year 2011 the growth rate of potato and ginger stood at −1.39% and −57.89%. In 2012, the yield of potato and ginger had increased and was 12.37 MT (2.74%), 6.54 (255.43%). Yield of potato in year 2013 was at12.69 MT (2.59%). Therefore, the yield of ginger remained constant after the period of 2013 which stood at 6.54 MT. But in year 2014 the yield of potato had declined by −12.29%. In 2015, yield of potato again declined by −10.15% (10 MT). In 2016, production of potato had increased by 30% (13 MT). In 2017 the production of potato had declined by −3.85%. In year 2018, the production of potato was 12.75 MT.

4 Conclusion

The analysis reveals significant fluctuations in rainfall and temperature over the years, impacting crop production and yields in district Sirmaur. Kharif season rainfall exhibited greater variability compared to Rabi, while an overall rise in temperature was observed, leading to increased evaporation and altered moisture levels. These climatic variations contributed to irregular crop production and yield trends, highlighting the vulnerability of agriculture to climate change. Despite technological advancements, such as the use of high-yield varieties (HYV), adverse weather conditions often negated productivity gains. Environmental factors like CO_2 emissions and deforestation exacerbate these challenges, underscoring the need for sustainable agricultural practices.

5 Policy Recommendations

5.1 Water Resource Management

- Develop robust irrigation infrastructure to reduce dependence on erratic rainfall.
- Promote rainwater harvesting and efficient water storage systems.

5.2 Climate-Resilient Agriculture

- Encourage the adoption of drought-resistant and heat-tolerant crop varieties.
- Provide incentives for climate-smart farming techniques, such as crop rotation and integrated pest management.

5.3 Environmental Conservation

- Implement afforestation programs to reduce CO2 levels and enhance local climate stability.
- Enforce stricter regulations on industrial emissions to mitigate the rise in temperature.

5.4 Research and Development

- Invest in research to develop crops resilient to climatic fluctuations.
- Establish localized weather monitoring systems for timely advisories to farmers.

5.5 Capacity Building and Awareness

- Conduct training programs for farmers on sustainable agricultural practices.
- Raise awareness about the impact of climate change and measures to adapt effectively.

5.6 Financial Support and Risk Mitigation

- Expand crop insurance schemes to cover losses due to weather extremes.
- Provide financial support for adopting advanced agricultural technologies.

By integrating these measures into agricultural policy, the district can enhance resilience against climate variability and ensure sustainable growth in the agriculture sector.

References

1. Thornton, P.K., Ericksen, P.J., Herrero, M., Challinor, A.J.: Climate variability and vulnerability to climate change: a review. Glob. Chang. Biol. **20**(11), 3313–3328 (2014)
2. Kumari, S.: High value cash crops agriculture in Himachal Pradesh: a study in documentation and valuation of environmental costs. Doctoral dissertation. Agricultural Economics, CSKHPKV, Palampur (2007)
3. Dore, M.H.: Climate change and changes in global precipitation patterns: what do we know? Environ. Int. **31**(8), 1167–1181 (2005)
4. Mahato, A.: Climate change and its impact on agriculture. IJSRP. **4**(4), 1–6 (2014)
5. Lal, M.: Implications of climate change in sustained agricultural productivity in South Asia. Reg. Environ. Chang. **11**(Suppl 1), 79–94 (2011)
6. Kim, G.C.: The Impact of Climate Change on the Agricultural Sector: Implications of the Agro-Industry for Low Carbon, Green Growth Strategy and Roadmap for the East Asian Region, pp. 1–50 (2009)
7. Partap, B., Partap, T.: Climate Change Impact on Hill Agriculture and Farmers Adaptive Strategies: A Case Study of Kullu Valley in H.P. (2007)

8. Himachal Pradesh Climate Change Cell (HPCCC): Impact of Climate Change Assessment on Agriculture Sector in Himachal Pradesh: District Shimla (n.d.). https://hpccc.hp.gov.in
9. Jreat, M.: Tourism in Himachal Pradesh. Indus Publishing (2004)
10. Panwar, S.: Vulnerability of Himalayan springs to climate change and anthropogenic impact: a review. J. Mt. Sci. **17**(1), 117–132 (2020)
11. Department of Environment, Science & Technology, Government of Himachal Pradesh: Climate Change Vulnerability Assessment for the Yamuna Basin (2024). https://dest.hp.gov.in
12. Research Gate: Impact of Climate Change on Agriculture, Horticulture, and Forestry in Himachal Pradesh: A Review (2024). https://www.researchgate.net
13. Anonymous: Statistical Abstract of Himachal Pradesh. Department of Economics and Statistics, Himachal Pradesh (2013)
14. Anonymous: Statistical Abstract of Himachal Pradesh. Department of Economics and Statistics, Himachal Pradesh (2015)
15. Anonymous: Statistical Abstract of Himachal Pradesh. Department of Economics and Statistics, Government of Himachal Pradesh (2016)
16. Anonymous: Statistical Abstract of Himachal Pradesh. Department of Economics and Statistics, Government of Himachal Pradesh (2017)
17. Anonymous: Statistical Abstract of Himachal Pradesh. Department of Economics and Statistics, Government of Himachal Pradesh (2018)

Optimization Techniques for Edge Deployment of Zero Shot Open Vocabulary Detection Methods for Plastic Waste Classification

B. Madhini and P. Supraja(✉)

Department of Networking and Communications, School of Computing, SRM Institute of Science and Technology, Chengalpattu, Tamil Nadu, India
{mb7992,suprajap}@srmist.edu.in

Abstract. Plastic waste has become a major environmental hazard for the past few decades. Automated waste segregation using Deep learning techniques has proven to be a very efficient way to identify and segregate the generated waste. Traditional and CNN-based machine learning models developed for plastic waste classification had pre-defined labels and static classes associated with them. In recent times, due to technological improvement in production methodology, different categories of plastic waste have been generated into the environment. A plastic-type identified under a particular category previously is now classified under a different category based on the recycling technology changes, property character, and its impact on the environment. To cater to this current need, the zero-shot open vocabulary capabilities of Vision Language Models (VLMs) can be used to identify unseen waste types or novel classes. However, the real-time zero-shot object detection on resource-constrained edge devices such as microprocessors, mobile devices, etc. poses several challenges such as memory constraints, computational limitations, and heterogenous hardware configurations. This paper provides a comprehensive understanding of the optimization techniques that can be employed in overcoming the shortcomings of edge deployment of zero-shot open-vocabulary object detection models and how to use them effectively for plastic waste classification.

Keywords: Plastic waste classification · Vision Language Model (VLM) · Edge Deployment · Optimization

1 Introduction

Smart Cities development project encompasses the integration of data sourced from edge devices, and sensors, with the backbone of communication networks that lead to a commitment to sustainability across various sectors such as waste management, energy usage, and modern infrastructure with smart buildings and transport systems. Smartphones with edge-AI enabled are pre-trained to capture the finest picture with the help of computer vision imaging technology [4] Automated plastic waste classification helps in sorting plastics at the primary segregation unit (like households, industry, construction

J. C. Casila et al. (Eds.): CCORE 2024, SPEES, pp. 148–159, 2025.
https://doi.org/10.1007/978-3-031-93887-0_15

sites, etc.), secondary segregation units (like plastic sorting utility areas, municipal plastic waste collection wards, etc.), and plastic waste recycling plants. Currently in India, in plastic sorting plants (secondary segregation sector), deep learning algorithms with computer vision technology have been used to identify the different plastic wastes. The plastic waste captured by RGB cameras on a conveyor belt helps identify its types based on visual characteristics such as shape, texture, and color. The identified plastic wastes are then sorted into respective recycling designated areas with the help of robotic arms which help in the physical sorting of the classified plastic waste. But at the primary segregation unit (source of waste origin), the plastic waste classification is carried out by the municipal and plastic waste vendor and the process is labor-intensive, time-consuming, and error-prone. By deploying automated waste sorting technology on edge devices, the plastic waste classification and segregation process will see a significant reduction in labor time and accurate predictions at the source segregation area.

With the breakthrough in Deep Neural Networks [DNN], the scope of object detection research in the computer vision area has seen a steep ascent in both industry and the educational field [6]. The edge device generated images are mostly unlabelled which presents a significant obstacle and mandates human annotations [3]. Zero-shot models identify novel classes by leveraging pre-trained visual and textual semantic embeddings that allow them to recognize and classify new plastic waste types without needing specific training. The textual description along with the visual embeddings aids in this process of classifying new classes without prior training. The encoders in the vision transformers do image embeddings and text embeddings into a common feature space to work across cross-modalities. Currently, the vision encoders are tuned towards RGB images predominantly.

Real-time analysis of images directly on the device where they get generated helps in reducing the latency time, improving privacy, reducing bandwidth, and minimal data transmission over the cloud. When employing zero-shot open vocabulary models, the need for huge datasets and retraining the models can be avoided. Zero-shot object Detection (ZSOD) models identify new unseen objects using textual descriptions linking with their visual embeddings. Zero-shot open vocabulary object detection models that can be used for waste classification are CLIP-based models, DINO, Vision Language and Distillation (ViLD), and YOLO-World.

Edge deployment of ZSOD is critical as it allows for real-time analysis of data on the place of its origin. Zero-shot models requires a fine-tuning on the downstream tasks inspite of their impressive detection performance. The large weights associated with them demands for model compression when deploying them at the network edge [4].The complex algorithms employed by ZSOD often pose a challenge in deploying on resource-constrained edge devices such as microcontrollers, small IOT devices, smartphones, industrial sensors, and wearable devices. Edge devices have restricted processing power, memory limitations, reduced battery life, power capacity, and network bandwidth. Optimization of the model is the key requirement when considering edge deployment. Hence designing a deployment framework for a real-time inference pipeline, choosing the right model, and edge hardware selection helps in overcoming the difficulties in edge deployment.

This paper provides firstly, a comprehensive survey of the optimization techniques that can be applied for edge deployment of zero-shot open vocabulary object detection models and secondly a novel proposal of the state-of-the-art ZSOD-PWM (Zero Shot Object Detection – Plastic Waste Management) model framework that addresses the limitation in deploying ZSOD for plastic waste classification.

The rest of this study is organized as follows. The existing related works on feasible edge deployment optimization techniques for plastic waste classification utilizing computer vision are surveyed in Sect. 2. In Sect. 3, this study analyzes the best-suited zero detection model, for plastic waste classification, challenges in real-time deployment, and the optimization techniques specific to the chosen models. Section 4 discusses the methodology adopted for evaluating the ZSOD-PWM model(s) performance. Finally, the conclusions of this study and future works identified are analyzed in Sect. 5.

2 Related Works

In paper [1] tHuang, et al. state that the resources from the waste need to be recycled to reduce its impact on the global carbon footprint. Manual sorting of waste proves to be less efficient and not cost-effective. Over recent years, Convolutional Neural Network (CNN) has become the choice for automated sorting at recycling plants. However, due to the limitation of CNN, the accuracy has set a constraint on implementing appropriate methods and systems. They have utilized a deep neural network model, the Vision Transformer that is based on a self-attention mechanism. This Vision transformer model proves to have higher prediction accuracy when compared to the previous models. The author proposes a portable device for image capturing and a server for classification. However, they have not discussed the latency issues that arise when communicating with the server.

In paper [2], Prova, N. N. I. argues that the VIT model chosen for waste segregation shows a significant accuracy rate of 95% that surpassed the rates of previous CNN models like VGG16 and MobileNetV2. The enhanced recycling process is achieved by choosing the self-attention VIT model. They used the VIT model to classify household wastes into twelve different types. However, this classification does not address the growing varieties of plastic types that get continuously ingested into the environment. The model needs to be retrained again for the new classes.

In paper [3], Cai et al. throw light on how the latest developments in Vision-Language models have triggered interest in edge device deployment. But at the same time, they mention handling visual modalities, annotating manually, and the constraints in computational efforts. They propose a novel framework named EdgeVL, that integrates knowledge distillation with dual-modality and contrastive learning embedded with quantization. Despite its innovative approach, EdgeVL faces challenges in cross-modal use that limit its scope of applications.

In paper [4], Kashef et al. discuss IoT's potential in transforming urban development. This paper explains a smart waste management system, a customized version called ViT-WM. The model utilizes an enhanced TrashNet dataset and uses 7 classes but fails to address the growing images and multi-label classification.

In paper [5], Y. Dong, et al., talk in detail about how large language models (LLM) have transformed into a global foundation model. With an impressive zero-shot ability, if

LLMs are fine-tuned with model compression techniques, the energy consumption, and operational costs can be significantly reduced. The higher level of data heterogeneity induces a more significant performance gap between fully and parameter-efficient fine tuning but compromises privacy.

In paper [6], Cai et al. propose a real-time object detection on mobile using a compression-compilation and co-design technique. A new pruning scheme called block-punch method is used for the kernel along with the compiler optimization. This method addresses the hardware aspects only when considering edge deployment and hence fails on the scalability aspect.

In paper [7], Xu J et al. develop a trim YOLO framework designed for edge device deployment named EdgeTrim-YOLO. The author proposed a customization of the YOLO framework, called Ghost-Conv, DFC Attention mechanism, and re-parameterization strategies into the native backbone. The model has shown a significant reduction in its complexity and reduced computational overhead. But this approach is specific to a particular zero-shot detection model namely YOLO and needs to be generalized for other models as well.

In paper [8], Liu et al. propose a state-of-the-art YOLO framework, that can be used on edge computing platforms. The model showcases low complexity and anchor-free object detection with enhanced data augmentation. The model also suppresses overfitting during the training phase and random loss function that improves the detection accuracy of tiny objects. The inference speed also improved with the development of the lighter model with fewer parameters. However, the detection accuracy needs improvement.

In paper [9], Shuvo et al. did research on efficient Deep Learning inferences on edge devices by proposing novel architecture and algorithm design, accelerator design, and optimization techniques. The research provides a comprehensive study on DL optimization in the processing of images on different hardware platforms and inference techniques. However, the state-of-the-art zero-shot object detection models are not considered.

In paper [10], Mittal et al. discuss the optimization technique of lightweight object detection models but fails to address the zero-shot model's optimization techniques.

3 Methodology

3.1 Zero Shot Open Detection Open Vocabulary Model

Zero-shot open vocabulary model on Plastic Waste Classification (ZSOD) aims to bridge the gap between traditional object detection and the need for generalization to unseen classes. It typically involves two components:

- The Visual Representation Learning component extracts features from images using backbone networks like ResNet or Vision Transformers.
- The Semantic Alignment components align visual features with semantic embeddings derived from textual descriptions, word embeddings, or external knowledge graphs.

State-of-the-art ZSOD models often rely on pre-trained vision-language models such as CLIP, which enable robust cross-modal representations. Despite their effectiveness, these models are computationally intensive, posing challenges for edge deployment.

Zero-shot open vocabulary detection proves to be an effective tool for plastic waste classification. Plastic waste sorting has issues such as that only certain types of plastic can be recycled and the other can only be upcycled. Till now, categorizing them based on the resin code was the only way of identifying them. Resin code classification helps in segregating plastic wastes into seven different types such as PETE, HDPE, PVC, LDPE, PP, PS, and others. Traditional and CNN-based machine learning algorithm identifies them when the images are captured on a white background and fail to identify real-time mixed plastic wastes. The biggest challenge in segregating different material types based on colors and shapes in real-time has been made easy with the development of a zero-shot open vocabulary mechanism. The text prompt along with the visual image are fed into the decoders where the tokenization happens and the embeddings are added. Later they are fed to the self-attention models which identify the segmented patches with the text coding and encode it into the final output. The output images consist of the bounding box, with the identified plastic waste type and threshold associated with it. The ZSOD models that are identified for this case study are 1) VILD, 2) OWL-ViT, 3) GDINO, and 4) YOLO-WORLD.

The challenges in using ZSOD over Real-Time are latency delay, increased power consumption, and inconsistency in the accurate prediction of novel classes. Mobile devices typically have limited processing power compared to desktop or server setups. Optimizing model size, and computation, and reducing inference time are critical for real-time performance. Intensive processing on mobile devices can drain the battery quickly, so it's essential to balance model complexity and energy efficiency. Zero-shot models may not be as accurate when recognizing unseen objects compared to models trained specifically on those objects. If possible, fine-tuning or retraining the model can improve accuracy.

3.2 Optimization Techniques for Edge Deployment of Vision Models

Zero-shot open vocabulary models that are considered for this case study include Vision and Language Knowledge Distillation (ViLD), Vision Transformer for Open-World Localization (OWL-ViT), Grounding DINO 1.5 Edge and YOLO-World.

3.2.1 Vision and Language Knowledge Distillation

ViLD follows an optimization technique called distillation. This method helps the model choose the teacher and student architecture generically. It also aids in transferring various models' expertise through different strategies. The strategies encompass response, feature, and relation. As the teacher models are improved, ViLD performances also increase. When the backbone CLIP model is replaced with an EfficientNetl2 ALIGN model, the Average Precision significantly increases (Fig. 1).

ViLD can be optimized by replacing the token classifier BERT with DistilBERT. DistilBERT has only half the layers of BERT and is trained on eight 16 GB V100 GPUs for 90 h. This reduces the model size by 40% but with a 97% accuracy rate and a 60% faster performance rate. After replacing it with DistilBERT, converting the model into a serialized format using the ONNX toolkit reduces the model size further.

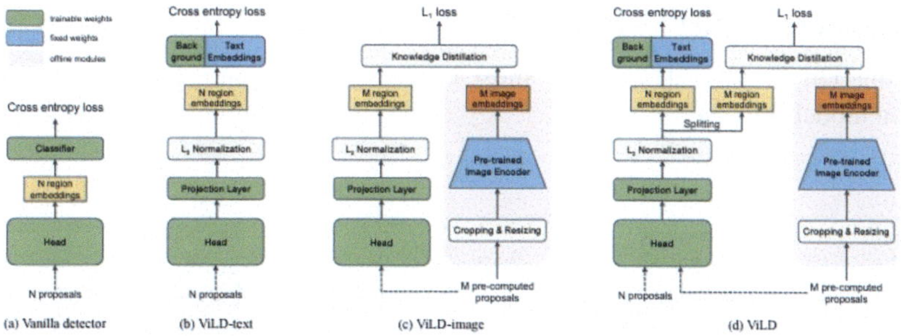

Model architecture and training objectives. **(a)** The classification head of a vanilla two-stage detector, *e.g.*, Mask R-CNN. **(b)** ViLD-text replaces the classifier with fixed text embeddings and a learnable background embedding. The projection layer is introduced to adjust the dimension of region embeddings to be compatible with the text embeddings. **(c)** ViLD-image distills from the precomputed image embeddings of proposals with an $\mathcal{L}_1$ loss. **(d)** ViLD combines ViLD-text and ViLD-image.

Fig. 1. ViLD Model Architecture.

3.2.2 OWL-ViT

OWL-ViT model uses a CLIP backbone and an open-world detection framework along with a lightweight classification and box prediction heads attached to each transformer output token. In order to deploy OWL-ViT on edge devices, the model can use the vanilla Transformers model and get exported to the ONNX model. The exported model further gets quantized and optimized using an ORTQuantizer and ORTOptimizer (Fig. 2).

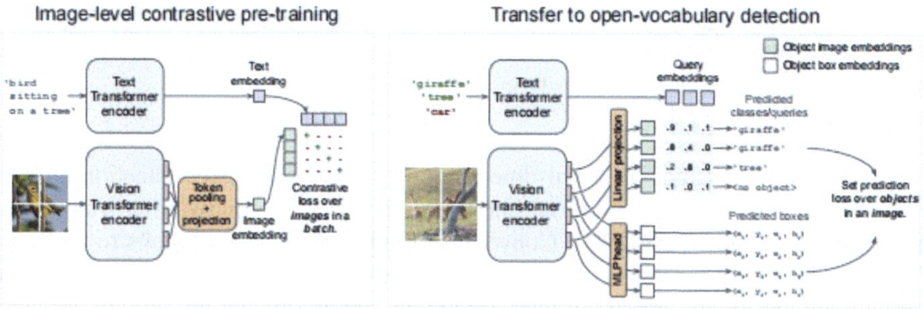

Fig. 2. OWL-ViT Model.

OWL-ViT can be fine-tuned according to the domain-specific input data. This improves performance according to the edge-specific applications. The threshold level between precision and recall value can also be adjusted based on the use case. OWL-ViT also stabilizes optimization by

(a) reducing the speed of the learning rate to prevent the destruction of representation learned during pre-training,
(b) biasing the predicted box coordinate,
(c) stochastic depth regularization, and
(d) shorter training schedules.

3.2.3 Grounding DiNO 1.5 Edge

Grounding DINO 1.5 Edge, is the most efficient, edge-optimized open-set object detection model (Fig. 3).

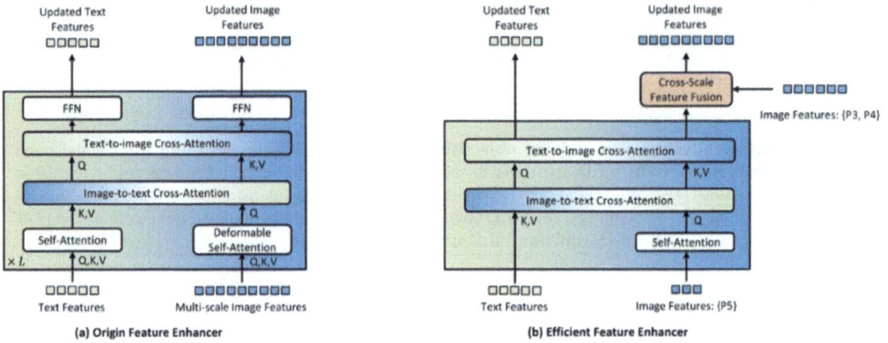

Fig. 3. Grounding DINO 1.5 Edge Feature Comparison.

The image backbone is replaced by EfficientViT-L1 for achieving rapid multi-scale feature extraction. The cross-modality fusion is limited to high-level image features only and a cross-scale feature fusion module is introduced to the low-level image feature. By adopting this methodology feature enhancement is balanced with computational efficiency. The speed of 72.5 FPS is achieved when TensorRT is used for optimization with a performance of 36.2 AP on the LVIS-minval dataset. This makes it suitable for edge computing applications.

3.2.4 Yolo-World

YOLO-World is described as a real-time model for zero-shot object detection model. However, the inference speed on multiple frames is possible only by employing high-performance T4 and V100 GPUs. Converting FP tensors or computations to INT8 or UINT8 tensors reduces the model storage and memory size. The computation speed on certain hardware greatly increases during the INT8 tensor conversion. Also, the latency and power consumption were significantly reduced. However, the consequence of this conversion contributes to accuracy loss to some extent (Table 1).

Table 1. Performance of YOLO model with applied quantization technique.

Model	mAP50–95 SiLU	mAP50–95 ReLU6
yolov8n F32	37.4	34
yolov8n INT8	33.9	31.4
yolov8n INT8 + F32	36.3	33.9

The various types of quantization employed are Float 16, Dynamic Range, and Full Integer.

$$f = (q - z) * s \tag{1}$$

f is the floating-point number.
q is the quantized value.
z is the zero-point.
s is the scale.

The utilization of an open-source toolkit such as ONNXRuntime helps in optimization and speeds up AI inference for various use cases including audio, video, and text.

3.3 Frameworks and Toolkits for Model Optimization

There are different toolkits available for Model optimization. TensorFlow Model Optimization Toolkit, NVIDIA TAO Toolkit, OpenCV, Intel OpenVINO Toolkit, TensorRT, and MLflow/TensorBoard ONNXRuntime are a few among them. In this paper, we used ONNX optimization to reduce the model size and measured the performance in detecting the plastic waste.

3.3.1 OpenVINO

OpenVINO toolkit helps maintain accuracy while providing optimization. It also helps in multi-device compatibility for running inference on multiple compute devices on a single system. OpenVINO runtime allows some layers to run in quant mode and other layers to run in float mode. While using this toolkit the post-processing part of the model is excluded from quantization.

3.3.2 ONNX

The ONNX format facilitates interoperability between different deep-learning frameworks. The performance of the ONNX format can be optimized for multiple hardware platforms and execution environments. Due to the compact nature of the ONNX format, the deployment has become easy on multiple devices such as mobile and embedded systems. ONNX format can be applied across various optimization tools and techniques for the model's performance, such as model size compression, faster inference rate, etc.

4 Implementation

4.1 Zero-Shot Object Detection Framework for Plastic Waste Classification (ZSOD-PWM)

The paper also proposes a novel Zero-shot object detection - plastic waste management (ZSOD-PWM) framework that can used for any type of image classification. ZSOD-PWM decouples the inference pipeline, model selection, post-processing, and visualizing platform. The model framework ZSOD-PWM (Fig. 4) can be generically applied to any type of classification and optimization techniques related to model compression is discussed in this paper.

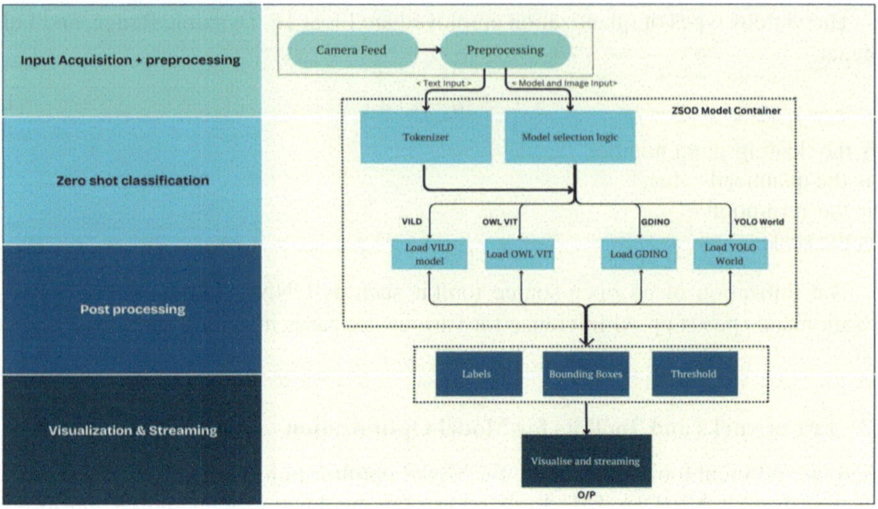

Fig. 4. Real-time Zero-Shot Object Detection Pipeline Framework for Plastic Waste (ZSOD-PWM).

The image classification goes through the pipeline which starts from a) Input Acquisition b) Preprocessing c) Zero-shot Classification d) Post-processing and e) Visualization and Streaming.

Input Acquisition – The image is obtained through the RGB cameras through mobile, webcam, or any camera attached to a microprocessor. The text input and the model selection is obtained through a User GUI.

Pre-processing – In this stage, the text input is sent to the tokenizer which splits into text strings. The image preprocessing is done at this stage with the regions of interest. The selected model input, along with the tokenized strings of text and visual inputs are passed to the respective model selected.

Zero-shot image classification – Based on the selected model and the text and image inputs the model performs the classification and generates the output.

Post-processing – The threshold, bounding boxes, and labels attached to the classified objects comprise the post-processing stage.

Visualization and Streaming – If the detected objects belong to an image, the corresponding classified image is displayed on the terminal of the edge device, or if it's a live stream inference, the corresponding video classification will be streamed on the terminal.

4.2 Zero-Shot Object Detection Model Size Compression

The experiment was performed on Google Collab with GPU T4 and CPU. Internet-sourced plastic waste images were used as input for plastic waste classification. The vision transformer models such as YOLO-World, OWL-ViT, and GDINO converted to the ONNX model. The converted ONNX model is again simplified by replacing the redundant operators with their constant outputs also known as constant folding.

This invariably reduced the model size. Refer Table 2. Further, the transformer models are quantized and optimized to suit edge deployment and also maintain 100 percent latency using Optimum. Dynamic quantization is currently only supported for CPUs. For Grounding DINO, the OpenVino platform offers a quantized version, thereby compressing the model size to perform on edge devices. The model's size was significantly reduced and the edge device was able to load the model and apply image classification on plastic waste images.

Table 2. Model size - Before and After Compression.

Model	ViLD	OWL ViT	GDINO	Yolo-World
Before Compression	5.25 MB	6.7 MB	661.85 MB	25.9 MB
After Compression	3.72 MB	5.97 MB	3.83 MB	4.08 MB (1.83 MB after Compacting)

5 Results and Discussion

Edge deployment has the limitation of processing power, memory limitation, reduced battery life, power capacity, and network bandwidth. By using the ZSOD-PWM framework the model size has been significantly reduced by compression, constant folding and fine-tuning (Fig. 5).

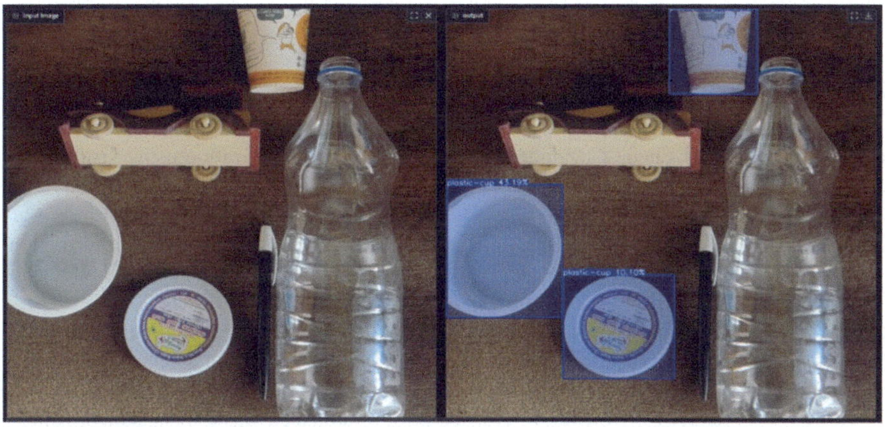

Fig. 5. Image classification - After Model compression on Edge device.

Real-time plastic images were taken as input to the compressed model and the classification was performed correctly despite the compression. The prediction accuracy was consistent. The experiment was carried out on YOLO-World, GDINO, OWL-ViT and ViLD.

6 Challenges and Future Direction

Zero-shot open vocabulary models impress with their performance and leveraging their ability across edge needs further improvements where real-time inference speed is not compromised. Future work in this area has to be in increasing the inference speed by building a common framework where the backbone, heads, and reparameterization can be plugged and played as per the need of the application.The future research should focus more on exploring the world of open vocabulary with zero shot detection as this gives the advantage of changing the text input prompt as per the requirement of the user. There is a big scope in scaling the hardware framework to adapt for the usage of ZSOD models.

7 Conclusion

The paper identifies the suitable zero-shot object detection model(s) that can be used for plastic waste classification. It also analyses the optimization techniques for deploying on edge devices and how model size reduction proves a significant factor in attaining the same. A generic real-time framework for zero-shot detection for plastic waste inference has also been proposed where the stages can be decoupled and optimization can be applied as per the requirement.Edge deployment gives the opportunity to implement zero shot open vocabulary object detection for plastic waste classification through mobile application,

Acknowledgments. We extend our heartfelt thanks to the support extended by the staff and faculty of Networking and Communications for carrying out the research project.

References

1. Huang, K., Lei, H., Jiao, Z., Zhong, Z.: Recycling waste classification using vision transformer on portable device. Sustainability. **13**(21), 11572 (2021). https://doi.org/10.3390/su132111572
2. Prova, N.N.I.: Garbage intelligence: utilizing vision transformer for smart waste sorting. In: 2024 Second International Conference on Intelligent Cyber Physical Systems and Internet of Things (ICoICI), pp. 1213–1219. IEEE (2024)
3. Cai, K., Duan, Z., Liu, G., Fleming, C., Lu, C.X.: Self-adapting large visual-language models to edge devices across visual modalities. In: Lecture Notes in Computer Science, pp. 301–318 (2024). https://doi.org/10.1007/978-3-031-73390-1_18
4. Kashef, R., Alegre, E.P., Prova, T., Aggarwal, S.: Automated waste management using a customized vision-based transformer model. In: 2022 IEEE World AI IoT Congress (AIIoT), pp. 300–309 (2024). https://doi.org/10.1109/aiiot61789.2024.10578946
5. Dong, Y., Zhang, H., Li, C., Guo, S., Leung, V., Hu, X.: Fine-Tuning and Deploying Large Language Models Over Edges: Issues and Approaches (2024)
6. Cai, Y., Li, H., Yuan, G., Niu, W., Li, Y., Tang, X., Ren, B., Wang, Y.: Yolobile: real-time object detection on mobile devices via compression-compilation co-design. Proc. AAAI Conf. Artif. Intell. **35**(2), 955–963 (2021)

7. Xu, J., Pan, F., Han, X., Wang, L., Wang, Y., Li, W.: Edgetrim-YOLO: improved trim YOLO framework tailored for deployment on edge devices. In: 2024 4th International Conference on Computer Communication and Artificial Intelligence (CCAI), pp. 113–118. IEEE (2024)
8. Liu, S., Zha, J., Sun, J., Li, Z., Wang, G.: EdgeYOLO: an edge-real-time object detector. In: In 2023 42nd Chinese Control Conference (CCC), pp. 7507–7512. IEEE (2023)
9. Shuvo, M.M.H., Islam, S.K., Cheng, J., Morshed, B.I.: Efficient acceleration of deep learning inference on resource-constrained edge devices: a review. Proc. IEEE. **111**(1), 42–91 (2022)
10. Mittal, P.: A comprehensive survey of deep learning-based lightweight object detection models for edge devices. Artif. Intell. Rev. **57**(9), 242 (2024)
11. Yang, M., Thung, G.: Classification of Trash for Recyclability Status (2016)
12. Cheng, T., Song, L., Ge, Y., Liu, W., Wang, X., Shan, Y.: YOLO-World: Real-Time Open-Vocabulary Object Detection (2024). (https://arXiv:2401.17270v3)

Eco-Waters: Climate Tales

Find the Future from the Past: Diatoms as Bioindicator of Water Quality

Tri Retnaningsih Soeprobowati[1,2](✉), Muhammad Hadi[2], Jumari Jumari[1,2], Rully Rahadian[1,2], Ni Kadek Dita Cahyani[1,2], Hernanda Afra Haniyyah[1,2], Muhammad Choirul[2], Marsya Abbel Iriyanti[2], and Peter Gell[1,3,4]

[1] Cluster for Paleolimnology (CPalim), Diponegoro University, Semarang, Indonesia
trsoeprobowati@live.undip.ac.id

[2] Departement of Biology, Faculty of Science and Mathematics, Diponegoro University, Semarang, Indonesia

[3] Diponegoro University, Semarang, Indonesia

[4] Future Regions Research Centre, Federation University, Ballarat, VIC, Australia

Abstract. Like nature's diary, sediment cores record an ecosystem's history layer by layer, preserving information on past climates, ecological changes, and human activities. These time capsules reveal environmental shifts over thousands of years through fossils like diatoms, single-celled algae that serve as powerful bioindicators of water quality. Diatoms are sensitive to changes in salinity, pH, and nutrient levels and provide insights into past and present aquatic conditions. Changes in species composition within sediment layers highlight ecological transitions, while chemical signals, such as heavy metals and nutrient levels, trace human impacts like industrialization and agriculture. Radiocarbon dating and isotopic techniques establish a timeline of these events, offering guidance for future ecosystem management. Mangrove ecosystems in Demak, Central Java, have faced severe degradation since the 1980s due to agricultural and aquaculture expansion, leading to coastal abrasion. Rehabilitation efforts since 2012 have supported natural regrowth, gradually restoring these critical habitats. This study focuses on Bedono Village, analyzing land-use changes, diatom distribution, and sediment records to understand environmental shifts and promote sustainable coastal management. Diatom analysis revealed freshwater species dominance (49%), followed by marine (38%), freshwater-marine (9%), and brackish (4%) taxa, reflecting adaptations to diverse salinity gradients. Conventional methods identified 140 species, while eDNA metabarcoding highlighted 20.3% of Bacillariophyta and other non-diatom taxa. These findings underscore diatoms' ecological plasticity and their critical role in monitoring environmental changes, offering valuable insights for the conservation and restoration of mangrove ecosystems.

Keywords: diatoms · lake · mangrove · water quality · sustainable water management

J. C. Casila et al. (Eds.): CCORE 2024, SPEES, pp. 163–173, 2025.
https://doi.org/10.1007/978-3-031-93887-0_16

1 Introduction

Aquatic ecosystems sustain biodiversity, human health, and economic activities worldwide [1]. However, the quality of water in lakes, rivers, and wetlands is increasingly compromised due to rapid industrialization, urbanization, and agricultural practices [2]. Maintaining water quality has become a pressing issue as pollutants accumulate, demanding innovative and practical monitoring tools. Among these, microscopic, single-celled algae diatoms stand out as robust bioindicators due to their sensitivity to environmental changes, extensive fossil record, and diversity [3]. This paper explores how diatoms, preserved in sediment cores, reveal past and present water quality and serve as predictors of future aquatic conditions.

Sediment cores extracted from lakebeds, mangroves, and riverbeds are natural archives that capture an area's ecological and environmental history layer by layer [4]. Much like pages in an ancient book, these cores preserve snapshots of the conditions at deposition. With each layer accumulating over centuries or even millennia, sediment cores record data on temperature fluctuations [5], pH levels, salinity [6], and nutrient concentrations [7]. These stratified layers act as nature's diary, logging natural changes and human impacts. Dating techniques, particularly radiocarbon dating, allow scientists to assign ages to each layer, constructing a chronological framework for environmental events. This ability to reconstruct the timeline of an ecosystem provides valuable context for understanding long-term ecological shifts and trends [8].

Diatoms, belonging to a group of single-celled algae, have emerged as some of the most effective bioindicators for water quality assessment. Found in nearly all aquatic environments, they are susceptible to changes in their surroundings, including shifts in nutrient levels, pH, temperature, and salinity [3]. Diatoms produce unique, silica-based cell walls that fossilize well, allowing their remains to persist in sediment layers for millions of years. Adaptability, sensitivity, and fossil preservation make diatoms invaluable for tracking historical environmental changes [9]. Different diatom species thrive under specific environmental conditions. For instance, some species are highly tolerant of acidic conditions [10], while others require high nutrient levels [11]. Researchers can identify fluctuations in water quality over time by analyzing the composition of diatom communities in sediment layers [12]. This way, diatoms reveal insights into past aquatic conditions, allowing scientists to understand how human activities and natural processes have influenced water ecosystems.

Mangrove ecosystems in Demak, Central Java, have faced significant degradation since the 1980s due to their conversion into agricultural land and aquaculture, leading to severe coastal abrasion [13]. However, since 2012, rehabilitation efforts by stakeholders and local communities, along with natural regrowth, have begun restoring these critical habitats. This study focuses on the mangrove ecosystem in Bedono Village, Demak Regency, analyzing land-use changes, the vertical distribution of diatoms, and sedimentary records to understand the impacts of environmental shifts and support sustainable coastal management. Examining diatom abundance, diversity, and community composition helps identify key drivers of ecological transformation, offering critical guidance for conserving and restoring coastal ecosystems like those in Bedono.

2 Methods

2.1 Study Area

Bedono Village, Demak is located in the Java, Indonesia, a volcanic highland characterized by active geothermal activity and nutrient-rich volcanic soils. The lake's catchment area spans agricultural and urban landscapes, making it a valuable case study for assessing the impacts of land-use change on freshwater ecosystems (Fig. 1). The two study locations were chosen based on proximity to contrasting environmental influences:

- **Site A:** Minimal human activity, located near undisturbed forested areas.
- **Site B:** Close to agricultural fields and urban development, subject to higher nutrient runoff.

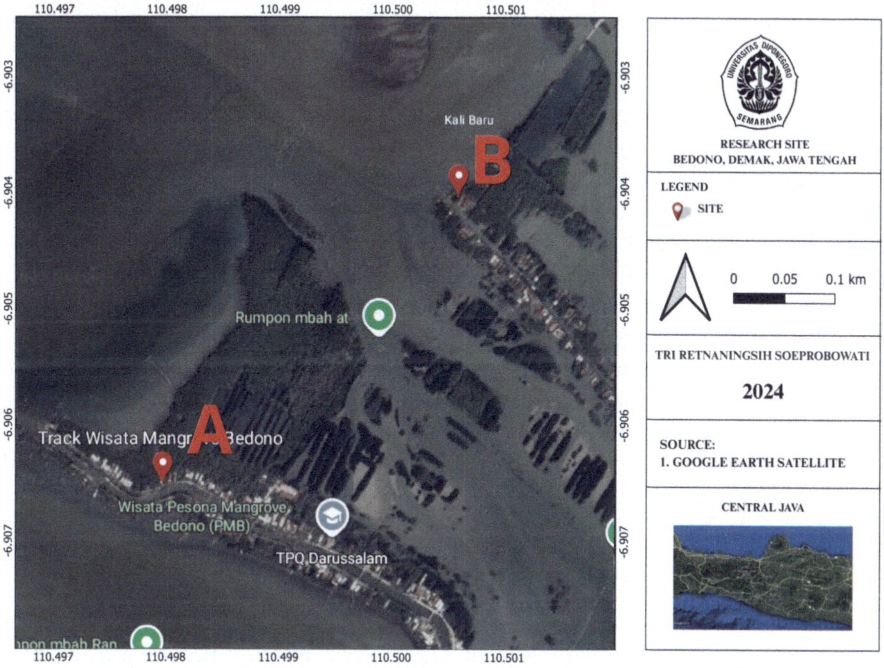

Fig. 1. Sampling sites for sediment in Bedono Village, Sayung, Demak Regency, Central Java.

2.2 Sediment Core Collection

Sediment cores were retrieved using a D'Section/Russian corer from the deepest points of the lake at Sites A and B. Each core reached a depth of 300 cm and was sectioned into 5 cm intervals. The samples were transported under controlled conditions to preserve their physical and chemical integrity.

Sediment cores were extracted from two sampling sites (see Fig. 1) using D'Section corer, which preserves sediment layers in sequence. Cores were taken from undisturbed areas with minimal human interference to ensure accurate historical records. The core length varied depending on sediment depth, and samples were collected to provide sufficient stratification for detailed analysis.

Each core was sectioned regularly (every 1 cm) to represent different sediment layers.

2.3 Diatom Preparation and Identification

From each sediment layer, diatoms were extracted using a series of chemical treatments to remove organic and mineral matter, leaving purified diatom frustules (silica-based cell walls). Samples were treated with hydrochloric acid (10% HCl) and hydrogen peroxide (10% H_2O_2) to clear organic material and then washed and centrifuged repeatedly to concentrate the diatom residue.

Processed samples were mounted on glass slides and examined under a high-resolution microscope. Using taxonomic keys and reference images [14–18], diatom frustules were identified until at least 300 valves were found at the species level. Diatom species composition and abundance were recorded for each layer, with special attention to indicator species sensitive to changes in pH, nutrient levels, and salinity.

Historical water quality trends were reconstructed by mapping diatom composition changes across dated layers, performed by PAST version 4.17. Significant shifts in species composition were examined to pinpoint potential events, such as eutrophication, pollution, or changes in salinity, associated with natural or anthropogenic factors.

The population and species found were processed into a diatom stratigraphic diagram using C2 version 1.8.0 [19] and cluster analysis using the PAST 4.17 program [20].

Reconstructed water quality data from diatom assemblages were compared with historical records and validated against modern water quality assessments from the same sites. This method provides a comprehensive approach to using diatoms as bioindicators of water quality through sediment core analysis, linking past conditions with current and future ecosystem management needs.

2.4 Land-Use Change Analysis

Land-use changes from 2003 to 2023 were analyzed using satellite imagery and GIS software. Changes in forest cover, agricultural expansion, and urbanization were quantif.

3 Result and Discussion

3.1 Reconstructing Historical Water Quality

The degradation of mangrove ecosystems in Bedono Village, Sayung, Demak Regency, Central Java, illustrates a stark environmental crisis. Mangrove coverage declined from 2.24 hectares in 2003 to just 24.49 hectares by 2023 (see Fig. 2). Rising sea levels, coastal erosion, and land subsidence drive this significant loss. The rapid disappearance of mangroves, which act as natural barriers against tidal surges and erosion, has left the area increasingly vulnerable to environmental and socio-economic challenges [21].

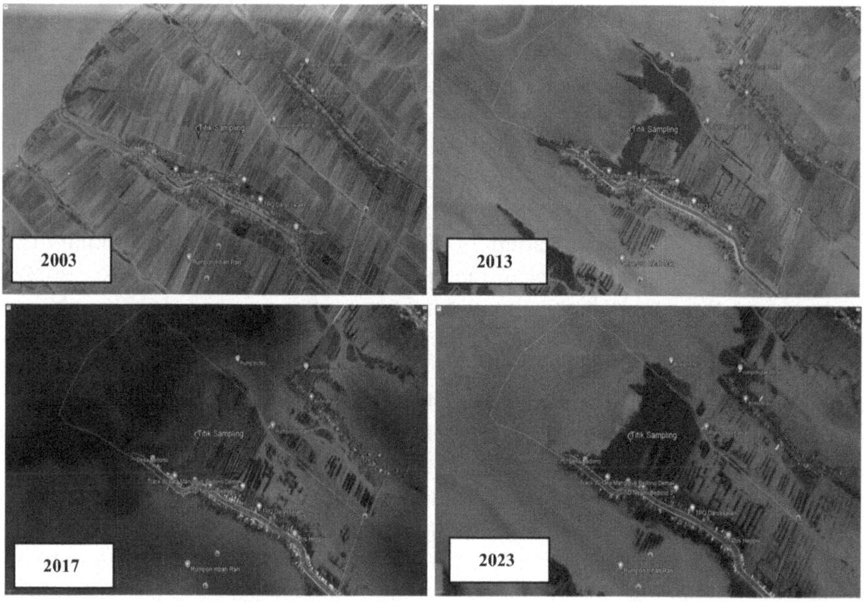

Fig. 2. Changes in Mangrove Coverage in Bedono Village from 2003 to 2023.

Understanding past water quality is essential for predicting future trends. Diatoms directly link to this historical perspective, offering insights into changes in salinity, pH, and nutrient levels that are otherwise challenging to reconstruct. For example, diatom species sensitive to nutrient overloads, such as those caused by agricultural runoff, can indicate periods of eutrophication—when nutrient levels become excessively high, leading to issues like algal blooms and oxygen depletion in aquatic ecosystems [22].

Through chemical analysis and isotope dating, scientists can correlate diatom assemblages with known environmental disturbances [23]. Heavy metals, pesticides, and increased nutrient levels leave chemical signatures within sediment layers that align with periods of industrialization, agricultural expansion, or urban development [24]. These signals allow scientists to identify the onset of anthropogenic impacts and assess the long-term effects on water bodies.

3.2 Diatoms and Modern Water Monitoring

Today, diatoms are at the forefront of water quality monitoring. Their rapid response to environmental changes allows real-time water quality tracking in lakes, rivers, and estuaries. The species composition within diatom communities can signal the onset of pollution events or nutrient shifts before they become severe, providing an early warning system for ecosystem managers. This predictive capability is precious in climate change, as diatoms respond quickly to temperature fluctuations and other climate-related stressors [25]. Monitoring diatom communities can also help manage water resources more sustainably. By identifying areas where water quality is declining, authorities can take preemptive actions to mitigate pollution sources, prevent habitat degradation, and maintain ecosystem health [26]. Diatom-based monitoring thus bridges the gap between historical records and contemporary water management, helping policymakers and environmentalists respond proactively to ecological threats [27].

The diversity of diatoms across different habitats reveals distinct ecological patterns influenced by environmental gradients. In the study, freshwater diatoms dominated the community, accounting for 49% of the identified taxa (see Fig. 3). This dominance reflects the widespread availability of freshwater systems and their nutrient-rich environments, which support a variety of diatom species adapted to low salinity [28]. Marine diatoms, contributing 38% of the diversity, underscore the ecological specialization of taxa thriving in high-salinity conditions, often associated with stable and nutrient-variable oceanic environments. Freshwater-marine diatoms, representing 9%, occupy transitional zones where freshwater meets marine systems. These habitats provide unique ecological niches characterized by fluctuating salinity, supporting taxa with high adaptability [29]. The presence of brackish diatoms, at 4%, highlights the specialized communities adapted to intermediate salinity levels found in estuaries and coastal lagoons. This habitat-based distribution of diatoms illustrates their ecological plasticity and adaptability to diverse environmental conditions [29]. Such diversity is crucial for understanding ecosystem dynamics, as diatoms play key roles in primary production and nutrient cycling [30]. Moreover, these findings reinforce the importance of conserving diverse aquatic habitats to maintain diatom communities' ecological functions and services.

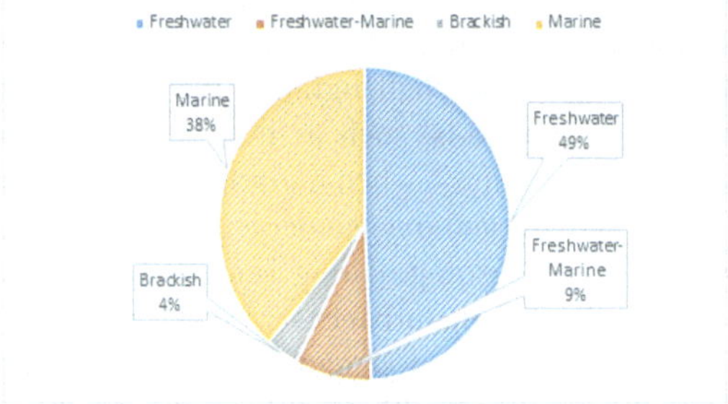

Fig. 3. Diversity of Diatoms Across Freshwater, Freshmarine, Marine, and Brackish Environments.

The analysis of diatom communities across freshwater, freshwater marine, and marine environments reveals significant similarities in composition (see Fig. 4), highlighting the ecological versatility of diatoms and their capacity to adapt to diverse aquatic habitats. Freshwater marine environments, as transitional zones, act as environmental bridges, supporting diatom species adapted to both freshwater and marine conditions. This overlap emphasizes the role of salinity gradients in shaping community composition. The persistence of specific diatom taxa across these environments further underscores their adaptive strategies, such as varying cell wall silicification [31] and osmoregulatory mechanisms, allowing them to thrive under fluctuating conditions [32]. The observed similarities also highlight the importance of diatoms as bioindicators. Their widespread distribution and sensitivity to environmental changes make them ideal for monitoring ecosystem health and detecting shifts in aquatic conditions. Understanding these patterns can aid in predicting ecosystem responses to environmental pressures, such as climate change and anthropogenic disturbances [33], particularly in regions where freshwater and marine systems interact.

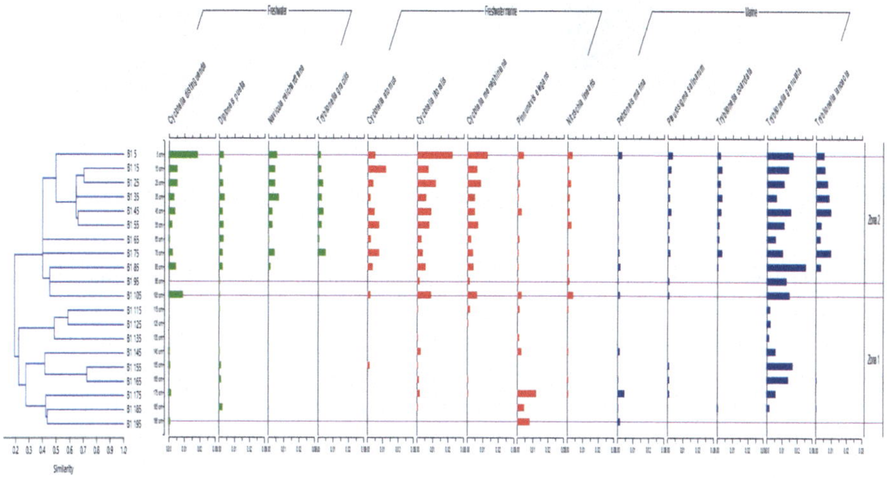

Fig. 4. Similarity of Diatom Communities Across Freshwater, Freshmarine-water and Marine Environments.

3.3 Linking Sediment Data to Land-Use Changes (2003–2023)

Figure 2 illustrates land-use changes over the past 20 years, highlighting agricultural expansion and urbanization as key drivers of alteration in the catchment landscape. These land-use changes coincide with significant shifts in diatom assemblages, as demonstrated by studies linking agricultural intensification to an increase in eutrophic indicators. Specifically, the abundance of *Nitzschia palea* (an indicator of eutrophic conditions) shows a marked increase between 2010 and 2015, aligning with periods of intensified agriculture [34]. Conversely, a notable decline in oligotrophic diatoms, such as *Aulacoseira spp.*, corresponds to the reduction in forest cover and increased nutrient runoff from agricultural lands [35, 36]. These shifts suggest that the landscape modifications have directly contributed to changes in water quality and ecosystem health.

3.4 Diatoms from eDNA Metabarcoding

The integration of conventional identification methods and eDNA metabarcoding has provided valuable insights into the biodiversity of aquatic ecosystems (see Fig. 5). Conventional methods identified 140 species, showcasing a comprehensive inventory of the observed taxa. However, the application of eDNA metabarcoding revealed a distinct subset of the community composition, with 20.3% of identified taxa belonging

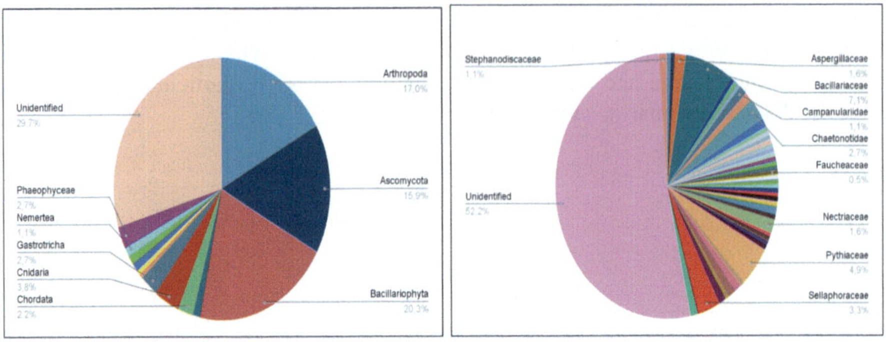

Fig. 5. eDNA metabarcoding analysis of sediment samples from Bedono mangrove ecosystem.

to Bacillariophyta, including families such as Stephanodiscaceae, Bacillariaceae, and Sellaphoraceae.

Additionally, the detection of Campanulariidae and Chaetonotidae highlights the diversity of non-diatom groups captured through eDNA analysis. The disparity between the two approaches underscores the complementary nature of these methods. Conventional techniques provide detailed morphological information, while eDNA metabarcoding offers rapid and sensitive detection of molecular signatures, capturing both abundant and cryptic species [37]. The dominance of Bacillariophyta in the eDNA results suggests their ecological significance and highlights the potential of molecular approaches for monitoring primary producers in aquatic systems [38]. This combination of methods enriches our understanding of biodiversity patterns and supports more robust ecosystem assessments. However, methodological limitations, such as biases in DNA reference databases and species detectability, must be addressed to maximize the reliability of eDNA metabarcoding in future studies.

3.5 Implications for Sustainable Water Management

As human activities continue to impact water ecosystems, diatoms will play an increasingly vital role in protecting water quality. Their presence across nearly all aquatic habitats and their resilience in responding to environmental shifts make them ideal for local and global water quality assessments. Diatoms are especially valuable for developing countries, where traditional water monitoring infrastructure may be limited; their analysis offers a low-cost, highly effective means of assessing ecosystem health [39]. Integrating diatom research into policy and management frameworks can strengthen efforts to maintain sustainable water quality. From guiding restoration projects to informing agricultural practices, diatoms provide critical insights to help governments, industries, and communities make informed decisions. Protecting water resources not only safeguards biodiversity but also ensures the continued availability of clean water for future generations [40].

4 Conclusion

Diatoms offer a window into water quality's past, present, and future. Their fossilized remains in sediment cores allow scientists to reconstruct environmental histories, revealing ecological resilience and vulnerability patterns. As bioindicators, diatoms have proven effective in tracking natural and human-induced changes in aquatic environments, offering a powerful tool for addressing today's water quality challenges. In a world facing escalating environmental pressures, diatoms stand as silent yet eloquent witnesses to the impact of human actions on aquatic ecosystems, guiding us toward sustainable water management practices.

References

1. Bernhardt, J.R., O'Connor, M.I.: Aquatic biodiversity enhances multiple nutritional benefits to humans. Proc. Natl. Acad. Sci. U. S. A. **118**(15), e1917487118 (2021)
2. Akhtar, N., et al.: Various natural and anthropogenic factors responsible for water quality degradation: a review. Water. **13**(19), 2660 (2021)
3. Taurozzi, D., Cesarini, G., Scalici, M.: Diatoms as bioindicators for health assessments of ephemeral freshwater ecosystems: a comprehensive review. Ecol. Indic. **166**, 1470–1160 (2024)
4. Ramírez, G.A., et al.: The Guaymas Basin subseafloor sedimentary archaeome reflects complex environmental histories. iScience. **23**(9), 101459 (2020)
5. Baumer, M.M., et al.: Climatic and environmental changes in the Yana Highlands of north-eastern Siberia over the last c. 57 000 years, derived from a sediment core from Lake Emanda. Boreas. **50**(1), 114–133 (2021)
6. Costa, L., et al.: Rare earth element distributions in salt marsh sediment cores reveal evidence of environmental lability during bioturbation and diagenetic processes. Chem. Geol. **584**, 120503 (2021)
7. Luo, M., et al.: Research on the geochemical background values and evolution rules of lake sediments for heavy metals and nutrients in the Eastern China Plain from 1937 to 2017. J. Hazard. Mater. **436**, 129136 (2022)
8. Palmisano, A., et al.: Long-term demographic trends in prehistoric Italy: climate impacts and regionalised socio-ecological trajectories. J. World Prehist. **34**(3), 381–432 (2021)
9. Bryłka, K., et al.: Uncertainties surrounding the oldest fossil record of diatoms. Sci. Rep. **13**(1), 8047 (2023)
10. Saxena, A., et al.: Diatoms recovery from wastewater: Overview from an ecological and economic perspective. J. Water Process. Eng. **39**, 101705 (2021)
11. Bi, R., et al.: Responses of marine diatom-dinoflagellate competition to multiple environmental drivers: abundance, elemental, and biochemical aspects. Front. Microbiol. **12**, 731786 (2021)
12. Shibabaw, T., et al.: Diatom community structure in relation to environmental factors in human influenced rivers and streams in tropical Africa. PLoS One. **16**(2), e0246043 (2021)
13. Rahim, A., Soeprobowati, T.R., Putranto, T.T.: Mangrove species structure at the Tuntang Estuary, Demak, Central Java. IOP Conf. Ser: Earth Environ. Sci. **1137**(1), 012053 (2023)
14. Krammer, K., Lange-Bertalot, H.: Bacillariophyceae. 2. Teil: Bacillariaceae, Epi-themiaceae, Surirellaceae In: Ettl, H., Gerloff, J., Heynig, H., Mollen-hauer, D. (eds.) Susswasserflora von Mitteleuropa. ustav Fisher Verlag, Jena (1988)

15. Krammer.: Die cymbelloiden Diatomeen. Eine Monographie der weltweit bekannten taxa. Teil 1. Allgemeines und Encyonema Part. In: Bibliotheca Di-atomologica, p. 36. J. Cramer, Berlin-Stuttgart (1997)
16. Krammer, K., Lange-Bertalot.: Bacillariophyceae: Teil 3: Centrales, Fragillaria-ceae, Eunotiaceae. In: Süßwasserf. Spectrum, Berlin (2004)
17. Krammer, K., Lange-Bertalot.: Bacillariophyceae: Teil 2: Bacillariphyceae, Ep-ithemiaceae, Surirellaceae. In: Süßwasserf. Spectrum, Berlin (2004)
18. Krammer, K., Lange-Bertalot.: Bacillariophyceae: Teil 1: Naviculae. In: Süßwas-serf. Spectrum, Berlin (2004)
19. Juggins, S.: C2 Software for Ecological and Palaeoecological, Data Analysis and Visualisation. Newcastle University (2022) Accessed 13 Sept 2024
20. Hammer, Ø., David, A.T.H., Ryan, P.D.: PAST: paleontological statistics software package for education and data analysis. Palaeontol. Electron. **4**(1), 9–18 (2024) Accessed 13 Sept 2024
21. Bhowmik, A.K., et al.: Global mangrove deforestation and its interacting social-ecological drivers: a systematic review and synthesis. Sustainability. **14**(8), 4433 (2022)
22. Wei, Y., et al.: Temperature and nutrients drive distinct successions between diatoms and dinoflagellates over the past 40 years: implications for climate warming and eutrophication. Sci. Total Environ. **931**, 172997 (2024)
23. Li, Y.-r., et al.: Effects of climate warming and declining water quality on the eco-environmental evolution of Jinmucuo Lake: evidence from sedimentary diatom assemblages. Heliyon. **10**(21), e39985 (2024)
24. Wang, Y., et al.: Synergetic loss of heavy metal and phosphorus: evidence from geochemical fraction and estuary sedimentation. J. Hazard. Mater. **416**, 125710 (2021)
25. Essa, D.I., et al.: Assessing phytoplankton populations and their relation to water parameters as early alerts and biological indicators of the aquatic pollution. Ecol. Indic. **159**, 111721 (2024)
26. Wolfram, J., et al.: Water quality and ecological risks in European surface waters—monitoring improves while water quality decreases. Environ. Int. **152**, 106479 (2021)
27. Kahlert, M., et al.: Biodiversity patterns of Arctic diatom assemblages in lakes and streams: current reference conditions and historical context for biomonitoring. Freshw. Biol. **67**(1), 116–140 (2022)
28. Stenger-Kovács, C., et al.: Diatom community response to inland water salinization: a review. Hydrobiologia. **850**(20), 4627–4663 (2023)
29. Glaser, K., Karsten, U.: Salinity tolerance in biogeographically different strains of the marine benthic diatom Cylindrotheca closterium (Bacillariophyceae). J. Appl. Phycol. **32**(6), 3809–3816 (2020)
30. Behrenfeld, M.J., et al.: Thoughts on the evolution and ecological niche of diatoms. Ecological monographs. **91**(3), e01457 (2021)
31. Kamakura, S., Bilcke, G., Sato, S.: Transcriptional responses to salinity-induced changes in cell wall morphology of the euryhaline diatom Pleurosira laevis. J. Phycol. **60**(2), 308–326 (2024)
32. Nakov, T., et al.: Transcriptional response of osmolyte synthetic pathways and membrane transporters in a euryhaline diatom during long-term acclimation to a salinity gradient. J. Phycol. **56**(6), 1712–1728 (2020)
33. Costa, A.P.T., Schneck, F.: Diatoms as indicators in running waters: trends of studies on biological assessment and monitoring. Environ. Monit. Assess. **194**(10), 695 (2022)
34. Smith, V.H., Sturm, B.S.M., Denoyelles, F.J., Billings, S.A.: The ecology of algal biodiesel production. Trends Ecol. Evol. **25**(5), 301–309 (2018)
35. William, D.M., et al.: Studies on Type Material from Kutzing's Diatom Collection IV: The Basionym, Author, and Type of Tetracyclus rupestre. Phytotaxa. **498**(1), 44–50 (2021)

36. Codex, Y.: Advancements in DNA barcoding for eDNA metabarcoding in freshwater ecosystems: unveiling cryptic species and enhancing biodiversity monitoring and conservation. Yubetsu Codex Ecol. **1**(2) (2023)
37. Kang, Y., Kwak, I.-S., Kang, C.-K.: Co-occurrence of Bacillariophyceae-based-and Cryptophyceae-based planktonic food webs in a temperate estuarine ecosystem revealed via eDNA. Estuar. Coast. Shelf Sci. **264**, 107661 (2022)
38. Apothéloz-Perret-Gentil, L., et al.: Monitoring the ecological status of rivers with diatom eDNA metabarcoding: a comparison of taxonomic markers and analytical approaches for the inference of a molecular diatom index. Mol. Ecol. **30**(13), 2959–2968 (2021)
39. Masouras, A., et al.: Benthic diatoms in river biomonitoring—present and future perspectives within the water framework directive. Water. **13**(4), 478 (2021)
40. Brown, T.A., et al.: Influence of nutrient availability on arctic sea ice diatom HBI lipid synthesis. Org. Geochem. **141**, 1–5 (2020)

Air Pollution Tolerance Index (APTI) Mangrove Plants in the Wonorejo Surabaya Mangrove Centre Area

Diana Putri Jumayu Andalbuti[1,2], Lilih Khotimperwati[1,2](✉), Tri Retnaningsih Soeprobowati[1,2], and Jumari Jumari[1,2]

[1] Biology Department, Faculty of Sciences and Mathematics, Diponegoro University, Semarang, Indonesia
lilihkhotimperwati@live.undip.ac.id

[2] Cluster for Paleolimnology, Diponegoro University, Semarang, Indonesia

Abstract. Surabaya is one of the centers of industrial areas in East Java, ranked as the 4th most polluted city in Indonesia according to the Air Quality Index (AQI) 2024; on the other hand, Surabaya City has great mangrove forest potential, particularly in the Wonorejo Mangrove Center located in Rungkut District. Due to its strategic position at the edge of the city and coastal areas, this mangrove forest holds promise as a natural solution to mitigate air pollution. Environmental pressures from industrial activities, population growth, and pond creation around the mangrove forests to the preservation of mangrove sustainability. The plants' assessment of the species diversity of mangroves and the evaluation of the tolerance levels of the plants were done using the Air Pollution Tolerance Index (APTI). The research was conducted from September 2023 to June 2024 in the Wonorejo Mangrove Center, Rungkut District, Surabaya City. The APTI assessment involved four biochemical parameters: relative water content, total chlorophyll content, ascorbic acid content, and leaf extract pH. The results showed that the APTI values of mangrove plants ranged from very tolerant to intermediate, with *Rhizophora mucronata* showing the highest value (107.24 ± 0.15) and *Anredera cordifolia* the lowest (18.53 ± 0.45). The results indicate that all mangrove species possess the potential as bioreactors for air pollution in Surabaya City.

Keywords: Mangrove · Air Pollution Tolerance Index (APTI) · Chlorophyll · Ascorbic acid · pH · Relative water content

1 Introduction

Urban areas face significant challenges in managing air pollution due to rapid urbanization, transportation, and industrialization. Surabaya, as the 4th most polluted city in Indonesia according to the Air Quality Index (AQI) [1] experiences severe air quality challenges. The contributors to this pollution are emissions from motorized vehicles, industrial activities, which release harmful pollutants such as SO_2, NO_2, CO, particulate matter ($PM_{2.5}$ and PM_{10}), and ground-level ozone [2, 3]. These pollutants have adverse effects on human health and environmental quality, making air pollution a critical problem in Surabaya [4].

J. C. Casila et al. (Eds.): CCORE 2024, SPEES, pp. 174–189, 2025.
https://doi.org/10.1007/978-3-031-93887-0_17

In response to these challenges, mangrove ecosystems in the Wonorejo Mangrove Centre, Rungkut Sub-district in Surabaya, hold great potential for mitigating air pollution [5, 6]. As one of the most extensive mangrove areas in the region, these ecosystems are strategically located on the city's edge, serving as a buffer zone for pollutant absorption. Mangroves play a significant role in absorbing carbon and air pollutants [7–9], making them a natural solution to reducing the impact of air pollution.

However, the process of absorbing these pollutants can affect the morphological, physiological, and biochemical structures of plants, especially their leaves, which act as the primary medium for interaction with air [10–12]. Air pollution is known to reduce chlorophyll levels, trigger leaf corrosion, and disrupt photosynthetic capacity [13]. Certain plants can adapt to these environmental stresses through biochemical changes, such as increased ascorbic acid levels, pH, and relative water content in leaves, which support their protective mechanisms against pollution [14, 15]. Understanding the tolerance levels of mangroves forests under such conditions is essential to optimize their role in pollution mitigation.

Although research on the Air Pollution Tolerance Index (APTI) has been conducted on various plant species in urban areas, the findings demonstrate that APTI values can effectively guide the selection of pollution-tolerant species suitable for urban greening, such as in Debrecen, Hungary [16]. For instance, researchers have utilized APTI values to assess the tolerance levels of plant species in green open spaces in the Kemiling sub-district, Bandar Lampung [17]. Furthermore, studies have shown that variations in APTI values among landscape plants reflect their physiological adaptations and resilience to environmental stress [18]. However, studies focusing on APTI in mangrove ecosystems are scarce. Given the strategic role of mangroves in mitigating air pollution, further in-depth research is crucial to identify pollution-tolerant mangrove species and evaluate their potential as environmental bioindicators.

This study aims to evaluate the APTI value of mangrove species in the Wonorejo Mangrove Center, Surabaya. Specifically, this research aims to identify highly tolerant mangrove species that can serve as natural bioreductors. The results of the study are expected to provide new contributions in the utilization of mangrove ecosystems as air pollution mitigation solutions, support sustainable mangrove management, and provide science-based policy recommendations in an effort to achieve SDGs 15 (life on land), SDGs 11 (sustainable cities and communities), and SDGs 13 (climate action).

2 Material and Method

2.1 Study Location

The Wonorejo Mangrove Center is located in the Rungkut sub-district of Surabaya City. This 200-hectare area encompasses two distinct ecotourism sectors: mangrove ecotourism and pond ecotourism. This research focuses on the mangrove plants in the area, and due to the diversity of species present, it allows for the comparison of individual species. The Rungkut sub-district serves as the center of the industrial zone in Surabaya City [19]. As a result, industrial activities in the area produce air pollutants such as SO_2, NO_2, and particulates, which can affect the ambient air quality, including in nearby residential areas. Additionally, this air pollution exerts ecological pressure on mangrove ecosystems, leading to a decline in leaf quality due to exposure to pollutants, which may disrupt photosynthesis and cause physiological and biochemical changes in plants.

Due to industrial activities and urbanization, the mangrove forests in Wonorejo have experienced a significant decline. However, since 2009, mangrove rehabilitation programs have been underway to restore the function of coastal ecosystems [20]. These efforts have succeeded in increasing vegetation cover and enhancing biodiversity. However, the continued exposure of mangroves to pollutants, such as SO_2, NO_2, and particulate matter, presents ongoing challenges to their ecological function. Mangroves in this location are critical not only for pollution mitigation but also for providing essential ecosystem services, such as carbon sequestration, coastal protection, and habitat for a variety of flora and fauna [7, 10, 28]. The reason for selecting this research site lies in its unique position as a meeting point of urban, industrial, and ecological dynamics. The proximity to industrial activities provides a valuable opportunity to study the direct impacts of air pollutants on mangrove physiology and the Air Pollution Tolerance Index (APTI). In addition, historical rehabilitation efforts offer a unique context to assess the resilience and adaptability of mangrove species under environmental stress [6, 28]. By evaluating the APTI of mangrove species in this area, this study will provide insights into their potential as natural bioreductors, contributing to local air pollution mitigation and global climate action goals.

2.2 Fieldwork

Based on its land function, the selected station is the Wonorejo Mangrove Center area as an ecotourism area, which is ±2.4 km from the estuary, with four sampling points as follows: SP1: 112°49′1.2″E 7°18′32.4″S; SP2: 112°49′12″E 7°18′36″S; SP3: 112°49′26.4″E 7°18′28.8″S; and SP4: 112°49′30″E 7°18′28.8″S. This research location was chosen due to its proximity to industrial and residential areas, which makes it highly susceptible to air pollution, including SO_2, NO_2, and other particulates known to exert ecological pressure on mangrove ecosystems. Identification of mangrove species was carried out directly at the research site using the plant net application and the Indonesian Mangrove Introduction Handbook.

Leaves from each species found at the research site were collected for further analysis. The leaves were taken from the 3rd, 4th, and 5th positions on the branches to ensure uniformity in leaf age. Subsequently, the leaves were placed in labeled zip-lock bags and stored in a cool box containing ice gel to maintain their freshness during storage (up to 1 week). This storage condition was designed to minimize physical and chemical changes in the leaves during transport and storage prior to analysis. These leaves will later be used to analyze four (4) APTI biochemical parameters, including, which include relative water content, total chlorophyll, ascorbic acid, and leaf pH.

The collected leaves were then brought to the Ecology and Biosystematics Laboratory, Department of Biology, Faculty of Science and Mathematics, Diponegoro University, to undergo a series of analyses carried out in three (3) replicates (Fig. 1).

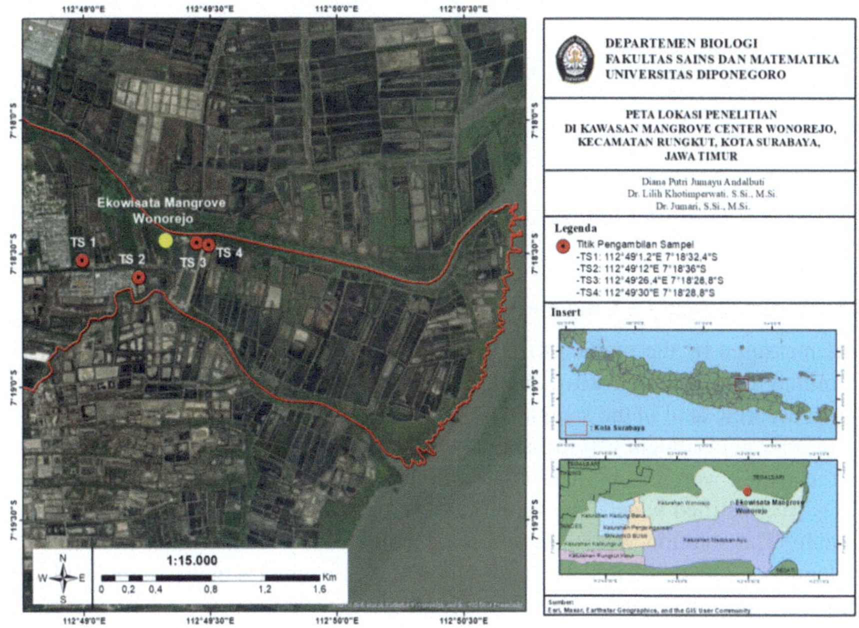

Fig. 1. Map of the research location in the Wonorejo Mangrove Center Area, Rungkut District, Surabaya City.

2.3 Measurement of Physiological and Biochemical Parameter

The selection of biochemical parameters in this study was based on their ability to reflect plant responses to air pollution. These parameters include leaf relative water content, total chlorophyll content, ascorbic acid, and leaf extract pH.

2.3.1 Leaf Relative Water Content (RWC%)

5 g of developed leaves from each plant (FW) were soaked with distilled water for 24 h. Then, they were dried with paper towels and weighed for turgid weight (TW). Next, they were dried in an oven at 60 °C for 48 h or until the weight was constant. RWC was calculated using the following equation [21, 22].

$$RWC\% = [((FW - DW))/((TW - DW))] \times 100\% \quad (1)$$

(FW: Fresh Weight, DW: Dry Weight, TW: Turgid Weight)

2.3.2 Total Chlorophyll Content (TCH mg.g−)

0.5 g of fresh leaves were pulverized using a mortar and then extracted with 10 mL of acetone. The solution was then filtered, homogenized, and put into a cuvette. The solution was analyzed using a spectrophotometer with absorbance at wavelengths of 645 nm and 663 nm [21, 23]. Then, the chlorophyll content was calculated with the following equation:

$$Chlorophyll\ a\ (mg/ml) = (12.7\ A663\text{--}2.69\ A645) \quad (2)$$

$$Chlorophyll\ b\ (mg/ml) = (22.9\ A645\text{--}4.68\ A663) \quad (3)$$

$$Total\ Chlorophyll\ (mg/ml) = (20.2\ A645 + 8.02\ A663) \quad (4)$$

2.3.3 Ascorbic Acid (AA mg.g−)

It was measured by the method of UV-Vis spectrophotometric [24, 25]. Starting with determining the maximum wavelength, making a standard curve, and determining the value of ascorbic acid in the sample.

Ascorbic acid content was measured by the spectrophotometric method. Samples as much as 0.25 g were taken, then pulverized using a mortar and extracted with 10 mL of CO_2-free distilled water. The solution was filtered and homogenized to be put into a cuvette. The solution was analyzed using a spectrophotometer with an absorbance wavelength of 290 to 300 nm. The absorbance value obtained is then extrapolated using the equation from the standard curve that has been received.

2.3.4 Leaf Extract pH

5 g of fresh leaves were pulverized using a mortar and then extracted with 50 mL of CO_2-free distilled water. The solution was filtered and homogenized to be measured with the help of a pH meter [21, 26].

2.3.5 Air Pollution Tolerance Index (APTI)

The APTI of each species was calculated using four parameters: moisture content, pH, chlorophyll value, and ascorbic acid content [21, 27]. Furthermore, the APTI value was calculated using the following equation:

$$APTI = \left(AA \times \left[(TCH + pH) + RWC\right]\right)/10 \quad (5)$$

(in which AA is in mg/g, TCH is in mg/g, and RWC is in percentage)

In the classification method, the tolerance level of each mangrove plant is obtained directly by comparing its APTI value with the index value described by [28], tolerance is divided into four levels, as shown in Table 1.

2.4 Data Analysis

ANOVA statistically analyzed data to determine differences in the mean values of biochemical components and APTI ($p < 0.05$). Then, Duncan's further test was conducted to determine which species pairs differed significantly ($p < 0.05$). A multiple correlation test was used to determine the relationship between biochemical parameters (independent) and APTI values (dependent), and graphs were made with MS Excel 2010.

Table 1. Plant Tolerance Level Criteria Based on APTI value.

APTI value range	Response
31–100	Very tolerant
21–30	Tolerant
11–20	Intermediate
1–10	Sensitive
<1	Very sensitive

3 Result and Discussion

The Air Pollution Tolerance Index (APTI) indicates the ability of plants to tolerate air pollution. Plants with higher APTI values can act as a solution to air pollution, while plants with low APTI values can be bioindicators of air pollution levels [27]. APTI values can also be used to determine the potential of plants as air pollution acceptors. Generally, when plants are exposed to pollutants, they will show symptoms of changes in the phytochemical content in their leaves [18]. The determination of APTI values is based on four parameters, including total relative water content, total chlorophyll content, ascorbic acid content, and pH of plant leaf extracts.

3.1 Leaf Relative Water Content (RWC)

Relative water content in mangrove species, both true mangroves and associated mangroves, has results that range from 31.35% to 96.32%, where the highest value is found in *Rhizophora apiculata* (96.32 ± 0.44%). The lowest value is found in *Excoecaria agallocha* (31.35 ± 20.51%) (Table 2 and Fig. 2).

Based on the single-factor ANOVA statistical test, it was found that species differences affected RWC values ($p = 0.046$). This indicates that all actual and associated mangroves have different abilities to maintain water content in the leaves when exposed to air pollution.

Based on Duncan's Multiple Range Test, it is known that several species pairs are significantly different: *Excoecaria agallocha*, *Hibiscus tiliaceus*, and Derris trifoliate with *Rhizophora mucronata* and *Rhizophora apiculata* (Table 2). Mangrove plants generally have thick, sturdy leaves to store water and prevent drought. According to a study by [29], true mangroves have thick leaves and can store significant amounts of water. Mangrove species store water in the sub-epidermal tissue (hypodermis) beneath the epidermal layer. Compared to other species, *Rhizophora apiculata* boasts the thickest hypodermis, which allows it to store more water and endure droughts brought on by environmental stressors like air pollution. Conversely, *Terminalia catappa*, although not a true mangrove, also exhibits high water content.

The plant's natural response to high light intensity involves reducing the size and number of stomata to minimize water loss through transpiration. Tensity, which includes reducing the size and number of stomata to minimize transpiration [30]. Further

Table 2. Biochemical parameters and APTI values, (Mean ± Standard error) in mangrove plants.

No	Species	RWC (%)	TCH (mg.g⁻)	AA (mg.g⁻)	pH	APTI
1	*Avicennia lanata*	57.27 ± 1.09^{ab}	4.86 ± 0.04^{d}	10.36 ± 0.01^{g}	6.08 ± 0.01^{k}	70.65 ± 1.10^{bcd}**
2	*Avicennia marina*	54.81 ± 3.70^{ab}	6.99 ± 0.11^{k}	11.65 ± 0.02^{k}	6.00 ± 0.01^{j}	79.03 ± 4.52^{bcd}**
3	*Bruguiera cylindrical*	56.32 ± 11.39^{ab}	3.73 ± 0.04^{b}	9.81 ± 0.03^{f}	5.90 ± 0.01^{h}	64.66 ± 11.02^{bc}**
4	*Bruguiera gymnorrhiza*	67.88 ± 4.10^{abc}	3.59 ± 0.02^{b}	8.03 ± 0.01^{b}	5.04 ± 0.00^{a}	61.44 ± 3.37^{bc}**
5	*Rhizophora apiculata*	96.32 ± 0.44^{c}	5.57 ± 0.02^{g}	9.06 ± 0.04^{c}	5.46 ± 0.01^{d}	43.53 ± 0.54^{ab}**
6	*Rhizophora mucronata*	82.98 ± 0.26^{cb}	4.28 ± 0.02^{c}	11.54 ± 0.02^{j}	5.68 ± 0.01^{e}	107.24 ± 0.15^{d}**
7	*Rhizophora stylosa*	65.18 ± 0.14^{abc}	6.36 ± 0.04^{i}	11.86 ± 0.03^{l}	5.72 ± 0.01^{f}	91.56 ± 0.38^{cd}**
8	*Sonneratia caseolaris*	55.82 ± 10.40^{ab}	5.05 ± 0.03^{e}	10.41 ± 0.03^{g}	5.16 ± 0.01^{b}	68.68 ± 10.66^{bcd}**
9	*Excoecaria agallocha*	31.35 ± 20.15^{a}	6.94 ± 0.05^{k}	11.35 ± 0.03^{i}	5.18 ± 0.01^{b}	49.24 ± 13.10^{abc}**
10	'*Acrostichum aureum*	52.29 ± 20.15^{ab}	5.61 ± 0.07^{g}	9.33 ± 0.02^{d}	6.24 ± 0.01^{m}	59.91 ± 18.93^{bc}**
11	'*Anredera cordifolia*	53.11 ± 0.22^{ab}	3.17 ± 0.01^{a}	2.97 ± 0.06^{a}	6.14 ± 0.00^{l}	18.53 ± 0.45^{a}*
12	'*Acanthus ebracteatus*	51.86 ± 11.98^{ab}	4.90 ± 0.03^{d}	9.29 ± 0.02^{d}	6.24 ± 0.00^{mn}	58.49 ± 11.07^{bc}**
13	'*Calophyllum inophyllum*	53.85 ± 2.92^{ab}	5.93 ± 0.05^{h}	9.68 ± 0.02^{e}	5.21 ± 0.00^{c}	62.90 ± 2.96^{bc}**
14	Derris trifoliate	41.06 ± 26.67^{a}	6.59 ± 0.06^{j}	12.34 ± 0.02^{m}	6.27 ± 0.01^{n}	66.46 ± 12.84^{bcd}**
15	'*Hibiscus tiliaceus*	32.58 ± 3.89^{a}	5.48 ± 0.03^{h}	12.43 ± 0.03^{m}	5.96 ± 0.01^{i}	55.15 ± 4.97^{abc}**
16	'*Pluchea indica*	54.17 ± 12.66^{ab}	6.47 ± 0.04^{ij}	10.76 ± 0.02^{n}	5.87 ± 0.00^{g}	71.56 ± 13.71^{bcd}**
17	'*Terminalia catappa*	65.20 ± 6.77^{abc}	5.21 ± 0.06^{f}	11.87 ± 0.02^{l}	5.16 ± 0.02^{b}	89.73 ± 8.21^{cd}**

(') Associated mangrove species (**) Tolerant (*) Intermediate, Consecutive letter differences indicate significant differences ($p < 0.05$) based on Duncan's test

explained that water deficits trigger physiological responses in plants, such as reducing transpiration, closing stomata, and decreasing the number of leaves.

Increased RWC in plants enhances their resistance to drought. However, plants may struggle to survive when the rate of transpiration in leaves decreases due to air pollution. This is because reduced transpiration limits the movement of water and dissolved minerals (1%–2%) from the roots to the leaves, where photosynthesis and biosynthesis occur. Consequently, plants fail to transport essential minerals, maintain leaf cooling, and sustain proper metabolic function [31, 32]. Changes in the RWC values in leaves are

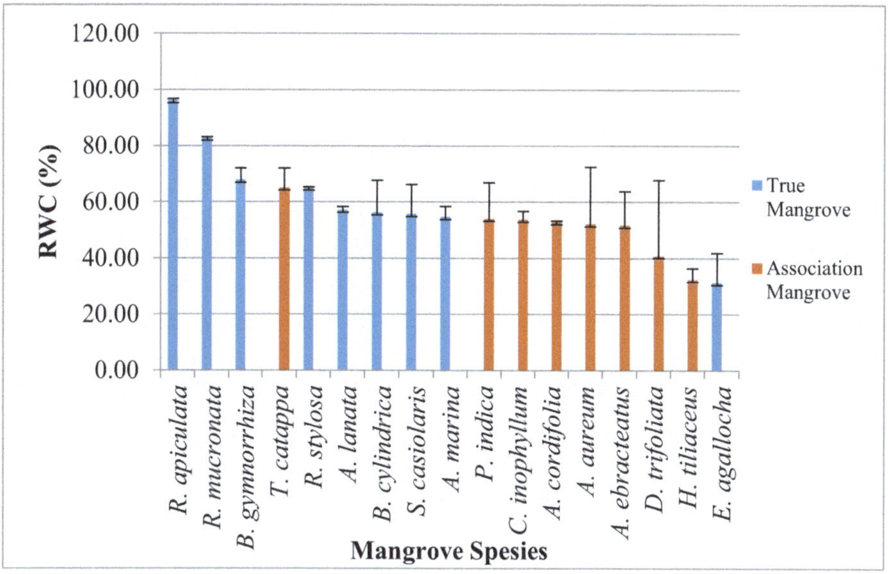

Fig. 2. Relative water content (mean) in mangrove plants.

one mechanism plants use to adapt to environmental conditions and mitigate the effects of air pollution.

3.2 Total Chlorophyll Content (TCH)

The total chlorophyll content of various mangrove species, both actual and associated mangroves, varied. The results showed that the chlorophyll content ranged from 6.99 mg.g^{-} to 3.17 mg.g^{-} . The highest chlorophyll content, 6.99 ± 0.11 mg.g^{-} was found in *A. marina* species, while the lowest value of 3.17 ± 0.01 mg.g^{-} was found in *A. cordifolia* species (Table 2).

Based on the single-factor ANOVA statistical test, it is known that species differences significantly affect the chlorophyll value ($p = 0.000$). This indicates that different mangrove species have different adaptation mechanisms to changes in chlorophyll content when exposed to air pollution. Based on Duncan's Multiple Range Test, several species pairs are not significantly different, namely, *B. gymnorrhiza* with *B. cylindrical, A. lanata* with *A. ebracteatus, R. apiculata* with *A. aureum, H. tiliaceus* with *C. inophyllum, C. inophyllum* with *D. trifoliata,* and *E. agallocha* with *A. marina* (Table 2).

Light, leaf age, and nutrients influence chlorophyll content in mangrove leaves. According to [32], pollutants are also responsible for macroscopic leaf damage and plant metabolism. Increased air pollution decreases leaf chlorophyll levels. Species with higher leaf chlorophyll content can be interpreted as pollution-tolerant plants (Fig. 3).

Chlorophyll measurement is an important parameter to determine the impact of air pollution on plants. Chlorophyll plays a vital role in plant metabolism. Exposure to pollution slowly reduces the chlorophyll content in leaves, resulting in the yellowing of leaves and reduced photosynthetic capacity [33, 34]. A decrease in chlorophyll content

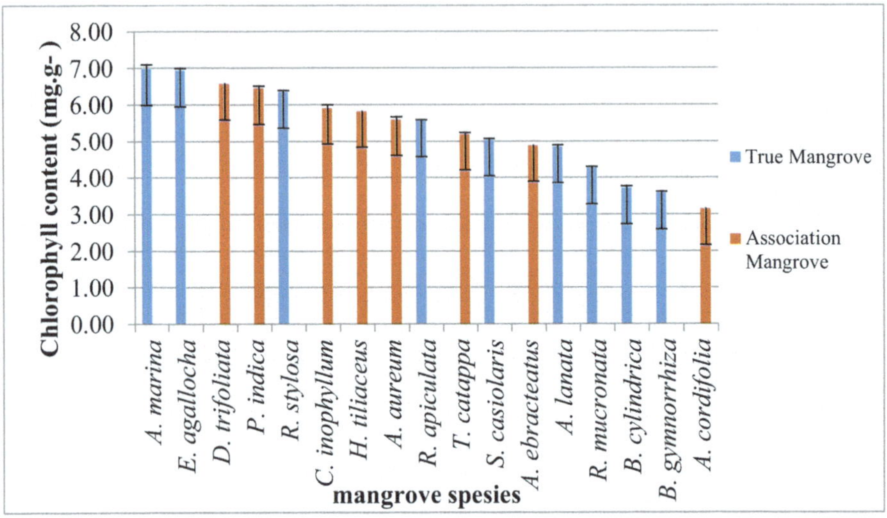

Fig. 3. Chlorophyll content (mean) in mangrove plants.

can be considered an indicator of increased pollutant content in plants [18]. Air pollution stress will reduce chlorophyll levels in plants. The loss of magnesium ions leads to the degradation of chlorophyll to pheophytin, which decreases the chlorophyll content in plants in industrial (polluted) areas [35].

3.3 Ascorbic Acid (AA)

The results showed that the ascorbic acid content ranged from 12.43 mg.g^{-} to 2.97 mg.g^{-}. The highest ascorbic acid content, 12.43 $\pm$ 0.03 mg.g^{-} was found in *H. tiliaceus* species, while the lowest value, 2.97 $\pm$ 0.06 mg.g^{-} was found in *A. cordifolia* species (Table 2 and Fig. 4).

Based on the single-factor ANOVA statistical test, it is known that species differences significantly affect the value of ascorbic acid ($p = 0.000$). This means that each mangrove species shows a different response to synthesizing ascorbic acid to adapt to environmental pressures, such as exposure to air pollution. Based on Duncan's Multiple Range Test, we know that several species pairs, such as *A. ebracteatus* with *A. aureum, A. lanata* with *A. caseolaris*, *R. stylosa* with *T. catappa*, and *D. trifoliate* with *H. tiliaceus*, have no significant effect. (Table 2).

Ascorbic acid content plays an essential role in synthesizing cell walls, plant defense, and cell division [21]. Additionally, this compound plays a crucial role in shielding chloroplasts from damage resulting from exposure to pollutants [36]. The adaptability of mangrove plants makes them tolerant of environmental stress (air pollution). According to a study, the ascorbic acid content directly correlates with the amount of pollutants in the air [37]. It can be seen that *H. tiliaceus*, *D. trifoliata*, *T. catappa*, *R. stylosa* and *A. marina* species have high ascorbic content (Fig. 4). These species have relatively many stomata and broad, round leaves in common.

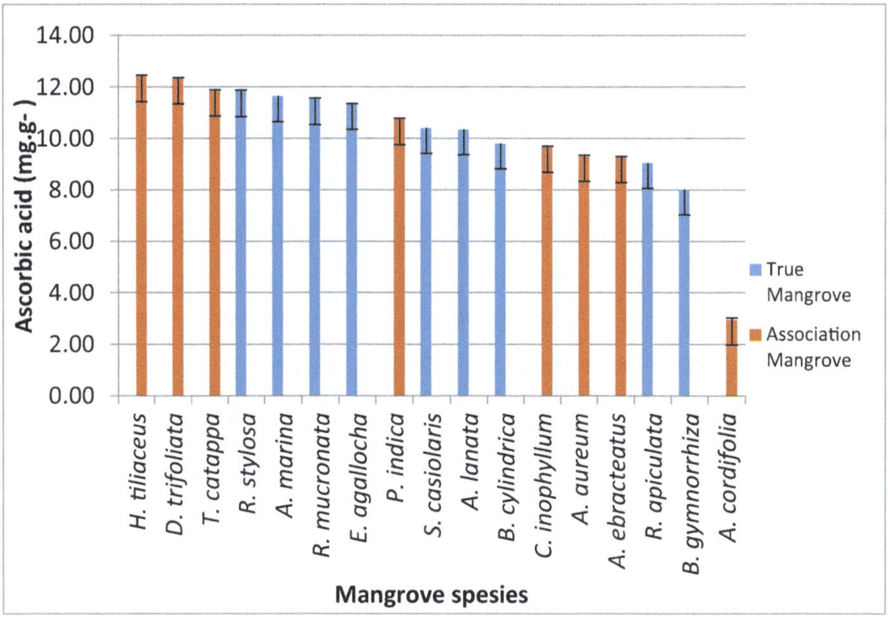

Fig. 4. Ascorbic acid content (mean) in mangrove plants.

Leaf surface area and the number of stomata influence plants' high ascorbic acid content. Dust pollutants trapped on the leaf surface can penetrate leaf cells through the stomata. When these pollutants induce oxidative stress, they trigger the production of reactive oxygen species (ROS), such as superoxide and hydrogen peroxide, altering the cellular environment. According to [15], an increase in reactive oxygen content can cause an increase in ascorbic acid to overcome environmental stress conditions. ROS can affect plant metabolism and activate enzymes involved in ascorbic acid biosynthesis, which is part of the plant's coping mechanism against environmental stress, including exposure to air pollution. Therefore, environmental conditions that lead to the accumulation of pollutants in leaves can trigger an increase in ascorbic acid as an adaptation response to oxidative stress. Thus, an increase in ascorbic acid content may indicate that the tolerance of plants to environmental stress (pollutants) is better [38].

3.4 pH Content

The pH of leaf extracts from different mangrove species, both actual and associated mangroves, varied. The results showed that the pH of leaf extracts ranged from 5.04 to 6.27. The highest pH value, 6.27 ± 0.01, was found in *D. trifoliata* species, while the lowest value, 5.04 ± 0.00, was found in *B. gymnorrhiza* species (Table 2 and Fig. 5).

Based on the single-factor ANOVA statistical test, it is known that species differences significantly affect the pH value ($p = 0.000$). This means there are significant differences in the acidity or basicity of leaf extracts between mangrove species in the face of environmental stress. Based on Duncan's Multiple Range Test, it is known that

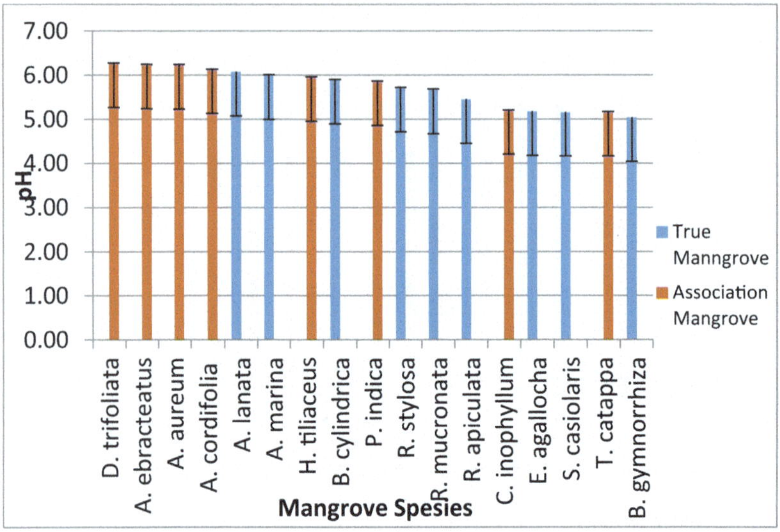

Fig. 5. Leaf extract pH content (mean) in mangrove plants.

some species that do not have a significant effect are *T. catappa, C. inophyllum with E. agallocha, A. aureum with A. ebracteatus, and A. ebracteatus* with *D. trifoliate* (Table 2).

Results show that mangrove plants such as *D. trifoliate, A. ebracteate, and A. aureum* have alkaline pH values. This is because these plants are mostly found growing in areas closer to land or in areas more affected by freshwater flow. The substrate in these areas tends to have a higher pH, or be closer to 7, than the more saline and inundated substrate where true mangroves grow.

Leaf pH acts as an indicator of air pollution sensitivity. Plants with low pH will be more sensitive to pollutants, while plants with a pH of around 7 will be more tolerant [18, 39]. Acidic pollutants will result in a decrease in leaf pH, and this decrease will be seen directly in plants that are sensitive to air pollution [38]. A high pH level will increase the efficiency of hexose sugar conversion to ascorbic acid and increase ascorbic acid reduction. So it can provide better resistance to pollutants [40], while low pH will cause the development of stress in mangrove plants [41].

3.5 Air Pollution Tolerance Index (APTI)

The Air Pollution Tolerance Index (APTI) shows a plant's response in tolerating pollutants in the air. Plant species found at the research site showed a variety of APTI values. According to [21], relative water content, total chlorophyll, ascorbic acid, and pH of leaf extracts used in the formulation of APTI values are plant parameters that can be used to assess the level of plant tolerance to air pollution. Determination of plant tolerance to air pollution based on criteria set by [28].

Mangrove species at the study site exhibited significant variation in APTI values. The mean APTI values among mangrove species ranged from 18.53 to 107.24. The highest mean APTI value, 107.24 ± 0.15, was recorded in *Rhizophora mucronata*, while the lowest mean value, 18.53 ± 0.45, was observed in *Acanthus cordifolia* (Table 2 and Fig. 6).

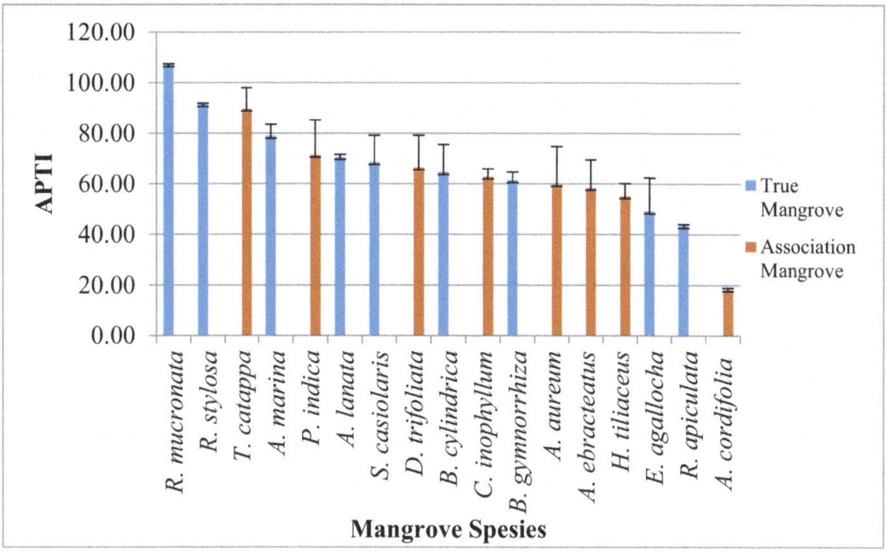

Fig. 6. Air Pollution Tolerance Index in Mangrove Plants.

The single-factor ANOVA statistical test revealed that species differences have a significant effect on APTI values ($p = 0.010$). This indicates that each species responds differently to air pollution. Duncan's Multiple Range Test (Table 2) showed that some pairs of species were significantly different. These pairs included *A. cordifolia* and *A. ebracteatus*, *A. aureum* and *B. gymnorrhiza*, *C. inophyllum* and *B. cylindrical*. Due to their high level of tolerance, almost all mangrove plants, both true and associated, can adapt to air pollution and serve as air pollution bioreductors (Table 2). True mangrove plants such as *R. mucronata, R. stylosa* and *A. marina*, as well as associated mangroves such as *T. catappa, P. indica*, and *D. trifoliate* show a very tolerant response with a high index value, so they have the potential to be used as air pollution bioreductors (Fig. 6). Bioreductor plants are plants that have the ability to reduce pollutants from the environment, one of which is air pollution [42], plants with high tolerance levels can be used as air pollution reductors. On the other hand, *Acanthus cordifolia* showed an intermediate response in its APTI value. Plants with an intermediate response are still tolerant to air pollution, although they are not fully adapted.

Based on previous studies, plants with low APTI values tend to exhibit visible signs of stress, such as chlorosis, necrosis, or leaf deformation, which provide clear indications of air quality [44]. As bioindicators, these plants play an important role in environmental monitoring, offering a natural and cost-effective method to assess the impact of pollution on ecosystems.

These findings have significant implications for urban planning and environmental management, particularly in coastal cities such as Surabaya, where industrial activities greatly contribute to air pollution. Mangrove species with high APTI values, such as *R. mucronata* (APTI: 107.24), can be strategically planted in industrial zones to act as air pollution bioreductors, helping to mitigate the negative effects of pollutants. Meanwhile, species like *A. cordifolia* (APTI: 18.53) can function as bioindicators, enabling authorities to monitor air quality more effectively.

Additionally, the use of APTI as a guiding tool for mangrove rehabilitation programs ensures the selection of resilient species, allowing for ecosystem restoration while addressing pollution challenges. This dual strategy, utilizing mangroves as both bioreductors and bioindicators, offers a sustainable and efficient approach to urban greening and environmental conservation [43].

4 Relationship Between Four Parameters and APTI Value

Based on the multiple correlation test, the relationship between the four parameters and the APTI value shows a strong correlation coefficient ($r = 0.876$), indicating a very strong relationship. This means that changes in these parameters are significantly related to changes in the APTI value (Table 3). Additionally, the coefficient of determination ($R^2 = 0.767$) shows that the four parameters strongly influence the APTI value.

Table 3. Multiple 4-parameter correlation coefficient with APTI.

Model	R	R Square	Adjusted R Square
1	0.876[a]	0.767	0.747

(a) Predictors: (constant), chlorophylll, pH, RWC AA, n (51)

In this regard, relative water content is related to a plant's ability to store water when exposed to pollutants [32, 45]. Additionally, total chlorophyll content plays a role in plant metabolism [32]. On the other hand, high ascorbic acid content increases plant resistance to air pollution [17], while low leaf pH (acidic conditions) makes plants more sensitive to air pollution [18, 39].

5 Conclusions

The mangroves found at the Wonorejo Mangrove Center have shown significant adaptability to the air pollution prevalent in Surabaya. This is evident from each mangrove species' APTI (Air Pollution Tolerance Index) values, categorized as either highly tolerant or intermediate. The presence of 17 species, including 9 true mangroves and eight associate mangroves, further supports the resilience of this ecosystem. These species have survived and thrived in conditions challenging for many plants, indicating a strong adaptive capacity to environmental stressors.

References

1. AQI: Ranking Kualitas Udara di Indonesia real-time. In: Indeks kualitas udara (AQI) (2024)
2. Setyo, G.A., Handriyono, R.E.: Analisis Penyebaran Gas Karbon Monoksida (Co) Dari Sumber Transportasi Di Jalan Tunjungan Surabaya. Semin. Nas. Sains dan Teknol. Terap. **9**(0), 360–369 (2021)
3. Rosyidah, M.: Polusi Udara dan Kesehatan. J. Tek. Ind. **1**(11), 5–8 (2016)
4. WHO: Air Pollution. World Health Organization, Geneva (2023)
5. Syamsu, I.F., Nugraha, A.Z., Nugraheni, C.T., Wahwakhi, S.: Kajian Perubahan Tutupan Lahan di Ekosistem Mangrove Pantai Timur Surabaya. J. Media Konserv. **13**(2), 122–131 (2018)
6. Ade, R., Rahmawati, D.: Identifikasi Pemanfaatan Kawasan Konservasi Mangrove di Wonorejo Surabaya Rizki. J. Tek. ITS. **6**(2), 604–606 (2017)
7. Dinilhuda, A., Akbar, A.A., Jumiati, J.: Peran Ekosistem Mangrove Bagi Mitigasi Pemanasan Global. J. UNTAN. **18**(2), 1–8 (2018)
8. Irawanto, R.: Keanekaragaman Vegetasi Mangrove di Pesisir Kota Surabaya Dan Potensinya Sebagai Fitoremediator Lingkungan. Pros. Semin. Nas. Biol. **6**(1), 413–422 (2020) http://journal.uin-alauddin.ac.id/index.php/psb/article/view/15889/9691
9. S. Rahim and D. W. K. Baderan, Hutan Mangrove dan Pemanfaatannya. (2017)
10. Mahrukh, M.: Air pollution mitigation potential of dominant landscape plants of an urban ecosystem (Lahore city) of pakistan: an air pollution tolerance index (APTI) assessment. Pol. J. Environ. Stud. **32**(4), 3233–3245 (2023). https://doi.org/10.15244/pjoes/161980
11. Nugrahani, P.: Air Pollution Tolerance Index (APTI). Deepublish Publisher, Yogyakarta (2014)
12. Megia, R., Ratnasari, Hadisunarso: Karakteristik Morfologi dan Anatomi, serta Kandungan Klorofil Lima Kultivar Tanaman Penyerap Polusi Udara Sansevieria trifasciata. J. Sumberd. Hayati. **1**(2), 34–40 (2016). https://doi.org/10.29244/jsdh.1.2.34-40
13. Sett, R.: Responses in plants exposed to dust pollution. Hortic. Int. J. Rev. **1**(2), 53–56 (2017). https://doi.org/10.15406/hij.2017.01.00010
14. Gabriela, M., Bereta, D., Eugenia, G., Celi, A., Lupino, P.: Plant physiology and biochemistry physiological and biochemical roles of ascorbic acid on mitigation of abiotic stresses in plants. Plant Physiol. Biochem. J. **202**, 1–13 (2023). https://doi.org/10.1016/j.plaphy.2023.107970
15. Joshi, N., Joshi, A., Bist, B.: Air pollution tolerance index of some tree species from the industrial area of Tarapur. Int. J. Life Sci. **2**(2), 173–182 (2016)
16. Molnár, V.É., Ozsér, D.T., Szabó, S., Tóthmérész, B., Simon, E.: Use of leaves as bioindicator to assess air pollution based on composite proxy measure (APTI), dust amount and elemental concentration of metals. Plan. Theory. **9**, 1–11 (2020)
17. Ergantara, R.I., Khikmawati, E.: Nilai APTI (Air Pollution Tollerance Index) Tanaman Pengisi RTH Privat di Kecamatan Kemiling, Kota Bandar Lampung. Pros. SNPBS (Semin. Nas. Pendidik. Biol. dan Saintek). **5**, 496–501 (2020) https://publikasiilmiah.ums.ac.id/handle/11617/12254
18. Salsabila, S.H., Nugrahani, P., Santoso, J.: Toleransi Tanaman Lanskap Terhadap Pencemaran Udara di Kota Sidoarjo. J. Lanskap Indones. **12**(2), 73–78 (2020). https://doi.org/10.29244/jli.v12i2.32533
19. Suprihatin, H., Ikhtisholiyah: Studi Kualitas Udara Ambien di Jalan Raya Industri Surabaya Industrial Estate Rungkut (SIER). Semin. Nas. Tek. Sipil. **1**(1), 1–8 (2023)
20. Wijaya, N.I., Huda, M.: Monitoring Sebaran Vegetasi Mangrove Yang Direhabilitasi Di Kawasan Ekowisata Mangrove Wonorejo Surabaya. J. Ilmu dan Teknol. Kelaut. Trop. **10**(3), 747–755 (2018)

21. Ghafari, S., Kaviani, B., Sedaghathoor, S., Allahyari, M.S.: Assessment of air pollution tolerance index (APTI) for some ornamental woody species in green space of humid temperate region (Rasht, Iran). Environ. Dev. Sustain. **23**(2), 1579–1600 (2020). https://doi.org/10.1007/s10668-020-00640-1
22. Turner, N.C.: Techniques and experimental approaches for the measurement of plant water stress. Plant and Soil. **58**(1), 339–366 (1981)
23. Lichtenthaler, H.K.: Chlorophylls and carotenoids: pigments of photosynthetic biomembranes. In: Methods in Enzymology, pp. 350–382 (1987)
24. Angraini, N., Desmaniar, P.: Optimasi penggunaan high performance liquid chromatography (HPLC) untuk analisis asam askorbat guna menunjang kegiatan Praktikum Bioteknologi Kelautan. J. Penelit. Sains. **22**(2), 69–75 (2020)
25. Putri, M.H., Septiyani, P., Aryani, W., Abriyan, E.: Penetapan Kadar Vitamin C Pada Buah Jambu Biji, Jeruk, Dan Nanas, Menggunakan Metode Spektrofotometri Uv-Vis. J. Ilm. Wahana Pendidik. **9**(4), 333–342 (2023). https://doi.org/10.5281/zenodo.7681039
26. Sahu, C., Basti, S., Sahu, S.K.: Air pollution tolerance index (APTI) and expected performance index (EPI) of trees in sambalpur town of India. SN Appl. Sci. **2**(8), 1–14 (2020). https://doi.org/10.1007/s42452-020-3120-6
27. Singh, S.K., Rao, D.N.: Evaluation of the plants for their tolerance to air pollution. In: Proceedings of the Symposium on Air Pollution Control (1983)
28. Kaur, L., Charan, P.D., Choyal, R.: Evaluation of air pollution tolerance index of tree species of Bikaner City (Rajasthan). Holist. Approach Env. **13**(4), 123–131 (2023)
29. Tihurua, E.F., Agustiani, E.L., Rahmawati, K.: Karakter Anatomi Daun sebagai Bentuk Adaptasi Tumbuhan Penyusun Zonasi Mangrove di Banggai Kepulauan , Provinsi Sulawesi Tengah. J. Kelaut. Trop. **23**(2), 255–264 (2020)
30. Pamungkas, S.S.T., Suwarto, S., Farid, N.: Drought stress: responses and mechanism in plants. Agric. Sci. **10**, 168–185 (2022)
31. Liu, Y., Ding, H.: Variation in air pollution tolerance index of plants near a steel factory: implications for landscape-plant species selection for industrial areas Beijing Centre for Physical and Chemical Analysis. WSEAS Trans. Environ. Dev. **4**, 14–16 (2007)
32. Allifah, A.N.A., Natsir, N.A., Mulyawati, N.Y., Surati: Kadar Klorofil Daun Mangrove Di Kawasan Pelabuhan Hurnala Maluku Tengah. Biotropic J. Trop. Biol. **6**(1), 20–28 (2022)
33. Joshi, P.C., Swami, A.: Physiological responses of some tree species under roadside automobile pollution stress around city of Haridwar, India. Environmentalist. **27**, 365–374 (2007)
34. Patidar, S., Bafna, A., Batham, A.R., Panwar, K.: Impact of urban air pollution on photosynthetic pigment and proline content of plants growing along the A.B road Indore City, India. Int. J. Curr. Microbiol. App. Sci. **5**(3), 107–113 (2016). https://doi.org/10.20546/ijcmas.2016.503.015
35. Giri, S., Shrivastava, D., Deshmukh, K., Dubey, P.: Effect of air pollution on chlorophyll content of leaves. Curr. Agric. Res. J. **1**(2), 93–98 (2013). https://doi.org/10.12944/carj.1.2.04
36. Gupta, R., Purohit, M.: Bio-monitoring of road side air pollution using ascorbic acid content of selected trees at Ujjain. Int. J. Sci. Res. Rec. Sci. **1**(1), 26–29 (2015)
37. Ergantara, R.I., Khikmawati, E.: Analisis Pemilihan Jenis Tanaman Penyerap Emisi Udara Dalam Mendukung Ruang Terbuka Hijau Privat Di Kecamatan Kemiling Kota Bandar Lampung. J. Rekayasa, Teknol. dan Sains. **4**(1), 7–12 (2020)
38. Uka, U.N., Hogarh, J., Belford, E.J.D.: Morpho-anatomical and biochemical responses of plants to air pollution. Int. J. Mod. Bot. **7**(1), 1–11 (2017). https://doi.org/10.5923/j.ijmb.20170701.01
39. Zouari, M., et al.: A comparative study of air pollution tolerance index (APTI) of some fruit plant species growing in the industrial area of Sfax, Tunisia. Pollution. **4**(3), 439–446 (2018)

40. S. Shrestha, "Responses of Some Crops to Air Pollution by Using Air Pollution Tolerance Index.," (2020)
41. Wong, W.E.N.Y.E.E., et al.: Water, soil and air pollutants' interaction on mangrove ecosystem and corresponding artificial intelligence techniques used in decision support systems—a review. IEEE Access. **9**, 105532–105563 (2021)
42. Enitan, I.T., Durowoju, O.S., Edokpayi, J.N.: A review of air pollution mitigation approach using air pollution tolerance index (APTI) and anticipated performance index (API). Atmos. **13**(374), 1–20 (2022)
43. Seyyednejad, S.M., Niknejad, M., Koochak, H.: A review of some different effects of air pollution on plants. Res. J. Environ. Sci. **5**(4), 302 (2011)
44. Waryanti, S., Irawan, D.: Angsana Sebagai Bioindikator untuk Polusi di Sekitar Lebak Bulus. Al Kauniyah J. Biol. **8**(1), 46–50 (2015)
45. Yan-Ju, L., Hui, D.: Variation in air pollution tolerance index of plants near a steel factory. implications for landscape-plant species selection for industrial areas. WSEAS Trans. Environ. Dev. **1**, 24–30 (2008)

Estimation of Soil Organic Carbon Stock at Marine Science Techno Park, Jepara, Central Java

Dihaven Abdullah Muchsinin Amran[1], Jumari Jumari[1,2](✉), Sri Widodo Agung Suedy[1,2], Deny Nugroho Sugianto[3], Tri Retnaningsih Soeprobowati[1,2], and Aulia Rahim[2,3]

[1] Department of Biology, Faculty of Science and Mathematics, Diponegoro University, Semarang, Indonesia
jumari@live.undip.ac.id
[2] Cluster for Paleolimnology (CPalim), Diponegoro University, Semarang, Indonesia
[3] Faculty of Fisheries and Marine Science, Diponegoro University, Semarang, Indonesia

Abstract. Carbon stocks are the carbon content stored below and above ground in a system. Soil is one of the places with the highest potential for carbon storage. Carbon stocks come from organic matter from the decomposition of animals, plants, and other organisms was his study aimed to analyze and estimate soil properties and soil carbon stocks in the Marine Science Techno Park (MSTP) area. The research was conducted at MSTP in Teluk Awur, Jepara. The research site was divided into two stations: terrestrial (Station 1) and mangrove (Station 2), with three sampling points at each station. Soil samples were taken to a depth of 40 cm and analyzed for bulk density (BD), soil organic carbon content, SOC stock, and CO_2 equivalent. Bulk density values obtained in terrestrial soil were lower (0.54 to 0.66 g/cm^3) compared to mangrove soil (0.72 to 0.83 g/cm^3). The organic carbon content obtained in terrestrial soil was 9.15%, and in mangrove soil (0.41% to 0.60%). The terrestrial SOC stock was significantly higher (113.76 tons/ha) than the mangrove (15.45 tons/ha). The total SOC stock in MSTP is 602.32 tons, with terrestrial lands making the most significant contribution. CO_2 equivalent storage potential is 2210.52 tons, mainly from terrestrial ecosystems. Terrestrial lands have more extensive SOC stocks due to higher organic matter inputs and differences in soil structure. This underscores the importance of effective management practices such as reforestation to increase carbon storage.

Keywords: soil organic carbon · carbon stock

1 Introduction

Global warming is one of the most pressing challenges faced by humanity today. The global average temperature is rising at an alarming rate of 0.3 °C per decade and is projected to increase by 1.5 to 4.5 °C between 2030 and 2050 if no significant mitigation efforts are undertaken [12]. One of the primary contributors to global warming is

J. C. Casila et al. (Eds.): CCORE 2024, SPEES, pp. 190–198, 2025.
https://doi.org/10.1007/978-3-031-93887-0_18

the rising concentration of greenhouse gases, particularly carbon dioxide (CO_2), in the atmosphere. Therefore, it is imperative to implement strategies to reduce CO_2 emissions and enhance carbon sequestration.

In this context, natural ecosystems such as terrestrial lands and mangrove forests play a critical role in carbon storage [1, 18]. These ecosystems have the capacity to sequester and store substantial amounts of carbon, both in vegetation biomass and in soil. Soil serves as a significant long-term carbon reservoir, holding more carbon than plant biomass and the atmosphere combined [1, 2]. The type of vegetation present in an area directly influences the amount of carbon accumulated in the soil layer, known as soil carbon stock. Different land types and vegetation cover have varying carbon storage potentials. Notably, mangrove ecosystems, with their intricate root systems, are particularly efficient carbon sinks, surpassing terrestrial forests in carbon storage capacity [13]. Terrestrial ecosystems, which include forests, grasslands, and farmlands, also contribute substantially to carbon sequestration efforts.

Understanding and comparing carbon stocks across different ecosystems is crucial for developing effective conservation and sustainable management strategies to combat climate change. This study aims to address this need by estimating soil organic carbon stocks in the Marine Science Techno Park (MSTP) located in Teluk Awur, Jepara. The MSTP, managed by Diponegoro University, spans an area of 51.83 hectares and serves as a maritime-themed education and recreation hub. The park includes diverse land uses such as lecture buildings, mangrove forest areas, terrestrial land, shrimp ponds, ports, irrigation facilities, and marine tourism attractions. The MSTP encompasses two distinct land types—mangrove ecosystems and terrestrial areas—making it an ideal location for comparative research on carbon stocks.

Despite previous research efforts focusing on mangrove diversity and surface carbon stocks in the MSTP area, there has been no comprehensive study examining soil carbon stocks. This study seeks to fill this research gap by analyzing soil organic carbon stocks within the Jepara MSTP region. The findings of this research will provide valuable insights for future management and conservation efforts in the area, ensuring that effective measures are implemented to enhance carbon sequestration and contribute to climate change mitigation.

2 Material and Methods

2.1 Research Location

Data and samples were obtained in the Awur Bay region of the Marine Science Techno Park (MSTP), Tahunan District, Jepara Regency, Central Java. The Ecology, Biosystematics Laboratory, Biology Department, Diponegoro University, and Cen-dekia Nanotech Hutama Laboratory. This study was conducted between October 2022 and March 2023.

2.2 Sampling Plot and Data Collection

Data was collected The MSTP area was divided into 2 stations based on vegetation type, namely station 1 terrestrial (6°37′8.63″S, 110°38′21.09′E) and station 2 mangrove

(6°37′18.19″S, 110°38′20.44″E). Each station is divided into 3 points, namely the front (near the beach), the middle, and the back (Fig. 1). Each point has a different history of land change, which can affect the supply of stored carbon. Station 1 is a coastal area covered by terrestrial vegetation, where the dominating plant species are Teak (*Tectona grandis*), Mahogany (*Swietenia macrophylla*), and Sea Spruce (*Casuarina equisetifolia*). In contrast, station 2 is a coastal area covered by mangrove vegetation, with the dominating mangrove species being *Rhizophora mucronata, Rhizopora apiculata,* and *Lumnitzera racemosa.*

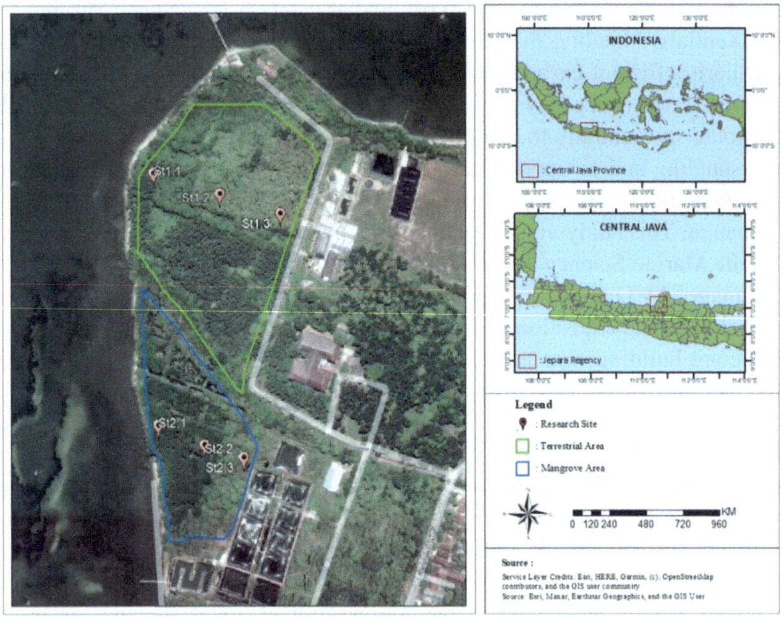

Fig. 1. Maps of research sites in MSTP, Jepara, Central Java.

Soil samples are taken using a soil corer, with a length of 50 cm and a diameter of 4.5 cm, made of galvanised steel, to a depth of 40 cm vertically. The samples are separated at 10 cm. The separated samples are placed into prepared zip plastic bags and labeled according to location points. The samples are analyzed in the Ecology and Biosystematics Laboratory, FSM, Diponegoro University, and the Cendekia Nanotech Hutama (CNH) Laboratory.

2.3 Data Analysis

2.3.1 Bulk Density

Soil bulk density is the mass density of soil or the amount of mass per unit volume of soil, describing how compact or loose the soil is. Soil bulk density can be measured with the equation [11]:

$$BD = (Ms\ (g))/Vs\ (cm^3)$$

where:

BD: Bulk Density (g/cm^3)
Ms: Mass of the dried sample (g)
Vs: Sample volume (cm^3)

2.3.2 Soil Organic Carbon Content

The organic carbon content of soil can be determined by the equation [19]:

C − organic (%C) = ppm curve x100/mg sample x fk.

where:

ppm curve: concentration of the sample obtained from the regression curve. The regression curve is established by plotting the concentration of a series of standards against their corresponding readings (after subtracting the blank reading)
Fk: water content correction factor = 100/(100 - % water content)

2.3.3 Soil Organic Carbon Stock

Total organic carbon is the total amount of carbon contained in organic matter in a system or sample, measured in tons per hectare. Total organic carbon can be calculated with [11]:

SOC = BD x Sample Depth Interval (cm) x % C x 100
SOCarea = SOC x Area.

where

KOT: Soil organic carbon stock value (Ton C Ha^{-1})
BD: Bulk density (g/cm^3)
%C: Organic Carbon Value
100: Conversion factor. It converts the units area from grams per square centimeter (g/cm^2) to tons per hectare (Ton/Ha)

2.3.4 CO_2 Equivalent

CO_2 equivalent is a metric unit used to measure the impact of greenhouse gases based on their potential to cause global warming, expressed in terms of CO_2 equivalent. CO_2 equivalent can be calculated with the equation [5]:

CO_2e = SOCarea x 3, 67

where:

3.67: Conversion factor to convert the amount of carbon (C) stored in the soil to its equivalent amount of carbon dioxide (CO_2)

3 Results and Discussion

3.1 Soil Properties

The soil bulk density at station 1 (terrestrial) ranges between 0.54–0.66 g/cm^3. The soil bulk density at station 1 increases with soil depth, with the highest density found in the deepest layer and the lowest density in the surface layer. The surface soil layer has lower density due to the high roots volume and stored organic matter [14]. The presence of high soil organic matter will affect the decrease in soil content weight. The presence of organic matter causes pore space in the soil, the higher the organic matter the higher the percentage of pore space in the soil. The increase in pore space in the soil shows a decreasing value of the content weight [7]. The deeper the soil layer, the denser the soil becomes. The minimal presence of organic material makes mineral material a more dominant component, as mineral material has a higher density compared to organic material (Table 1).

Table 1. Bulk density in the Marine Science Techno Park (MSTP) Teluk Awur, Jepara.

Soil depth (cm)	Bulk density (g/cm^3)									
	Terrestrial					Mangrove				
	Fr	Md	Bk	Avg	Dvt	Fr	Md	Bk	Avg	Dvt
10	0.46	0.61	0.55	0.54	0.08	0.68	0.80	0.88	0.79	0.10
20	0.57	0.62	0.64	0.61	0.04	0.48	0.91	0.77	0.72	0.22
30	0.66	0.63	0.54	0.61	0.06	0.55	0.80	0.85	0.73	0.16
40	0.71	0.63	0.63	0.66	0.05	0.61	0.99	0.89	0.83	0.19

Abbreviation: Fr: Front Area; Md: Middle Area; Bk: Back Area; Avg: Average; Dvt: Deviation

On the other hand, soil bulk density at station 2 (mangrove) has higher values than station 1 (terrestrial). The soil bulk density at station 2 ranges between 0.72–0.83 g/cm^3. The mangrove land density varies, with the highest density in the deepest layer and the lowest density in the 20 cm layer. This could be due to the soil type structure at the research location, where the soil is sandy clay. Soil with coarser texture is more dense than finer-textured soil [3]. The presence of sand in soil structure can increase soil bulk density.

The organic carbon content at station 1 ranges from 2.4% to 9.15%, with the highest value found in the surface layer and the lowest value in the deepest layer. The organic carbon content at station 1 tends to decrease with increasing soil depth (Fig. 2). The 10 cm depth layer has the highest average value, with a value of 9.15%. Organic content tends to be high in the surface layer because most of the organic matter input into the soil comes from litter on the soil surface and roots located in the surface layer [16]. In contrast, at the 40 cm depth, the source can be attributed to the decrease in the distribution of soil organic matter in each layer. Soil organic carbon originates from the degradation of necromass and plant roots on its surface. The deeper the soil layer, the higher its density. High soil bulk density results in less space for storing organic carbon [8] (Table 2).

Meanwhile, station two's soil organic carbon content shows shallow values. The soil organic carbon content at station two ranges from 0.41% to 0.6%, with the highest value found in the deepest layer and the lowest at the 20 cm depth. Litter or necromasa is very difficult to see on the surface of mangrove land. This can occur due to the condition of mangrove land directly affected by the tides [8]. Carbon stored in leaves and stems will be channeled to the soil layer through litter and necromass. These two things' absence

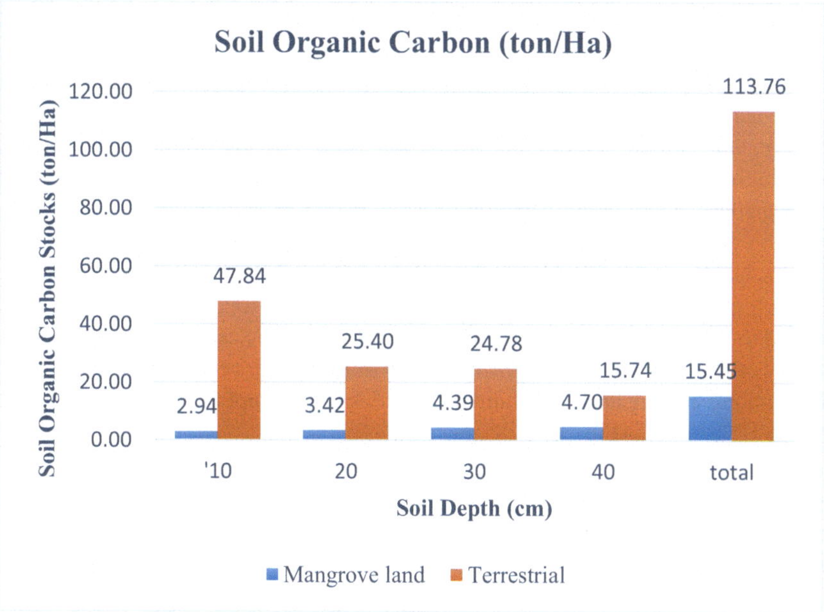

Fig. 2. Comparison of soil organic carbon stock estimates in terrestrial and mangrove.

Table 2. Carbon organic content in the Marine Science Techno Park (MSTP) Awur Bay, Jepara.

Soil depth (cm)	Organic carbon content (%)									
	Terrestrial					Mangrove				
	Fr	Md	Bk	Avg	Dvt	Fr	Md	Bk	Avg	Dvt
10	12.31	5.98	9.17	9.15	3.17	0.62	0.28	0.40	0.43	0.17
20	4.61	4.79	3.55	4.32	0.67	0.60	0.18	0.44	0.41	0.21
30	3.64	4.39	3.73	3.92	0.41	0.63	0.20	0.97	0.60	0.39
40	2.40	1.54	3.25	2.39	0.85	0.80	0.09	0.81	0.57	0.41

Abbreviation: Fr: Front Area, Md: Middle Area, Bk: Back Area, Avg: Average, Dvt: Deviation

greatly reduces carbon input to the soil layer [15]. So that carbon input to the soil layer becomes limited. The movement of rainwater and river flow from the back zone to the front zone can transport organic carbon from the previous zones to the beach. Tides further enhance this process. Teluk Awur Beach, with its flat and steep rock formations, has slower water flow, allowing organic carbon to accumulate in the front zone.

3.2 Estimated Soil Organic Carbon Stock

Based on the results of the study, it is known that soil organic carbon reserves in terrestrial are higher than mangroves (Fig. 2). Terrestrial soil organic carbon reserves have a value of 15.74–47.84 tons/ha with the highest value at the surface and the lowest at the deepest layer. The value of terrestrial soil organic carbon reserves tends to decrease with increasing soil depth. The total soil organic carbon reserves in terrestrial have a value of 113.76 tons/ha.

Soil organic carbon stock in mangroves has values ranging from 2.94–4.7 tons / ha, with the highest value in the deepest layer and the lowest value in the surface layer. The value of mangrove soil organic carbon reserves increases with increasing soil depth. It is based on (Fig. 2). Total soil organic carbon reserves in mangroves have a value of 15.45 tons/ha. The value of soil organic carbon stock is shallow compared to the results of research in Baturapa Village, Lolak District, Bolaang Mongondow Regency, which amounted to 398.82 tons/ha [13]. Apart from the low organic carbon content, this can be caused by other variables such as the condition of mangrove vegetation, the presence of ponds, land use change and other human activities that disrupt the mangrove ecosystem in the MSTP area [17] (Fig. 3).

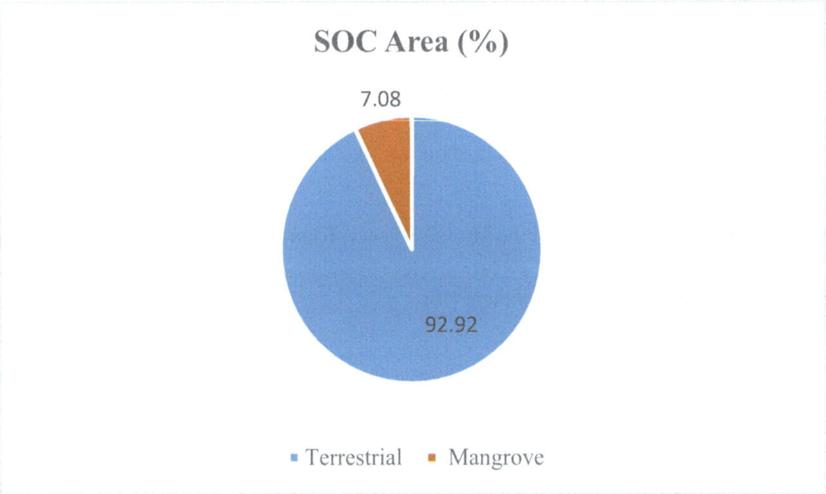

Fig. 3. Comparison of SOC area and CO equivalent in terrestrial and mangrove.

Carbon stored in soil can be converted as CO_2 equivalent. Based on data above, MSTP terrestrial and mangrove land can have a CO_2 equivalent value of 2210.52 tons. This value is dominated by terrestrial land with a value of 2054.04 tons. The CO_2 equivalent can show how much CO_2 is reduced from the atmosphere and stored in the soil [13]. Increased carbon stock in terrestrial ecosystems (forests, soils) and oceans helps to reduce the concentration of CO_2 in the atmosphere. This is crucial because CO_2 is a major greenhouse gas that traps heat and contributes to global warming [9]. By reducing atmospheric CO_2, carbon stock helps to stabilize the climate system, mitigating the impacts of climate change such as rising temperatures, sea-level rise, and extreme weather events [6]. MSTP, with its CO_2 equivalent value, has potential as a carbon sink. However, this requires more attention to its management. Many ways that can be done to increase the CO_2 equivalent value, such as reforestation with vegetation with high carbon uptake potential or vegetation optimization tailored to land conditions. Increasing storage in the soil can be done by increasing carbon input and reducing carbon expenditure [14].

4 Conclusion

Based on research of the Marine Science Techno Park (MSTP) area in Awur Bay, Jepara, it can be concluded that terrestrial soils in the MSTP area have much higher SOC reserves compared to mangrove soils. Terrestrial soils showed SOC reserves ranging from 15.74–47.84 tons/ha, with a total of 113.76 tons/ha, much higher than the 15.45 tons/ha found in mangrove soils. The total SOC stock in the MSTP area is 602.32 tons, primarily stored in terrestrial soils (559.68 tons) compared to mangrove soils (42.63 tons). When converted to CO_2 equivalent, the MSTP area has the potential to store 2210.52 tons of CO_2, with terrestrial soils contributing 2054.04 tons. Carbon stored in the soil can reduce the concentration of carbon in the atmosphere. In addition, carbon stored in the soil can improve the quality of the soil and the ecosystems within it. therefore, continuous management and monitoring is needed to keep track of the state of the carbon stocks at the site.

Acknowledgments. This research was supported by Riset Pengembangan dan Penerapan (RPP), contract number 753-05/UN7.D2/PP/IV/2023.

References

1. Afrina, N., Hidayat, M., Nisa, K.: Simpanan Karbon Pada Tanah Di Kawasan Geothermal Ie Brok Seulawah Agam Desa Meurah Kecamatan Seulimeum Kabupaten Aceh Besar. Prosiding Seminar Nasional Biologi, Teknologi dan Kependidikan. **8**(1), 71–77 (2020) [in Indonesia]
2. Ambarsari, N., Tedjasukmana, B.S.: Kajian Perkembangan Teknologi Sounding Untuk Mengukur Konsentrasi (CO_2) Di Atmosfer. Berita Dirgantara. **12**(1) (2011) [Indonesia]
3. Ayu, S.M., Rosdayati, A., Nadjib, N.N.: Simpanan Karbon Tanah Pada Ekosistem Mangrove Kelurahan Songka Kota Palopo. J. TABARO Agric. Sci. **4**(2), 484–489 (2020) [in Indonesia]
4. Bahari, A.S., Ganis, R.E., dan Muhammad, F.R.: Komposisi Jenis Mangrove di Mangrove Education Center of KeSEMaT (MECoK). Yayasan IKAMaT, Semarang (2018). [Indonesia]
5. Bossio, D.A., Cook-Patton, S.C., Ellis, P.W., Fargione, J., Sanderman, J., Smith, P., Wood, S., Zomer, R.J., von Unger, M., Emmer, I.M., Griscom, B.W.: The role of soil carbon in natural climate solutions. Nat. Sustain. **3**(5), 391–398 (2020)
6. Carbon Collective: Carbon stock. https://www.carboncollective.co/sustainable-investing/carbon-stock. Accessed 23 Jan 2025
7. Damayanti, A.S.: Beberapa sifat fisika kimia tanah yang berpengaruh terhadap model kecepatan infiltrasi pada tegakan mahoni. Berkala Penelitian Hayati. **17**(1), 85–91 (2012) [in Indonesia]
8. Eluozo, S.N.: Predictive model to monitor the rate of bulk density in fine and coarse soil formation influenced variation of porosity in Coastal Area of Port Harcourt. Am. J. Eng. Sci. Technol. Res. **1**(8), 115–127 (2013)
9. European Environment Agency: Carbon stocks and sequestration rates in ecosystems in Europe. https://www.eea.europa.eu/publications/carbon-stocks-and-sequestration-rates. Accessed 23 Jan 2025

10. Farhaby, A.M., Utama, A.U.: Analysis of mangrove litter production at Mang Kalok Beach, Bangka Regency. Jurnal Enggano. **4**(1), 1–11 (2019) [in Indonesia]
11. Howard, J., Hoyt, S., Isensee, K., Telszewski, M., Pidgeon, E.: Coastal blue carbon: methods for assessing carbon stocks and emissions factors in mangroves, tidal salt marshes, and seagrasses, Conservation International, Intergovernmental Oceanographic Commission of UNESCO, International Union for Conservation of Nature, Arlington, VA (2014)
12. Kikstra, J.S., Nicholls, Z.R., Smith, C.J., Lewis, J., Lamboll, R.D., Byers, E., Gidden, M.J., Rogelj, J., van Vuuren, D.P., Riahi, K.: The IPCC Sixth Assessment Report WGIII climate assessment of mitigation pathways: from emissions to global temperatures. Geosci. Model Dev. **15**(24), 9075–9109 (2022)
13. Marbun, A., Rumengan, A.P., Schaduw, J.N.W., Paruntu, C.P., Angmalisang, P.A., Dan Manopo, V.E.: Analisis Stok Karbon Pada Sedimen Mangrove Di Desa Baturapa Kecamatan Lolak Kabupaten Bolaang Mongondow. Jurnal Pesisir Dan Laut Tropis. **8**(1), 20 (2020) [in Indonesia]
14. Sari, T., Rafdinal, R.L., Tanah, H.K.: Karbon Organik Tanah dan Cadangan Karbon Organik Tanah Di Kawasan Agroforestri Tembawang Nanga Pemubuh Sekadau Hulu Kalimantan Barat. Jurnal Protobiont. **6**(3) (2017) [in Indonesia]
15. Sierra, C.A., Crow, S.E., Heimann, M., Metzler, H., Schulze, E.D.: The climate benefit of carbon sequestration. Biogeosciences. **18**(3), 1029–1048 (2021)
16. Siringoringo, H.H.: Peranan Penting Pengelolaan Penyerapan Karbon Dalam Tanah. Jurnal analisis kebijakan kehutanan, 29285 (2014) [in Indonesia]
17. Sulistyorini, I.S., Edwin, M., Imanuddin, I.: Estimasi Stok Karbon Tanah Organik pada Mangrove di Teluk Kaba dan Muara Teluk Pandan Taman Nasional Kutai. Agrifor: Jurnal Ilmu Pertanian dan Kehutanan. **19**(2), 293–302 (2020) [in Indonesia]
18. The Potential Of Indonesian Mangrove Forests For Global Climate Change Mitigation. Bogor (Indonesia): Center for International Forestry Research (CIFOR). https://www.cifor-icraf.org/publications/pdf_files/infobrief/3773-infobrief.pdf. Accessed 16 Jun 2024. [in Indonesia]
19. Walkley, A., Black, I.A.: An examination of the degtjareff method for determining soil organic matter, and a proposed modification of the chromic acid titration method. Soil Sci. **37**(1), 29–38 (1934)

Diatom Community Structure in Balekambang Lake Sediment Stratigraphy

Barra Madini Noor[1,2], Tri Retnaningsih Soeprobowati[1,2], Jumari Jumari[1,2(✉)], Lilih Khotimperwati[1,2], Riche Haryati[1,2], Mirza Hanif Al Falah[2,3], and Henk Heijnis[4]

[1] Biology Department, Faculty of Sciences and Mathematics, Diponegoro University, Semarang, Indonesia
jumari@live.undip.ac.id
[2] Cluster for Paleolimnology, Diponegoro University, Semarang, Indonesia
[3] PhD Program of Environmental Science, School of Postgraduate Studies, Diponegoro University, Semarang, Indonesia
[4] School of Postgraduate Studies, Diponegoro University, Semarang, Indonesia

Abstract. Balekambang Lake, one of the lakes on the Dieng Plateau, is a lake that has banks of organosol soil that continue to grow, causing lake size shrinkage. The identical phenomenon occurs in several other lakes on the Dieng Plateau, making it a threat to the existence of the lakes and a concern for the ecosystem within them. Diatoms are microalgae (Bacillariophyta) often used as indicators due to their sensitivity to environmental changes. The inorganic nature of their silica structure makes them well preserved in the depth of sediment. This study was conducted to determine the diatom community structure in the sediment stratigraphy of Balekambang Lake banks. A total of 250 cm of sediment was sampled vertically using a dissection corer. Each 10 cm of soil in the sample was digested using 10% H_2O_2 and 10% HCl and then prepared using the Battarbee method. The sample was observed under a 1,000-magnification student microscope. A total of 86 diatom species from 31 different genera were found in Balekambang Lake; Diversity and the diversity and evenness index values of the diatom community were relatively moderate to high, indicating good ecosystem stability. A high abundance of eutrophic indicating diatoms such as *Cylotella meneghiniana*, *Nitzchia palea*, *Navicula cryptocephala*, and *Staurosira construens* were found across sediment stratigraphy. Domination of *Staurosira pinnata*, an alkaline tolerant species, was found at a depth of 230 cm, indicating unstable of the ecosystem caused by ecological variability such as pH fluctuations. The indicated high level of nutrients in Balekambang Lake that is contributing to the ongoing eutrophication could result from Dieng's land use for settlement and agriculture that produce organic pollution to the lake.

Keywords: Balekambang Lake · Diatom stratigraphy · Eutrophication

J. C. Casila et al. (Eds.): CCORE 2024, SPEES, pp. 199–206, 2025.
https://doi.org/10.1007/978-3-031-93887-0_19

1 Introduction

1.1 Eutrophication of Balekambang Lake

Balekambang Lake is a lake located on the Dieng Plateau, Batur District, Banjarnegara Regency, Central Java Province [1]. It is located close to the Arjuna Temple site and surrounded by agricultural land. Residents use this lake as a water reservoir for agrarian irrigation [2, 3]. The soil around Balekambang Lake is a floating organosol soil formed from swamp plant litter, humus, and igneous rock material [4]. Year to year, the lake continues to experience a narrowing phenomenon that threatens the ecosystem and residents who depend on the water in this lake [2].

Eutrophication is one of many possible causes of the lake-narrowing phenomenon [5, 6]. The rapid growth of biomasses induced by the high availability of nutrients in eutrophic lakes creates a buildup of organic particles that can be sedimented and form organosol soils [7]. This result leads to siltation and narrowing lake water bodies in the long run [8]. Uncontrolled nutrient enrichment in lakes itself can be a result of natural variability or the impact of pollution from anthropogenic activities [9, 10].

Protective measures against the threat of lake narrowing in Balekambang Lake must be planned to preserve its ecosystem and maintain its use for residents. Through land management planning, a strategy to sustain Balekambang Lake may be developed to mitigate the impacts of induced eutrophication-induced lake narrowing [11, 12]. One of several assessments that could support this development is lake environmental change monitoring, which can provide data on natural variability and human activities that contribute to the lake-narrowing phenomenon [13].

1.2 Diatom as Past Environmental Indicator

Microfossil community structure analysis is one of many methods to determine the history of changes in the lake environment [14–16]. In this approach, researchers interpret environmental conditions based on the structure of the stratigraphically preserved microfossil community that formed during the sedimentation process [17]. Along with the development of technology and science, various types of microfossils are used in this approach, ranging from pollen, ostracods, radiolarians, and microalgae [18, 19].

Diatoms, one of the organisms that contribute microfossils to lake sediments, have long been recognized as effective bioindicators in environmental reconstruction studies [20, 21]. This is due to the inorganic nature of their silica frustule that makes them well preserved in sediment layers and their sensitivity to environmental changes [22]. Several studies related to the study of diatom community structure in lake sediment stratigraphy have been conducted in Indonesian waters [15, 23, 24]. Some of them are lakes in the Dieng Plateau, which include Penggilon, Cebong, and Menjer Lakes [25–28].

The closest research about diatom stratigraphy research on this lake was pollen stratigraphy analysis by Pudjiarianto *et al.* (2001) and epipelic diatom community structure observation by Pamrayoga *et al.* (2017). To deepen the understanding of changes in environmental conditions in this lake and the structure of diatom communities in general, this research was conducted to analyze the structure of diatom communities in the sediment stratigraphy of Balekambang Lake. The data results of this research could hopefully be used as a reference in land management planning around Balekambang Lake and as a reference for research on diatom stratigraphy and environmental changes in general.

2 Material and Methods

Balekambang Lake is coordinated at 7°12′27.4″S 109°54′29.7″E; located on the Dieng Plateau, Karangsari, Banjarnegara Regency, Central Java Province. A road connecting the lake with the Arjuna Temple site complex. Agricultural land and dense wild shrub vegetation can be seen around the road to the lake.

2.1 Field Working

The dissection coring method was carried out on the lake shore area to obtain sediment samples while maintaining the stratigraphic structure. Coring was carried out to a depth of 250 cm, the maximum depth achievable. Each 10 cm depth from a 250 cm sample was used as a sample.

2.2 Lab Working

Digestion is performed to remove organic matter and impurities in the sediment. A total of 5 g of sample was heated for 3 h by adding 50 mL of 10% HCl. The sample was precipitated for 6 h to separate the debris from the supernatant. The supernatant was removed, and distilled water was added until the sample reached a neutral pH. The second digestion was performed on the sample using 50 mL of 10% H_2O_2. The treatment of removing the supernatant and adding distilled water was performed on the sample again. Sample residues of 400 μmL and 800 μmL were prepared on glass objects using the Battarbee method with Naphrax© adhesive. Observations were made with a Yazumi XSZ-107 binocular microscope at 1000× magnification with immersion oil.

Observation, identification, and counting of individuals were performed on each sample. A total of 300 diatoms obtained were used as a database for community structure analysis. Identification was made by comparing morphological matches with the reference book for diatom identification, Kramme and Lange-Bertalot [29–32], and the official diatom identification site (Algaebase [33], Diatom of North America [34]).

2.3 Data Analysis

The number of individuals of diatom species found in each sediment layer was collected as a database for calculating community structure indices. The indices used include diversity, evenness, and dominance indices. To obtain accurate values, index values were calculated using PAST 4.0 software [35]. The calculation results are presented of bar graphs to facilitate the interpretation of environmental conditions indicated by community structure.

3 Results and Discussion

Based on the coring results at the sampling point, a 250 cm long sediment sample was obtained from Balekambang Lake, Dieng, Central Java. Through the identification and counting of diatom species in sediment samples with a depth of 10 cm multiples, 86 species from 31 genera were found. The identified genus includes *Achnanthes, Adlafia, Aulacoseira, Caloneis, Cyclotella, Cymbella, Denticula, Diploneis, Encyonema, Encyonopsis, Epithemia argus, Eunotia bilunaris, Fragilaria, Frustulia, Geissleria, Gomphonema, Lemnicola, Navicula, Neidium, Nitzschia, Pinnularia, Planothidium, PseudoStaurosira, Rhopalodia, Sellaphora, Stauroneis, Staurosira, Staurosirella,* and *Ulnaria.* One of the species that was also found with high abundancy in previous research by Pramyoga *et al.* (2017) [36] was *Nitzchia palea.*A total of 7 species from all diatoms found have an average relative abundance percentage value >3%, which includes *Cyclotella meneghiniana* Kützing (5%*), Fragilaria gracilis* Østrup (4%), *Navicula cryptocephala* Kützing (4%), *Nitzchia palea* (Kützing) W.Smith (4%), *Staurosira construens* Ehrenberg (4%), *Staurosira pinnata* Ehrenberg (17%), and *Ulnaria ulna* (Nitzsch) Kützing (7%). The species with an average relative abundance percentage > 3% were referenced in this study as the primary environmental indicator, as it is significantly more able to indicate ecological conditions.

Environmental indices score on every 10 cm depth of Balekambang Lake sediment were shown in Fig. 1. An increase or decrease in the values of the calculated indices indicates a change in community structure, indirectly indicating a change in the environment [37, 38]. Diatoms' sensitivity and tolerance level to their environment affects their population size. Diatom species with high adaptability can form large populations and vice versa [39, 40].

The diversity index value in the sediments of Balekambang Lake ranges from 0.03 to 3.3, with an average of 2.73. The range of values indicates relatively stable environmental conditions, but dominance also appears at some depths with poor environmental conditions.

The trend of diversity index values decreased from the most resounding, most profound sample start (250 cm) until it reached its lowest point at 230 cm with a value of 0.03. A significant increase in S. pinnata species caused this decrease. The abundance of the species increased the dominance index to 0.8. The dominance intensity of the *S. pinnata* population is supported by the low evenness value of 0.15, which shows that the distribution of *S. pinnata* is centered at this depth. Low diversity and evenness index values and high dominance values indicate an unstable habitat. The ability to tolerate *S. pinnata* in waters with alkaline pH (8–9) can be one of the causes of high dominance [41]. This assumption is based on the number of other diatoms in the same depth, including species from the genus Fragilaria, Ulnaria, and Nitzchia, which tend to be found in neutral or slightly acidic waters [42, 43].

The diversity index value increased after 230 cm depth. The increase in index value indicates the recovery of habitat stability from damage where diatom variation re-emerges and the dominance of *S. pinnata* decreases. The diversity index peaked to >3 at 170–130 cm depth. Diversity values that exceed 3 indicate a high level of diversity and a stable habitat [44, 45]. High species variability itself is a sign that the habitat can accommodate many species with a wide tolerance to stressors [46]This may indicate

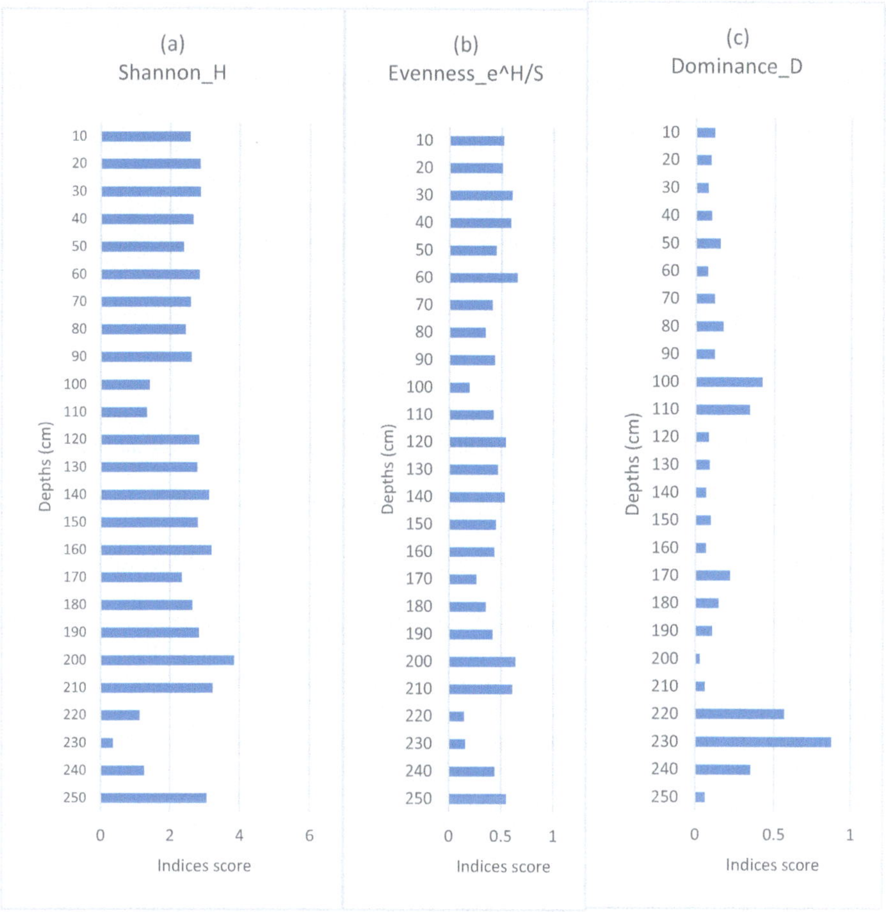

Fig. 1. Environmental indices score of Lake Balekambang Diatoms. (a) Diversity index; (b) Evenness index; (c) Dominance index.

that most species living in Balekambang Lake waters have acidic preferences and low alkaline pH tolerance.

After reaching a depth of 110 cm, the diversity index value increased once more with the value surpassing three. Until it reaches the surface layer, a constant high diversity index value is displayed. At 80 cm depth, a peak with a value of 3.3 was discovered. At this depth range, a number of species from the genera *Aulacoseira*, *Encyonema*, *Gomphonema*, *Lemnicola*, *Planothidium*, and *Pinnularia* started to show up in small numbers. Meanwhile, large population of diatoms were found in *C. meneghiniana*, *N. palea*, *F. gracilis*, and *S. pinnata*. These four species were indicators of eutrophic waters [47–49].

4 Conclusion

The diatom community of Balekambang Lake, Dieng comprises 86 species from 31 diatom genera. The species composition obtained included Cyclotella meneghiniana (5%), *Fragilaria gracilis* (4%), *Navicula cryptocephala* (4%), *Nitzchia palea* (4%), *Staurosira construens* (4%), *Staurosira pinnata* (17%), *Ulnaria ulna* (7%), and other species with abundance ≤3% (55%). Most species are evenly distributed and show a stable habitat. An abundance of eutrophic condition diatom indicator species was found in some depths, mainly at the top of the core.

References

1. Harijoko, A., et al.: Geochronology and magmatic evolution of the Dieng Volcanic Complex, Central Java, Indonesia and their relationships to geothermal resources. J. Volcanol. Geotherm. Res. **310**, 209–224 (2016). https://doi.org/10.1016/j.jvolgeores.2015.12.010
2. Harriyadi, H.: Pertimbangan Pemilihan Lokasi Kompleks Candi Dieng. Amerta. **37**(2), 123–138 (2020). https://doi.org/10.24832/amt.v37i2.123-138
3. Rusiah, Satya, M.N., Wahyudin, A.: Wisata Dataran Tinggi Dieng. Pelita-Jurnal Penelit. Mhs. UNY. **1**(1), 5–11 (2005)
4. Setyowati, D.L., Hardati, P.: Fenomena Dataran Dieng, pp. 1–140 (2009)
5. Chang, Y., et al.: Trends and causes of raw water quality indicators in the five most famous lakes of jiangsu province, china. Int. J. Environ. Res. Public Health. **19**(3), 1580 (2022). https://doi.org/10.3390/ijerph19031580
6. Qin, B.Q., et al.: Lake eutrophication and its ecosystem response. Chin. Sci. Bull. **58**(9), 961–970 (2013). https://doi.org/10.1007/s11434-012-5560-x
7. Sedyaaw, P., Bhatkar, V.R.V., Sawant, A.N.: A review on effects of eutrophication in aquatic ecosystem. Int. J. Dev. Res. **14**(04), 65362–65369 (2024). https://doi.org/10.37118/ijdr.28143.04.2024
8. Sonarghare, P., Masram, S., Sonparote, U., Khaparde, K., Kharkate, S.: Causes and effects of eutrophication on aquatic life (a review). Int. J. Environ. Rehab. Conserv. **XI**(SP 2), 213–219 (2020)
9. Pudjoarinto, A., Cushing, E.J.: Pollen-stratigraphic evidence of human activity at Dieng, Central Java. Palaeogeogr. Palaeoclimatol. Palaeoecol. **171**(3–4), 329–340 (2001). https://doi.org/10.1016/S0031-0182(01)00252-8
10. Eastwood, N., et al.: 100 Years of anthropogenic impact causes changes in freshwater functional biodiversity. Elife. **12**, 1–28 (2023). https://doi.org/10.7554/elife.86576
11. Li, T., Chu, C., Zhang, Y., Ju, M., Wang, Y.: Contrasting eutrophication risks and countermeasures in different water bodies: assessments to support targeted watershed management. Int. J. Environ. Res. Public Health. **14**(7), 695 (2017). https://doi.org/10.3390/ijerph14070695
12. Preisner, M., Neverova-Dziopak, E., Kowalewski, Z.: Mitigation of eutrophication caused by wastewater discharge: a simulation-based approach. Ambio. **50**(2), 413–424 (2021). https://doi.org/10.1007/s13280-020-01346-4
13. Akinnawo, S.O.: Eutrophication: causes, consequences, physical, chemical and biological techniques for mitigation strategies. Environ. Challenges. **12**, 100733 (2023). https://doi.org/10.1016/j.envc.2023.100733
14. Reavie, E.D., et al.: Paleolimnology of the Lake of the Woods southern basin: continued water quality degradation despite lower nutrient influx. Lake Reserv. Manag. **33**(4), 369–385 (2017). https://doi.org/10.1080/10402381.2017.1312648

15. Soeprobowati, T.R.: The challenges of paleolimnology Indonesia. J. Phys. Conf. Ser. **1943**(1), 012100 (2021). https://doi.org/10.1088/1742-6596/1943/1/012100
16. Soeprobowati, T.R., Hidayat, J.Y., Baskoro, K.: Epipelic diatoms as a bioindicator of lake rawa pening water quality. J. Sains dan Mat. **19**(4), 107–118 (2011)
17. Del Mouro, L., et al.: Organic walled microfossils in wet peperites from the early Cretaceous Paraná-Etendeka volcanism of Brazil. Sci. Rep. **13**(1), 1–10 (2023). https://doi.org/10.1038/s41598-023-42483-6
18. Burge, D.R.L., Edlund, M.B., Frisch, D.: Paleolimnology and resurrection ecology: the future of reconstructing the past. Evol. Appl. **11**(1), 42–59 (2018). https://doi.org/10.1111/eva.12556
19. Sremac, J., Bošnjak, M., Firi, K.F., Šimičević, A., Aščić, Š.: Marine microfossils: tiny archives of ocean changes through deep time. AIMS Microbiol. **10**(3), 644–673 (2024). https://doi.org/10.3934/microbiol.2024030
20. Hattikudur, R.S., Alakananda, B., Ramachandra, T.V.: Diatoms as bioindicators. In: LAKE 2014: Conference on Conservation and Sustainable Management of Wetland Ecosystems in Western Ghats, 13–15 November, 2014, pp. 2012–2015 (2014) http://ces.iisc.ernet.in/energy
21. Khalil, S., et al.: Exploration and determination of algal role as Bioindicator to evaluate water quality—probing fresh water algae. Saudi J. Biol. Sci. **28**(10), 5728–5737 (2021). https://doi.org/10.1016/j.sjbs.2021.06.004
22. Li, Q., et al.: Morphology and mechanical properties of fossil diatom frustules from genera of ellerbeckia and melosira. Nanomaterials. **11**(6), 1615 (2021). https://doi.org/10.3390/nano11061615
23. Nastuti, R., Soeprobowati, T.R., Utomo, S.: Land use changes toward water quality status of lake maninjau west sumatera. E3S Web Conf. **448**, 03046 (2023). https://doi.org/10.1051/e3sconf/202344803046
24. Al Falah, M.H., Soeprobowati, T.R., Hadiyanto: Epipelic diatom assemblages to determine water quality in the tuntang river downstream, demak, central java. IOP Conf. Ser. Earth Environ. Sci. **1137**(1), 012003 (2023). https://doi.org/10.1088/1755-1315/1137/1/012003
25. Putri, H.A., Soeprobowati, T.R., Hariyati, R., Purnaweni, H., Jumari, J., Heijnis, H.: An interpretation of diatom community for environmental record in 5-65 cm of Cebong Lake sediment. Pol. J. Environ. Stud. **32**(5), 4781–4788 (2023)
26. Soeprobowati, T.R., Purnaweni, H., Jumari, J., Sari, K.: The relationship of water quality to epipelic diatom assemblages in Cebong lake, Dieng Indonesia. Pol. J. Environ. Stud. **31**(1), 281–295 (2022). https://doi.org/10.15244/pjoes/137084
27. Soeprobowati, T.R., Suedy, S.W.A., Hadiyanto: Find the future from the past: paleolimnology in Indonesia. E3S Web Conf. **31**, 1–8 (2018). https://doi.org/10.1051/e3sconf/20183108002
28. Geyga, P., Soeprobowati, T.R.: Komunitas Fitoplankton di Telaga Menjer, Dieng, Jawa Tengah. In: Prosiding Seminar Nasional Hasil-Hasil Penelitian Pascasarjana, SPS UNDIP, pp. 1–23 (2016)
29. Krammer, K., Lange-Bertalot: Bacillariophyceae: Teil 3: Centrales, Fragillariaceae, Eunotiaceae, Süßwasserf. Spectrum, Berlin (2004)
30. Krammer, K., Lange-Bertalot: Bacillariophyceae: Teil 1: Naviculae, Süßwasserf. Spectrum, Berlin (2004)
31. Krammer, K., Lange-Bertalot: Bacillariophyceae. 3. Teil: Centrales, Fragilariaceae, Eunotiaceae. In: Süsswasserflora von Mitteleuropa 2, Süsswasser. Jena, Germany (1991)
32. K. Krammer and Lange-Bertalot, Bacillariophyceae 5. English and French translation of the keys, Süßwasserf. 2000
33. Guiry, M. D, G., Guiry: AlgaeBase. World-wide electronic publication, University of Galway.," https://www.algaebase.org. searched on June 12 (2024)
34. Spaulding, S.A., et al.: Diatoms.org: Supporting Taxonomists, Connecting Communities. https://diatoms.org/. Accessed 12 Jun 2024

35. Hammer, Ø., David, A.T.H., Ryan, P.D.: PAST: paleontological statistics software package for education and data analysis. Palaeontol. Electron. **4**(1), 1–9 (2001)
36. Pamrayoga, G., Soeprobowati, T.R., Putro, S.: Diatom Epipelik di Telaga Balekambang, Dieng, Jawa Tengah. In: Prosiding Seminar Nasional (2017)
37. Ehrlén, J., Morris, W.F.: Predicting changes in the distribution and abundance of species under environmental change. Ecol. Lett. **18**(3), 303–314 (2015). https://doi.org/10.1111/ele.12410
38. Rashit, E., Bazin, M.: Environmental fluctuations, productivity, and species diversity: an experimental study. Microb. Ecol. **14**(2), 101–112 (1987). https://doi.org/10.1007/BF02013016
39. Masithah, E.D., Islamy, R.A.: Checklist of freshwater periphytic diatoms in the midstream of Brantas River, East Java, Indonesia. Biodiversitas. **24**(6), 3269–3281 (2023). https://doi.org/10.13057/biodiv/d240621
40. Passy, S.I.: Diatom ecological guilds display distinct and predictable behavior along nutrient and disturbance gradients in running waters. Aquat. Bot. **86**(2), 171–178 (2007). https://doi.org/10.1016/j.aquabot.2006.09.018
41. Pisera, A., Siver, P.A., Mandic, O.: Miocene siliceous microfossils from the open cast coal mine Gračanica (Bugojno paleolake, Bosnia and Herzegovina) and their significance: a preliminary report. Palaeobiodivers. Palaeoenviron. **100**(2), 507–517 (2020). https://doi.org/10.1007/s12549-019-00378-3
42. Kahlert, M., et al.: Co-occurrence, ecological profiles and geographical distribution based on unique molecular identifiers of the common freshwater diatoms Fragilaria and Ulnaria. Ecol. Indic. **141**, 109114 (2022). https://doi.org/10.1016/j.ecolind.2022.109114
43. Sevindik, T.O., Hamilton, P.B., Solak, C.N., Yilmaz, E., Güzel, U.: Three New Nitzschia (Bacillariophyceae) Species from Highly Acidic Artificial Lakes in Çanakkale, Türkiye. Water. **15**(21), 3784 (2023). https://doi.org/10.3390/w15213784
44. Cardinale, B.J., et al.: Biodiversity simultaneously enhances the production and stability of community biomass, but the effects are independent. Ecology. **94**(8), 1697–1707 (2013). https://doi.org/10.1890/12-1334.1
45. Odum, E.P.: Dasar-Dasar Ekologi Edisi Ketiga, 3rd edn. Gadjah Mada University Press, Yogyakarta (1994)
46. Kireta, A.R., et al.: Planktonic and periphytic diatoms as indicators of stress on great rivers of the United States: testing water quality and disturbance models. Ecol. Indic. **13**(1), 222–231 (2012). https://doi.org/10.1016/j.ecolind.2011.06.006
47. Kociolek, J.P., Spaulding, S.A.: 15. Centric Diatoms. Elsevier, Amsterdam (2003). https://doi.org/10.1016/B978-0-12-741550-5.50016-7
48. Kock, A., Pheiffer, W., Wepener, V., Smit, N.J., Taylor, J.C.: Using confocal microscopy and pigment analyses to detect adverse insecticide effects in non-target freshwater diatom species—a proof-of-concept using Nitzschia palea. Bull. Environ. Contam. Toxicol. **110**(6), 1–6 (2023). https://doi.org/10.1007/s00128-023-03741-5
49. Rzdokiewicz, M., Zawiska, I.: Diatoms as paleoecological indicators of environmental change in the Lake Czechowskie catchments ecosystem (Northern Tuchola Pinewoods, Poland). Inst. Geol. Sci. **18**(1), 321–340 (2016)

Land Use Change and Mangrove Dynamics in Tambakbulusan and Morodemak, Central Java: A Spatial Analysis from 1973 to 2021

Aulia Rahim[1,2](✉), Tri Retnaningsih Soeprobowati[3,4], and Muhammad Helmi[2,4]

[1] Doctoral of Environmental Science, Department of Environmental Science, School of Postgraduate Studies, Diponegoro University, Imam Bardjo SH Street No. 5, Semarang 50241, Indonesia
auliarahim063@gmail.com
[2] Department of Biology, Faculty of Science and Mathematics, Diponegoro University, Prof Soedarto, SH Street, Semarang, Indonesia
[3] Department of Oceanography, Faculty of Fisheries and Marine Science, Diponegoro University, Semarang 50275, Indonesia
[4] Cluster for Paleolimnology (CPalim), Diponegoro University, Imam Bardjo SH Street No. 5, Semarang, Indonesia

Abstract. Land use change in coastal areas has become a significant phenomenon, influencing the dynamics of mangrove ecosystems and their critical role in maintaining coastal resilience and biodiversity. This study aims to analyze land use changes in Tambakbulusan and Morodemak, Central Java, from 1973 to 2021, focusing on the impacts of anthropogenic activities and rehabilitation efforts on mangrove ecosystems. Using spatial analysis, the results reveal that mangrove areas experienced a substantial decline between 1989 and the early 2000s due to the expansion of aquaculture ponds and settlements. However, from 2003 to 2021, mangrove coverage showed a positive trend, indicating the success of rehabilitation efforts. Meanwhile, aquaculture ponds expanded steadily, becoming the dominant land use by 2021, while bare land and scrubland exhibited fluctuating trends. Settlement areas also increased, further altering the landscape. These findings highlight the dual impact of human activities: the degradation of natural ecosystems and the potential for successful restoration efforts. This study underscores the importance of sustainable coastal land use management and continued mangrove rehabilitation to balance ecological and socio-economic needs in the region.

Keywords: Land Use Change · Mangrove Dynamics · Spatial Analysis

1 Introduction

Coastal areas are dynamic regions where natural ecosystems and human activities interact intensively, often resulting in significant land use changes. These changes, driven by urbanization [1], aquaculture development [2], and agricultural expansion [3], have profound implications for the environment [4, 5], particularly for mangrove ecosystems [6].

J. C. Casila et al. (Eds.): CCORE 2024, SPEES, pp. 207–216, 2025.
https://doi.org/10.1007/978-3-031-93887-0_20

Mangroves are critical coastal ecosystems that provide essential services such as shoreline stabilization [7], biodiversity support [8], and carbon sequestration [9]. However, their vulnerability to anthropogenic pressures has made them one of the most threatened ecosystems globally [10].

In Indonesia, which hosts the largest mangrove coverage in the world, mangrove loss has been accelerated by land conversion for aquaculture and settlement development [11]. Tambakbulusan and Morodemak, located in the coastal area of Demak Region, Central Java, exemplifies this phenomenon. Over the past decades, the region has undergone significant land use transformation, driven primarily by the expansion of aquaculture ponds and settlements [12]. These changes have led to substantial mangrove degradation, which threatens the ecological balance and coastal protection provided by these forests.

Despite historical losses, mangrove rehabilitation efforts in recent years have shown encouraging outcomes. Government and community-led programs to restore mangroves have gradually reversed the decline, as observed in the increasing mangrove coverage since the early 2000s. This indicates the potential for effective restoration strategies to mitigate the adverse impacts of past land use changes and enhance the resilience of coastal ecosystems.

This study explores the spatial and temporal dynamics of land use change in Tambakbulusan and Morodemak, Central Java, from 1973 to 2021, focusing on the interplay between anthropogenic influences and mangrove rehabilitation efforts. By analyzing historical data and spatial patterns, the research provides insights into the drivers of land use change and their implications for sustainable coastal management and ecosystem conservation.

2 Material and Methods

2.1 Research Location

This study was conducted in the mangrove ecosystems of Tambakbulusan and Morodemak Villages, located in Demak Regency, Central Java Province (Fig. 1). Demak has a mangrove ecosystem area of 980.1 hectares, making it the second-largest in Central Java; however, much of this area is degraded. The primary cause of mangrove damage in that area is land conversion for aquaculture, industrial development, and port construction. Geographically, Tambakbulusan and Morodemak Villages are situated at 6°50′10.19″S and 110°31′31.47″E, with a tropical climate characterized by an average temperature of 27.4–29.2 °C and peak rainfall in December [13]. The tidal regime along the northern coast of Demak is mixed semidiurnal, and surface sediments primarily consist of sand (97–99%), with smaller amounts of silt and clay [14, 15].

Tambakbulusan and Morodemak Villages lie in the downstream area of the Tuntang River Basin, which spans four regencies and one city. The Tuntang River originates from Lake Rawa Pening and flows into the Java Sea along the northern coast of Demak. Land use in the downstream region has shifted significantly since 2006, with much of the protected area converted into built-up land, reducing water absorption capacity and increasing the risk of flooding [16]. Despite these changes, the western coastal areas of Tambakbulusan and Morodemak Villages still host significant mangrove ecosystems,

Fig. 1. Map of study location at Tambakbulusan and Morodemak, Demak Regency, Central Java, Indonesia.

covering 167.36 hectares and 87.20 hectares, respectively [17]. These ecosystems play a critical role in providing ecological services, mitigating coastal erosion, and supporting biodiversity, highlighting their importance for sustainable coastal management.

2.2 Data Collection

The data used in this study includes Landsat satellite imagery, specifically Landsat 1 MSS (1973), Landsat 5 TM (1989, 1995, 2003), and Landsat 8 OLI/TIRS (2013, 2021), covering the areas of Tambakbulusan and Morodemak Villages. These satellite images were sourced from the U.S. Geological Survey (USGS) website. Landsat imagery was selected due to its capability to provide consistent data over the 1973–2021 period. In addition, spatial planning maps were obtained from the Regional Development Planning Agency of Demak Regency, and the Indonesian Topographic Map was used to support the analysis of land use changes in the study area.

2.3 Data Analysis

The data analysis process in this study involved several systematic steps to examine land use changes in the study area. Preprocessing of Landsat satellite imagery from 1973, 1989, 1995, 2003, 2013, and 2021 was conducted to enhance data quality, including radiometric and geometric corrections to reduce distortions and standardize the datasets [18]. The processed imagery was then analyzed through visual interpretation and on-screen digitization to classify various land use types, with particular emphasis on mangrove ecosystems and other coastal land cover [19]. Ground truth data collected from the field were integrated to refine the classification process and ensure accuracy.

To validate the results, an accuracy assessment with the confusion matrix method was conducted, with adjustments made iteratively until the classification accuracy exceeded 75% [20]. Following validation, an overlay analysis was performed to integrate the classified maps from different time periods, enabling a detailed comparison of land use changes over time. The final output was a comprehensive map illustrating the spatial and temporal dynamics of land use transformation in the Tambakbulusan and Morodemak coastal areas, emphasizing key trends such as mangrove loss, aquaculture expansion, and other coastal modifications.

3 Results and Discussion

Based on the satellite image analysis, land use in Tambakbulusan and Morodemak Villages comprises aquaculture ponds, mangroves, settlements, bare land, and shrubs. The land-use change mapping results clearly demonstrate the dominance of aquaculture ponds in these areas. This dominance can be attributed to the fact that the majority of the local population relies on aquaculture farming as their primary livelihood [21]. The land-use mapping results for Tambakbulusan and Morodemak Villages from 1973 to 2021 are presented in Fig. 2.

Referring to Fig. 3, it can be observed that compared to other land-use types, mangroves, bare land, and aquaculture ponds in the study area have undergone significant changes in the area between 1973 and 2021. Notably, aquaculture ponds experienced rapid expansion during the 1973–2003 period but showed a considerable decline between 2003 and 2021. Similarly, bare land increased significantly during the early period (1973–1995) but decreased again between 2003 and 2021.

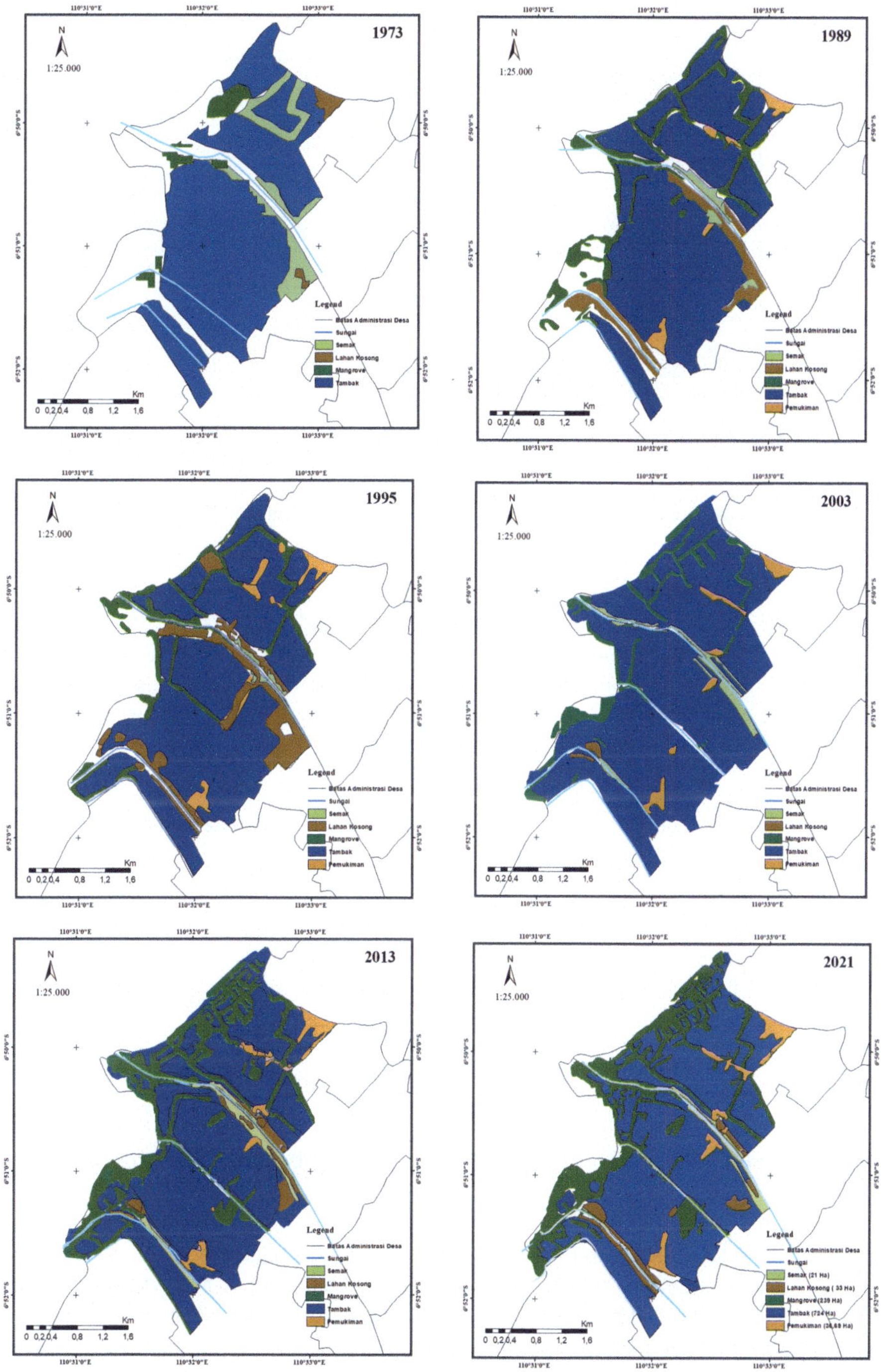

Fig. 2. Land use map in Tambakbulusan and Morodemak Villages (1973–2021).

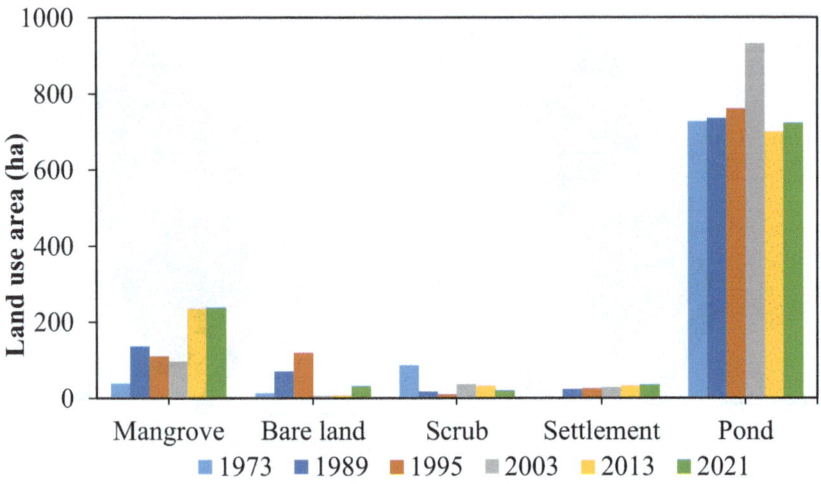

Fig. 3. Land use area changes in Tambakbulusan and Morodemak Villages (1973–2021).

In contrast, mangrove areas tended to experience a significant decline during the early period, particularly from 1989 to 2003. This trend suggests that the expansion of aquaculture ponds and bare land during this period likely encroached on mangrove ecosystems, indicating deforestation and land-use conversion of mangroves into aquaculture areas. Further details on the changes in area and percentage of land use during the 1973–2021 period are presented in Table 1.

Table 1. Changes in land use area in Tambakbulusan and Morodemak Villages.

Land use type	1973		1989		1995		2003		2013		2021	
	Ha	%	Ha	%	Ha	%	Ha	%	Ha	%	Ha	%
Mangrove	39	4.49	136	13.8	111	10.8	96	8.8	237	23.3	239	22.7
Bare land	14	1.61	71	7.21	120	11.7	6	0.5	9	0.9	33	3.1
Scrub	87	10	18	1.83	11	1.07	36	3.3	35	3.4	21	2.0
Settlement	–	–	24	2.44	26	2.52	28	2.6	35	3.4	36.69	3.5
Pond	728	83.9	736	74.7	762	74	931	85	702	69.0	724	68.7
Total	868	100	985	100	1030	100	1097	100	1018	100	1053.7	100

The calculation of area changes and percentages for each land use type, as presented in Table 1, reveals significant land expansion in Tambakbulusan and Morodemak Villages from 1973 (868 ha) to 2021 (1053.7 ha). This increase is likely due to accretion or sediment deposition in the area, which added to the landmass. Additionally, the results indicate relatively stable growth in settlement areas and more significant changes in other land use types during the 1973–2021 period. Settlement areas showed a steady increase, growing from 24 ha in 1989 to 36.69 ha in 2021, corresponding to a rise in percentage from 2.44% to 3.5% during this period. However, settlement data from 1973 could not be identified due to low-quality imagery.

Aquaculture ponds demonstrated substantial growth during the early period (1973–2003), expanding from 728 ha to 931 ha. This represents an increase in the percentage of aquaculture area from 83.9% to 85% by 2003. However, this trend reversed during the later period (2003–2021), with aquaculture area drastically declining from 931 ha (85%) to 724 ha (68.7%) in 2021. Similarly, bare land initially showed a significant increase, rising from 1.61% in 1973 to 11.7% in 1995 before decreasing sharply to just 3.1% in 2021. This pattern suggests substantial deforestation and aquaculture expansion during the early period (1973–2003).

In contrast, mangrove land exhibited an opposing trend to aquaculture ponds and bare land. Mangrove areas initially increased significantly between 1973 and 1989, growing from 39 ha (4.49%) to 136 ha (13.8%). However, it declined sharply, reaching 96 ha (8.8%) in 2003. Interestingly, the period from 2003 to 2021 saw rapid growth in the mangrove area, expanding from 96 ha to 239 ha. This represents an increase in mangrove coverage from 8.8% to 22.7% of the total land area in 2021. These findings reflect a dynamic interplay between deforestation, aquaculture expansion, and mangrove restoration efforts over the study period.

The significant surge in aquaculture ponds and bare land between 1989 and 2003 is predicted to be a consequence of deforestation and the conversion of mangrove ecosystems driven by the booming black tiger shrimp farming industry during that period. The rise in global shrimp prices spurred this boom [22]. Generally, this phenomenon occurred across most coastal areas in Indonesia as an outcome of the government's program to expand aquaculture ponds by 30,000 hectares, which was mandated under the Minister of Agriculture Decree No. 5/SK/Mentan/Bimas/VI/1984 [11]. This initiative aimed to offset the ban's impact on fish capture using trawls. It is suspected that a significant portion of the targeted areas for aquaculture development during this period were mangrove ecosystems [11]. According to [23], the largest-scale mangrove conversion into aquaculture ponds in Indonesia occurred between 1998 and 2001.

The coastal areas of Demak Regency, located along the North Coast of Java, also experienced a significant expansion of aquaculture ponds during this period. According to [24], the rapid growth of aquaculture ponds in Demak Regency occurred notably between 1980 and the early 2000s. By 1987, aquaculture pond coverage in one of Demak Regency's districts, Sayung, had already reached 1271.4 hectares and continued to expand into the early 2000s, resulting in extensive land conversion into aquaculture ponds. This phenomenon was driven by the economic dependency of coastal communities in the region on fishing and aquaculture, which were considered to have high economic value during the 1980s.

Based on local information, it revealed that extensive mangrove conversion to black tiger shrimp ponds occurred in Tambakbulusan and Morodemak from the mid-1980s to early 2000s, leading to significant mangrove loss. This conversion contributed to flash floods in the early 2000s, which damaged settlements and most ponds. Many abandoned the damaged ponds, causing a significant decline in aquaculture activities. Mangrove rehabilitation efforts by the community and government have since been implemented to mitigate tidal flooding and coastal erosion. The findings suggest that mangrove loss during 1989–2003 resulted from uncontrolled land conversion, while the increase in coverage from 2003–2021 reflects the success of rehabilitation efforts.

However, tidal flooding remains a critical threat to coastal communities in Tambakbulusan and Morodemak. According to the Wetlands Internasional data [25], tidal flooding in Morodemak damaged aquaculture embankments, inundated approximately 100 hectares of ponds, and caused losses of around IDR 72,540,000. Similarly, in Tambakbulusan, tidal flooding resulted in crop failures in shrimp and milkfish ponds and caused extensive damage to public facilities, roads, and settlements, with significant social and economic impacts [26].

Based on Demak Regional Regulation No. 1 of 2020, which amended the 2011–2031 Spatial Plan, parts of Demak's coastal area are designated for both mangrove conservation and industrial or aquaculture zones. Proper planning and monitoring are necessary to prevent pollution and mangrove ecosystem degradation caused by industrial and aquaculture activities [27].Given the critical role of mangroves in preventing tidal flooding in Tambakbulusan and Morodemak, sustainable management plans are essential. These plans must balance ecological preservation with the social and economic needs of local communities [28], integrating mangrove conservation with sustainable aquaculture practices.

4 Conclusion

The present study showed significant land use changes have occurred in Tambakbulusan and Morodemak from 1973 to 2021, with rapid expansion of aquaculture ponds, particularly from the mid-1980s to early 2000s, leading to widespread mangrove deforestation. This conversion contributed to environmental degradation and flash floods that damaged settlements and aquaculture areas. However, mangrove rehabilitation efforts initiated in the early 2000s have successfully restored mangrove coverage. Despite these efforts, tidal flooding remains a significant threat to the region, impacting aquaculture and infrastructure. The findings emphasize the need for sustainable land use planning that integrates mangrove conservation with aquaculture development, ensuring the long-term ecological stability and resilience of coastal communities. The success of the rehabilitation efforts demonstrates that continued restoration and effective management strategies can mitigate flooding impacts and preserve vital mangrove ecosystems.

Acknowledgement. This research was funded by the PMDSU Research Grant program by the Ministry of Education, Culture, Research, and Technology of the Republic of Indonesia through Diponegoro University with Number 345-42/UN7.D2/PP/V/2024, and International Publication Quality Improvement For PMDSU Scholarship Recipients Grant by the Ministry of Education, Culture, Research, and Technology of the Republic of Indonesia with Number 250.5/E4.4/KU/2024.

References

1. Huang, F., Huang, B., Huang, J., Li, S.: Measuring land change in coastal zone around a rapidly urbanized bay. Int. J. Environ. Res. Public Health. **15**(6), 1–16 (2018). https://doi.org/10.3390/ijerph15061059
2. Jayanthi, M., Thirumurthy, S., Muralidhar, M., Ravichandran, P.: Impact of shrimp aquaculture development on important ecosystems in India. Glob. Environ. Chang. **52**, 10–21 (2018). https://doi.org/10.1016/j.gloenvcha.2018.05.005
3. Periyasamy, R., Roy, P.D., Chokkalingam, L., Natarajan, L., Sundar, S., Moorthy, P., Gowrappan, M.: Transformation analysis on landuse/land Cover changes for two decades between 1999 and 2019 CE with reference to aquaculture—Nagapattinam Coast, Southeast India. J. Indian Soc. Remote Sens. **49**(11), 2831–2845 (2021). https://doi.org/10.1007/s12524-021-01432-4
4. Ai, B., Ma, C., Zhao, J., Zhang, R.: The impact of rapid urban expansion on coastal mangroves: a case study in Guangdong Province, China. Front. Earth Sci. **14**(1), 37–49 (2020). https://doi.org/10.1007/s11707-019-0768-6
5. Wang, Y., Chao, B., Dong, P., Zhang, D., Yu, W., Hu, W., Ma, Z., Chen, G., Liu, Z., Chen, B.: Simulating spatial change of mangrove habitat under the impact of coastal land use: coupling MaxEnt and Dyna-CLUE models. Sci. Total Environ. **788**, 147914 (2021). https://doi.org/10.1016/j.scitotenv.2021.147914
6. Osland, M.J., Feher, L.C., Lopez-Portillo, J., Day, R.H., Suman, D.O., Menendez, J.M.G., Rivera-Monroy, V.: Mangrove forests in a rapidly changing world: global change impacts and conservation opportunities along the Gulf of Mexico coast. Estuar. Coast. Shelf Sci. **214**, 120–140 (2018). https://doi.org/10.1016/j.ecss.2018.09.006
7. Soper, F.M., MacKenzie, R.A., Sharma, S., Cole, T.G., Litton, C.M., Sparks, J.P.: Non-native mangroves support carbon storage, sediment carbon burial, and accretion of coastal ecosystems. Glob. Chang. Biol. **25**(12), 4315–4326 (2019). https://doi.org/10.1111/gcb.14813
8. Rahman, M.M., Zimmer, M., Ahmed, I., Donato, D., Kanzaki, M., Xu, M.: Co-benefits of protecting mangroves for biodiversity conservation and carbon storage. Nat. Commun. **12**(1), 3875 (2021). https://doi.org/10.1038/s41467-021-24207-4
9. Rovai, A.S., Coelho-Jr, C., de Alemida, R., Cunha-Lignon, M., Manghini, R.P., Twilley, R.R., Cintron-Molero, G., Schaeffer-Novelli, Y.: Ecosystem-level carbon stocks and sequestration rates in mangroves in the Cananéia-Iguape lagoon estuarine system, southeastern Brazil. For. Ecol. Manage. **479**, 118553 (2021). https://doi.org/10.1016/j.foreco.2020.118553
10. Hayashi, S.N., Souza-Filho, P.W.M., Nascimento, W.R., Fernandes, M.E.B.: The effect of anthropogenic drivers on spatial patterns of mangrove land use on the Amazon coast. PLoS One. **14**(6), 1–20 (2018). https://doi.org/10.1371/journal.pone.0217754
11. Ilman, M., Dargusch, P., Dart, P., Onrizal, O.: A historical analysis of the drivers of loss and degradation of Indonesia's mangroves. Land Use Policy. **54**, 448–459 (2016). https://doi.org/10.1016/j.landusepol.2016.03.010
12. Rahim, A., Soeprobowati, T.R., Putranto, T.T., Al Falah, M.H., Gell, P.: Contribution of mangrove forest carbon stocks on climate change mitigation: a case study at Tuntang Estuary, Central Jawa. J. Coast. Conserv. **28**(4), 65 (2024). https://doi.org/10.1007/s11852-024-01059-w
13. BPS-Statistic of Demak Regency: Demak Regency in Figures, pp. 1–234. BPS-Statistic of Demak Regency (2021)
14. Alkautsar, M.D., Suryono, C.A., Pratikto, I.: Correlation between non-sand sediment grain size and organic material content in Morodemak Waters, Demak Regency. J. Mar. Res. **11**(3), 391–398 (2022). https://doi.org/10.14710/jmr.v11i3.35020. [In Indonesia]

15. Muskananfola, M.R., Supriharyono, Febrianto, S.: Spatio-temporal analysis of shoreline change along the coast of Sayung Demak, Indonesia using Digital Shoreline Analysis System. Reg. Stud. Mar. Sci. **34**, 101060 (2020). https://doi.org/10.1016/j.rsma.2020.101060
16. Sriyana: Characteristic study of Tuntang Watershed and integrated watershed management model. Teknik. **32**(3), 180–186 (2011) [In Indonesia]
17. Faturrohmah, S., Marjuki, B.: Identification of spatial dynamics of mangrove resources in coastal areas of Demak Regency, Central Jawa. Majalah Geografi Indonesia. **31**(1), 56 (2017). https://doi.org/10.22146/mgi.24234. [In Indonesia]
18. Young, N.E., Anderson, R.S., Chignell, S.M., Vorster, A.G., Lawrence, R., Evangelista, P.H.: A survival guide to Landsat preprocessing. Ecology. **98**(4), 920–932 (2017). https://doi.org/10.1002/ecy.1730
19. Mas, J.F., Lemoine-Rodríguez, R., González-López, R., López-Sánchez, J., Piña-Garduño, A., Herrera-Flores, E.: Land use/land cover change detection combining automatic processing and visual interpretation. Eur. J. Remote Sens. **50**(1), 626–635 (2017). https://doi.org/10.1080/22797254.2017.1387505
20. Wiggers, M.J., Nuarsa, I.W., Putra, I.D.N.N.: Analysis of coastal land use changes in Batu Layar District, West Lombok Regency in 2002 and 2019. J. Mar. Res. Technol. **3**(2), 68 (2020). https://doi.org/10.24843/jmrt.2020.v03.i02.p02. [In Indonesia]
21. Djamal, R.: Institutional analysis of pond farmers' groups in the development of farming business. J. Litbang Provinsi Jawa Teng. **12**(1), 55–70 (2014) [In Indonesia]
22. Pramudji: Mangrove status in the north coastal area of West Java (Larawnag and Indramayu) and management efforts. Oseana. **9**(2), 43–52 (2015) [In Indonesia]
23. Dutrieux, E., Proisy, C., Fromard, F., Walcker, R., Ilman, M., Pawlowski, F., Ferdiansyah, H., Ponthieux, O.: Mangrove restoration in the vicinity of oil and gas facilities: lessons learned from a large scale project. In: SPE International Conference on Health, Safety and Environment. 2014 Journey Contin.(2), pp. 1103–1120 (2014). https://doi.org/10.2118/168449-ms
24. Munasikhah, S., Wijayati, P.A.: From mangrove forest to fish pond: ecological crisis in Sayung area, Demak Regency 1990-1999. J. Indones. Hist. **10**(2), 129–140 (2022) [In Indonesia]
25. Wetlands Internasional: Impact of Tidal Floods in Demak Regency, p. 10 (2020) [In Indonesia]
26. Ward, P.J., Marfai, M.A., Yulianto, F., Hizbaron, D.R., Aerts, J.C.J.H.: Coastal inundation and damage exposure estimation: a case study for Jakarta. Nat. Hazards. **56**(3), 899–916 (2011). https://doi.org/10.1007/s11069-010-9599-1
27. Irsadi, A., Anggoro, S., Soeprobowati, T.R., Helmi, M., Khair, A.S.E.: Shoreline and mangrove analysis along semarang-demak, Indonesia for sustainable environmental management. J. Pendidik. IPA Indones. **8**(1), 1–11 (2019). https://doi.org/10.15294/jpii.v8i1.17892
28. Bagarinao, T.U.: Biodiversity in mangrove-derived aquaculture ponds in Dumangas, Iloilo, Philippines. Philipp. J. Sci. **150**(1), 153–169 (2021)

Trophic Status Changes in the Lower Tuntang River, Demak, Based on Diatom Stratigraphic Evidence

Mirza Hanif Al Falah[1,6], Tri Retnaningsih Soeprobowati[1,2,6](✉), Hadiyanto Hadiyanto[1,3], Marta Wojewódka-Przybył[4,6], Łukasz Kruszewski[4,6], and Paweł Rydelek[5,6]

[1] Department of Environmental Science, School of Postgraduate Studies, Diponegoro University, Semarang 50241, Indonesia
trsoeprobowati@live.undip.ac.id
[2] Department of Biology, Faculty of Science and Mathematics, Diponegoro University, Semarang 50241, Indonesia
[3] Department of Chemical Engineering, Faculty of Engineering, Diponegoro University, Semarang, Indonesia
[4] Institute of Geological Sciences, Polish Academy of Sciences Twarda 51/55, Warsaw 00-818, Poland
[5] Faculty of Geology, University of Warsaw, Warsaw 02-089, Poland
[6] Cluster for Paleolimnology (CPalim), Diponegoro University, Semarang 50275, Indonesia

Abstract. Rivers are essential ecosystems, yet anthropogenic impacts, including agricultural runoff, industrial discharges, and urban wastewater, have significantly degraded water quality. Eutrophication, driven by excessive nitrogen and phosphorus inputs, poses severe ecological threats by triggering algal blooms and depleting oxygen levels. To address these challenges, it is crucial to investigate ecological indicators that can provide insights into past and present water quality. This study explores diatom assemblages as bioindicators to reconstruct historical changes in water quality and environmental conditions. Sediment cores were collected using a Russian corer from two sites along the river and analyzed for diatom species diversity and abundance. A total of 100 and 105 diatom species were identified in the Agricultural (TA) and Mangrove (TM) areas, respectively, revealing taxa that were both overlapping and site-specific influenced by local conditions. The analysis showed a dominance of benthic diatoms, indicative of nutrient-rich conditions and eutrophication, along with planktonic species affected by hydrological fluctuations. Cluster and stratigraphic analyses using indices like Shannon-Wiener and Bray-Curtis uncovered temporal and spatial variations in species composition, reflecting the impacts of agriculture, land-use changes, and pollution. Key diatom species, such as *Nitzschia palea, Fragilaria crotonensis, Melosira varians,* and *Cyclotella meneghiniana*, served as indicators of eutrophic conditions and anthropogenic pressures. The findings another valuable insights for future restoration efforts, emphasizing the importance of diatom-based monitoring in managing and improving riverine ecosystems.

Keywords: Diatom · River · Eutrophication · Environmental Sciences · Pollution

J. C. Casila et al. (Eds.): CCORE 2024, SPEES, pp. 217–233, 2025.
https://doi.org/10.1007/978-3-031-93887-0_21

1 Introduction

Rivers are vital ecosystems that support diverse life forms and play crucial role in human survival [1]. They provide essential services, including drinking water, irrigation, energy production, and recreational opportunities [2]. However, rising pollution levels in river systems pose a serious threat to these ecosystems and the critical functions they perform. Pollutants from agricultural runoff, industrial discharges [3], and urban wastewater introduce harmful substances into rivers, causing significant water quality degradation [4]. Maintaining river water quality is essential for preserving environmental balance and safeguarding the well-being of humans and animals. There is a well-established connection between the condition of river water and its impact on public health, highlighting the importance of rivers as a indispensable natural resource [5].

One of the most alarming consequences of river pollution is eutrophication, a process driven by the excessive accumulation of nutrients such as nitrogen and phosphorus [6]. These nutrients often originate from agricultural fertilizers, untreated sewage, and other anthropogenic sources. Eutrophication accelerates the growth of algae and cyanobacteria, leading to the formation harmful algal blooms (HABs). The decay of these blooms depletes oxygen levels in the water column, creating hypoxic or anoxic conditions that threaten aquatic life. This cascade of ecological disturbances significantly impacts biodiversity and disrupts ecosystem balance [7].

Eutrophication is a severe environmental problem affecting Indonesian water ecosystems, such as Rawapening Lake [8], Cirata Reservoir [9], Bangkau Peatland [10], and Citarum River [11]. Issues related to water quality degradation are also evident downstream of the Tuntang River, Demak Regency. The growing population in Demak has contributed to increased water pollution in the Tuntang River. Data from the BBWS Pemali Juana (Central Java Water Service) for the period 2017–2022 indicates that the Tuntang River's conditions range from light to heavily polluted [12]. The source of this pollution includes industrial waste, inorganic agricultural fertilizers, and the domestic household waste. In recent years, human activities have further exacerbated channel narrowing, wetland degradation, and salinization in the estuarine region, which represents the river's terminal point [13].

A fundamental requirement for effective water quality management is reliable monitoring. Biological monitoring, or biomonitoring, typically involves the systematic use of living organisms and their responses to evaluate environmental conditions or changes [14]. Biological communities provide a more comprehensive reflection of both current and historical conditions, as well as the impact of multiple potential stressors, compared to chemical and physical measurements, which only capture a narrow set of factors and provide a temporary snapshot of the current state [15, 16]. Various biological metrics, based on different groups of organisms, have been developed to evaluate the health of freshwater ecosystems [17].

Diatoms, a group of photosynthetic microorganisms, have emerged as powerful indicators of river health and water quality [18]. These microscopic algae are highly sensitive to changes in environmental conditions, particularly nutrient concentrations and pollution levels. Different diatom species thrive under specific water quality conditions, making them effective bioindicators for monitoring eutrophication and pollution [19]. Benthic diatoms, a key component of river phytobenthos, are widely used as biological quality elements (BQEs) in surface water monitoring. They have effectively identified river ecosystems' eutrophication, organic pollution, and acidification [20]. Furthermore, using diatoms aligns with global efforts to adopt sustainable monitoring practices emphasizing ecosystem health. This approach not only aids in detecting pollution sources but also helps track the success of restoration initiatives to reduce eutrophication and improve water quality [21].

The use of diatoms to monitor water quality in river systems has been applied in various studies. In the Austral River system [22], diatom analysis confirmed the occurrence of eutrophication. In the Marshyangdi River [23], findings revealed that the physical and chemical qualities of the water significantly influenced the composition of diatom species. In the Gogor River [24], diatom studies demonstrated that the river is only lightly polluted. Research on diatoms from surface sediments in the lower reaches of the Tuntang River [25] indicated that the waters are currently eutrophic. Additionally, a stratigraphic study confirms that tidal water is still frequent along an 8 km stretch near the sea [26].

Based on these observations, conditions downstream of the Tuntang River have undergone annual changes due to variations in in water quality, sedimentation, and land conversion. This research uses diatom stratigraphy to reconstruct conditions downstream of the Tuntang River, providing valuable insights that can inform future river restoration efforts.

2 Materials and Method

2.1 Study Area

Demak Regency is located between 6°43′26″–7°09′43″ S and 110°27′58″–110°48′47″ E, covering an area of approximately 1149.07 km^2, of which about 897.43 km^2 consists of land and 252.34 km^2 comprises ocean. The Tuntang River originates from Lake Rawa Pening and flows into the Java Sea, passing through the Demak Regency in Central Java. It traverses three sub-districts: Karang Tengah, Demak, and Bonang. The Tuntang River plays an essential role in the community, providing water for agricultural irrigation and household needs such as washing and bathing. Additionally, as the Tuntang River flows through Salatiga Regency before reaching Demak Regency, it serves as a vital source of drinking water for the local population. The research was conducted in 2022, utilizing the purposive sampling method to determine two sampling points (Fig. 1). Land conversion in downstream areas primarily involves agriculture and pond development. Therefore, the selected sampling locations taken were in river flows adjacent to agriculture areas (TA) and river flows near mangroves (TM).

2.2 Sediment Sampling, Preparation, and Analysis

Sediment core sampling was conducted using a Russian corer until the maximum depth of sediments was reached. This study collected sediment core samples reaching150 cm in TA and 200 cm in TM. The sediment was placed into the pipe blade covered with plastic wrap, put into a cooler for transport to the laboratory, and then kept in a freezer

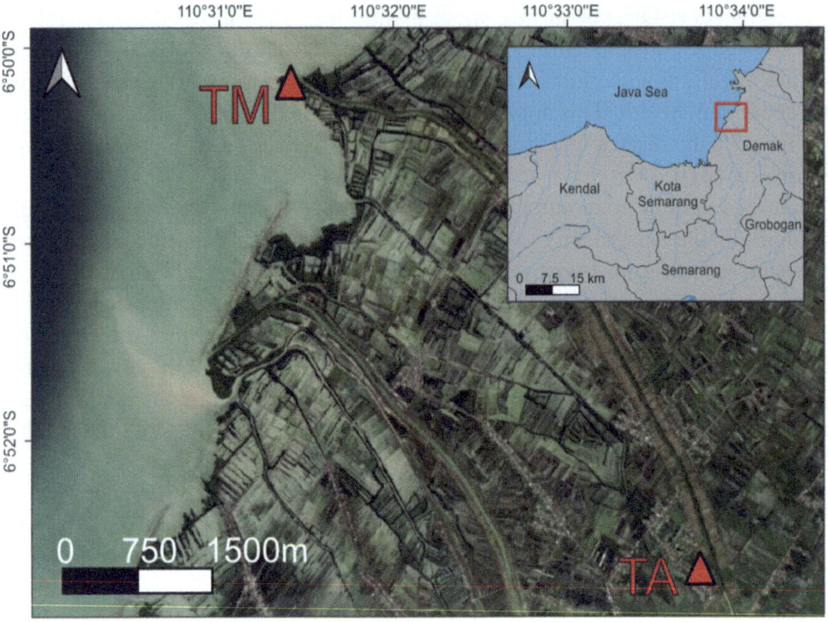

Fig. 1. Sampling locations of sediment cores: red triangles indicate the points of sampling. Inlet map: location of the study area (red frame) within Demak Regency. Abbreviations: TA: agricultural area, TM: mangrove area. Source: ESRI.

at −4 °C. The sediment cores were sectioned at a 1 cm resolution. For analysis, samples were examined at 10 cm depth intervals to capture significant environmental changes. Sediment digestion followed the method of Soeprobowati et al. [27], which is a modification of Battarbee [28]. This process involved using 10% HCl and H_2O_2 on 5 g of sediment. The purpose of diatom digestion was to remove organic and inorganic materials, allowing diatoms to be observed under a microscope [29].

After digestion, diatom preparations were made by mounting 22 × 22 mm coverslips on a hotplate. A volume of 400 μL of diatom residue was added to each coverslip and fixed onto slides using Naphrax, a mounting medium with a refractive index of 1.73. At least 300 diatom valves were counted to determine abundance using a microscope at 1000× magnification. Identification was carried out concerning standard literature and photomicrographs, which were cross-checked on Diatoms.org [30] and AlgaBase.org [31].

2.3 Data Analysis

The analysis of diatom counts was further performed using **PAST 5.0.2** to evaluate community structure metrics, including the Shannon-Wiener diversity index, Evenness, and Dominance [32]. Key species were identified and analyzed by selecting those with a relative abundance greater than 2% ($x > 2\%$) [33]. Stratigraphic visualization was done using C2 1.8.0 [34].

The Shannon-Weiner Species Diversity Index is calculated by determining the proportion of each species relative to the total number of individuals, multiplying this proportion by its natural logarithm for each species, and then summing these values. Since the result is negative, the absolute value (negative of the sum) is taken. A higher index value indicates greater species diversity, reflecting a more balanced ecosystem. For accurate comparisons, the populations being evaluated should ideally have similar numbers of individuals [35]. The formula is as follows:

$$H = -\sum_{T=1}^{S} Pi \ln Pi$$

where H′ is the species diversity index, s is the number of species, and Pi is the proportion of individuals of each species belonging to the ith species of the total number of individuals [36]. The Shannon-Wiener diversity index has three categories in its analysis:

(a) $H' < 1$ = Biota community is unstable, or water quality is heavily polluted;
(b) $1 < H' < 3$ = Medium biota community stability or moderately polluted water quality;
(c) $H' > 3$ = Stability of the biota community in prime condition (stable) or clean water quality.

Species evenness measures the relative abundance of different species within a community. It accounts for both the number of species and their proportions, providing insight into the community's diversity and the balance of species dominance [37]. The formula is as follows:

$$e = \frac{H'}{\ln s}$$

where e is the evenness index, S is the number of species, and H′ is the species diversity index. The evenness index is categorized into three levels for analysis [38]:

(a) 0–0.4 = low evenness, high dominance of species
(b) 0.4–0.6 = medium evenness
(c) 0.6–1.0 = high evenness, low dominance of species

The Simpson Index (D) quantifies the dominance within a multispecies community and measures the likelihood that two individuals randomly selected from the community belong to the same species [37]. The original formulation of the Simpson Index is as follows:

$$D = -\log \sum\nolimits_{i}^{s} = 1\ Pi^2$$

In this formula, s represents the species richness of a community, and pi denotes the relative proportion of species i. The index can be adjusted to 1–D, where the value increases as diversity rises (reflecting a decrease in the dominance of a few species) [39].

3 Results and Discussion

3.1 Diatom Composition

The analysis and identification of diatoms at both stations revealed d the presence of 100 species at TA and 105 species at TM found s, with 77 species common to both locations (Fig. 2). Diatom abundance ranged from 13 to 6890 individuals per gram (ind/gr).

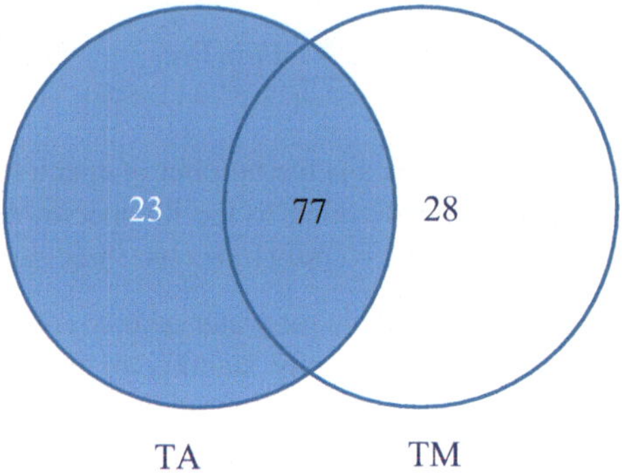

Fig. 2. The proportion of diatom abundance found at both research stations.

The difference in species composition is attributed to the presence of three types of diatoms in TM that thrive in distinct habitats: freshwater, brackish, and marine environments [31]. This finding can be compared with two case studies: research on the Morosari mangrove ecosystem, which identified 137 species [40], and the Ashi River, where 129 species were recorded at its peak [41]. Variations in nutrient levels between the two stations also contribute to differences in species diversity and abundance. Mangroves support the colonization of a diverse range of benthic diatoms, making them a valuable tool for monitoring and assessing estuarine conditions [42, 43]. However, human-induced nutrient enrichment in estuaries elevates the levels of dissolved inorganic nitrogen (DIN) and phosphorus (DIP) [44], which significantly impact diatom assemblages.

Both stations are dominated by benthic diatom species, based on their modes of life (Fig. 3). Planktonic diatoms primarily inhabit the water column, drifting downstream with the current. In contrast, benthic diatoms grow in relatively stable environmental conditions, often attaching themselves to substrates such as rocks and plants, making them inable to move or escape unfavorable conditions [45, 46]. Some species, however, can exists as either planktonic or benthic, including *Cylindrotheca closterium*, *Diadesmis contenta*, and *Denticula kuetzingii* [31].

However, planktonic and benthic diatoms are not entirely distinct in river ecosystems. Benthic diatoms can become resuspended in the water column, behaving as planktonic due to water agitation or invertebrate feeding. Conversely, planktonic diatoms can settle and function as benthic diatoms during calm river conditions [47, 48]. Both types compete for abiotic resources such as light and nutrients. Planktonic diatoms benefit from increased light availability due to reduced penetration to benthic zones, while benthic

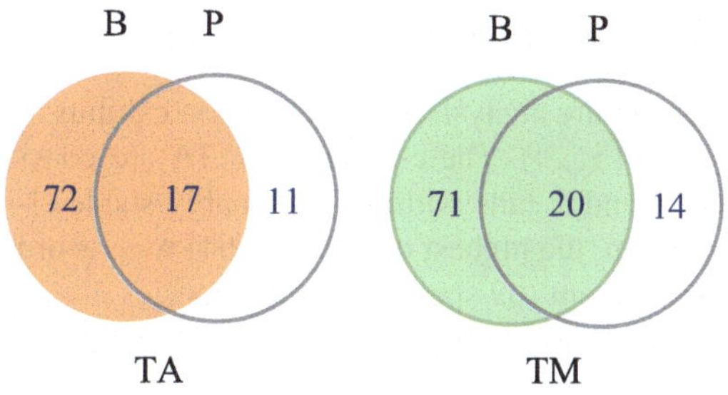

Fig. 3. The proportion of diatom type found at both research stations. (B: Benthic, P: Planktonic).

diatoms gain an advantage in nutrient acquisition by absorbing nutrients from both the sediments and the water column [45, 47, 49].

Ma et al. [41] observed the dominance of benthic diatoms in the Ashi River across three distinct seasons: wet, dry, and regular. A similar trend was also noted by Nunes et al. [50] in estuaries where benthic diatoms dominate due to the nutrient support in the mangrove ecosystem. Mangrove sediments provide an important niche for taxonomically diverse benthic communities [51]. These communities are often dominated by diatoms, primarily composed of pennate species. These benthic diatoms are typically either epipsammic, attaching to sand grains, or epipelic, consisting of motile forms that inhabit sediments [42, 43, 52, 53].

The genus Nitzschia is diverse, with 15 species identified, including *Nitzschia palea*, which is the most commonly found and abundant species within this genus. *Nitzschia* species are well associated with aquatic sediments, particularly in ecosystem with relatively high sedimentation rates [54, 55]. These diatoms exhibit a movement toward the sediment surface, where there light is available and to avoid being buried in the sediment [56].

Several species were found to occur exclusively in either the TA and TM areas. *Amphora pediculus* is one such species, identified only in TA. This taxon thrives oligosaprobic to mesosaprobic environments with high nutrient conditions [57], and is tolerant of organic pollution [58]. It is commonly recognized as a typical pioneer species in both intermittent and perennial rivers. Previous research has demonstrated that *A. pediculus* tends to be more abundant in areas influenced by human activities [59]. This pattern is likely reflected in the downstream region of the Tuntang River, where agricultural activities and fish ponds exert significant influence on the ecosystems.

Coscinodiscus centralis was found exclusively in the TM site. Research conducted by Sarker et al. in Bangladesh [60] indicates that *C. centralis* is a dominant estuarine species. Fluid mudflats are often primarily dominated by marine planktonic species, such as *C. centralis,* carried by oceanic waters [61]. Other studies have shown that *C. centralis* is well-adapted to a wide range of temperatures and thrives in low-light environments [62, 63]. Typically found in oceanic habitats, this species is considered cosmopolitan in its distribution [63]. Furtado et al., in their research on the Amazon Macrotidal Mangrove Coast [64], highlighted *Thalassiosira subtilis, C. centralis, Skeletonema costatum,* and *Pseudo-nitzschia pungens* as harmful species capable of causing Harmful algal blooms (HABs).

3.2 Community Structure of Diatom

Based on community structure analysis, the TA location exhibits a diversity index (H′) ranging from 2.03–3.64 (Fig. 4). The conditions in TA are considered favorable for diatoms, as the diversity index falls within a reasonably stable (1–3) and stable (>3) range. A a depth of 150 cm, the highest diversity (3.64) was recorded, attributed to the large variety of species present (69 species). Higher species diversity indicates a more balanced ecosystem, while lower species diversity suggests that the aquatic ecosystem is under pressure [65, 66].

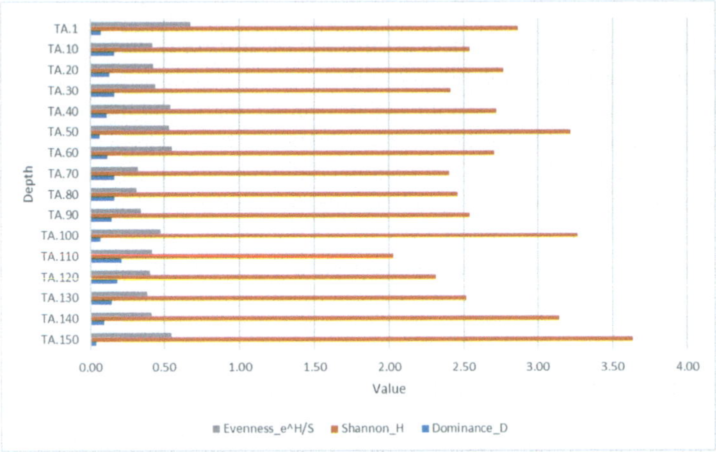

Fig. 4. The community structure of the TA area shows that the river is still in tolerance condition.

The graph's diversity decrease at a depth of 110 cm (2.03) indicates a reduction in species variety and abundance, though the environmental conditions remain relatively good. The variety and abundance of genera *Nitzschia* and *Sellaphora* decreased at this depth. *Sellaphora pupula* is a freshwater diatom that is a bioindicator of water quality. This species is notably tolerant of organic materials present in the substrate. According to Castillejo et al. [67], the diatom species *S. pupula* is closely associated with sampling locations characterized by poor water quality and has a notable tolerance for eutrophication. Among the genus *Nitzschia,* only *N. palea* does not exhibit a decline. This species is highly tolerant of a wide range of environmental condition [68]. The community structure of the TA area suggests that the river remains in a healthy condition.

TM area has a diversity index ranging from 1.89 to 3.4, indicating conditions similar to station 3, categorized as relatively stable to stable (Fig. 5). The decrease observed at a depth of 70 cm is attributed to an increased abundance of *C. centralis* and *Fragilaria crotonensis*. The blooming of algae species, such as *C. centralis* and *Rhizosolenia bergonii suggests* that additional nutrients are being introduced into the water runoff, likely facilitated by rainfall [69].

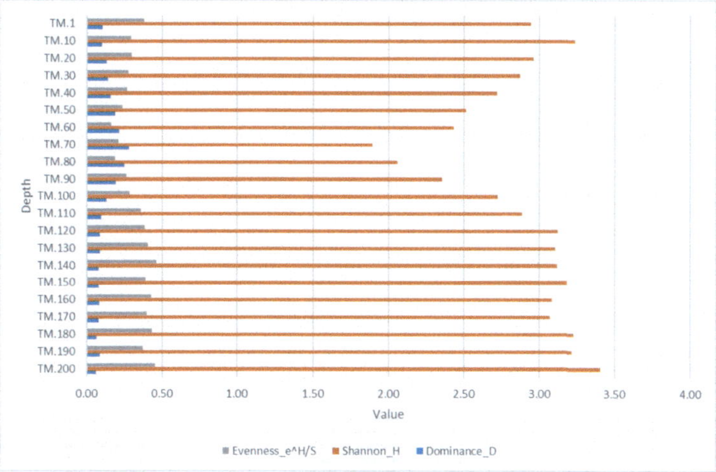

Fig. 5. The community structure of the TM area shows that the river is still in tolerance condition.

3.3 Diatom Stratigraphy for Reconstruct Past Environment

To reconstruct past conditions, a cluster analysis was performed using the Bray-Curtis index also known as Bray-Curtis dissimilarity. This statistical tool is widely used in ecology and biology to assess the similarity between two datasets. It compares species compositions across different environments by incorporating both species abundance and their presence or absence. This index produces a a dissimilarity value ranging from 0 (completely identical) to 1 (entirely different) [70].

Cluster analysis in TA divides the 150 cm of stratigraphic results into three distinct zones of diatoms community development: Zone 1 (150–140 cm), Zone 2 with subzones -2a (130–100 cm), Zone 2b (90–70 cm), Zone 2c (60–50 cm), and Zone 3 (40–1 cm) (Fig. 6). Three species alternately dominate in terms of abundance: *N. palea* (150–100 cm and 60–50 cm), *Fragilaria capucina* (90–70 cm), and *F. crotonensis* (40–1 cm). *N. palea* is a freshwater diatom that serves as an indicator of moderately polluted waters containing organic N and P [71]. This is supported by research from Han et al. (2016), which highlights that *N. palea* requires nitrogen, as it is crucial for the production of carbohydrates and proteins.

N. palea dominates diatom assemblages at the 150–100 cm depth range; however, cluster analysis differentiates the 150–130 cm and 130–100 cm zones because the species variation at a depth of 140 cm is more similar to 150 cm than 130–100 cm. *D. kuetzingii* was found in zone 1, indicating conditions characterized by conductivity and alkaline conditions [73]. The continued abundance of *N. palea* and *F. crotonensis* in the TA area, suggests that the water conditions are undergoing a eutrophication process.

Cluster analysis made for TM data reveals that stratigraphy is divided into three zones, similar to the TA area with divisions: Zone 1a (200–190 cm), Zone 1b (180–110 cm), Zone 2 (100–60 cm), and Zone 3 (50–1 cm) (Fig. 7). The appearance of *Cocconeis placentula* indicates meso-eutrophic conditions in the waters [74]. The genus

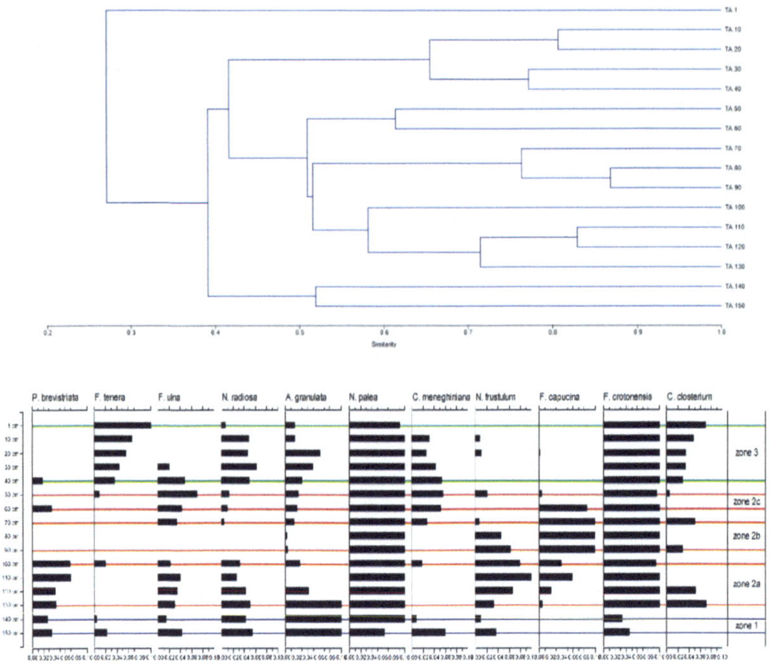

Fig. 6. Diatom cluster analysis (upper) and stratigraphy (lower) in the TA area show the area's dominance of eutrophic species.

Cocconeis has not demonstrated specific, as it has been observed under varying conditions in multiple studies. For example, *C. placentula*, has been found: a) in flowing waters in Italy [75]; b) sensitive to organic pollution in Poland [76]; c) in less polluted waters or even clean (not polluted) in Taiwan [77] [78]; d) in high-flow environments in Argentina [54]; e) under meso-eutrophic to hyper-eutrophic conditions in Finland [79] and Estonia [80]; f) in clean water containing high Ca^{2+} in Zimbabwe [81]; g) in clean water in India [82]; h) and in agriculturally impacted locations in Canada [83, 84].

This broad adaptability makes it challenging to use *C. placentula* as an indicator of specific environmental conditions. Furthermore, it is also known that this species exhibits diverse morphological forms, which give rise to several varieties within the species.

Melosira varians serves as a bioindicators for eutrophic habitats in lowland rivers influenced by agriculture activities in South Africa [85]. The genus *Encyonema* is commonly found in freshwater environments and is associated with eutrophic conditions. Specifically, *E. minutum* has been identified in polluted waters [86]. Previous studies have shown that *E. minutum* is closely linked to elevated nutrient concentrations resulting from agriculture runoff and wastewater in Andean river basins [87]. Another species indicating eutrophication in this research is *Cyclotella meneghiniana*, which was found in both the TA and TM areas. The presence of *C. meneghiniana* in the sediments suggests that the lake environment had elevated nutrient levels and increased turbidity [88, 89]. This lake remains in the eutrophication stage, likely driven by the expansion of

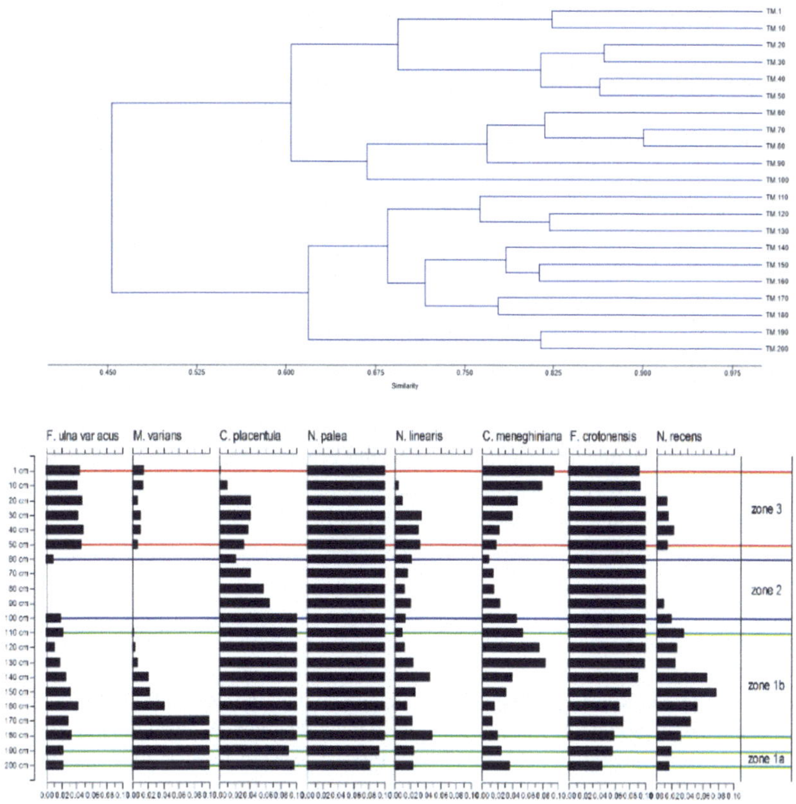

Fig. 7. Diatom cluster analysis (upper) and stratigraphy (lower) in the TM area show the area's dominance of eutrophic species.

aquaculture, poultry farming [90], and the combined impacts of shipping, agriculture, and ponds.

The fluctuating graph at each station indicates that the river is undergoing eutrophication due to land conversion, waste discharge, and agriculture activities. These factors cause the diatom composition to change continuously, particularly affecting species that are less adaptive to environmental changes. This is evident from the presence of *C. meneghiniana, F. crotonensis, N. palea,* and *M. varians.*

These anthropogenic activities have accelerated changes in river systems through pollution, habitat destruction, the introduction of alien species, hydrological manipulation and other physical disturbances. Beyond the direct physical impacts, human activities also have numerous adverse effects on water quality and the biotic communities that inhabit rivers [91]. Physiographic variables, such as drainage area, geology, slope, and land use type, significantly influence the composition of the diatom community. Among biotic communities, benthic diatom assemblages respond more rapidly to environmental changes making them highly sensitive indicators of ecological shifts [92].

4 Conclusion

This study highlights the utility of diatom stratigraphy in assessing and reconstructing past and present environmental conditions in the downstream region of Tuntang River. The dominance of eutrophic species such as like *N. palea, F. crotonensis, M. varians,* and *C. meneghiniana* confirms that the river has transitioned into the eutrophic stage. Shifts in diatom community structure reflect significant anthropogenic impacts, including nutrient enrichment from agriculture and land-use changes. The findings emphasize the importance of implementing sustainable water management practices and incorporating diatom-based biomonitoring as a tool for ecological restoration. By addressing the drivers of eutrophication and promoting land-use planning can help preserve the ecological health of the downstream Tuntang River and similar ecosystems.

Acknowledgement. This research was funded by the Ministry of Education, Culture, Research, and Technology of the Republic of Indonesia in the Program Magister menuju Doktor untuk Sarjana Unggul (PMDSU) program through Diponegoro University with contract number 345-43/UN7.D2/PP/V/2024 and the Peningkatan Kualitas Publikasi Internasional (PKPI) Program from PMDSU with contract number 250.6/E4.4/KU/2024.

References

1. Wang, H., He, G.: Rivers: linking nature, life, and civilization. River. **1**(1), 25–36 (2022)
2. Ashong, G.W., Kwaansa-Ansah, E.E., Ababio, B.A., Yagra, E.A., Antwi, G.: Pollution profiling and quality assessment of Bonsa River, Tarkwa Nsuaem, Ghana; toxic element, ecotoxicology, health risk assessment, and multivariate analysis. Environ. Challenges. **18**, 101078 (2025)
3. Khusanov, A., Sabirov, O., Frank, Y., Vorobev, D., Vorobev, E., Rakhmatullina, S., Tashbaev, S., Mamatkarimova, S., Yakhyoyev, A., Juraev, M., Isakov, I., Karimjonov, D., Zokirov, I., Abdullaev, I.: Microplastic pollution of the Zrafshan river tributary in Samarkand and Navoi regions of the Republic of Uzbekistan. Green Anal. Chem. **12**, 100200 (2025)
4. Liang, H., Zhang, Y., Du, S., Cao, J., Liu, Y., Zhao, H., Ding, T.: Heavy metals in sediments of the river-lake system in the Dianchi basin, China: their pollution, sources, and risks. Sci. Total Environ. **957**, 177652 (2024)
5. Atif, A., Kumar, G.: Evaluation of physicochemical, heavy metal pollution and microbiological indicators in water samples of Ganges at Uttarakhand India: an impact on public health. Int. J. Environ. Rehabil. Conserv. **4**, 445–466 (2020)
6. Khan, M., Mohammad, F.: Eutrophication: challenges and solutions. In: Ansari, A., Gill, S. (eds.) Eutrophication: Causes, Consequences and Control. Springer, Dordrecht (2014)
7. Lan, J., Liu, P., Hu, X., Zhu, S.: Harmful algal blooms in eutrophic marine environments: causes, monitoring, and treatment. Water. **16**(17), 2525 (2023)
8. Nugroho, N.P.: Spatial distribution of nutrient export from the catchment area of lake Rawapening. In: Yulihastin, E., Abadi, P., Sitompul, P., Harjupa, W. (eds.) Proceedings of the International Conference on Radioscience, Equatorial Atmospheric Science and Environment and Humanosphere Science, 2021. Springer Proceedings in Physics, vol. 275. Springer, Singapore (2022)
9. Pratiwi, N.T.M., Imran, Z., Ayu, I.P., Iswantari, A., Wulandari, D.Y.: The phosphorus load and the variation of the trophic states of Cirata Reservoir (West Java, Indonesia) from 1988 to 2017. Biodiversitas. **21**, 4176–4183 (2020)

10. Siswanto, S., Ainun, M.A.: Eutrophication of Danau Bangkau peatland based on nitrate-phosphate concentrations and fish diversity. Russ. J. Agric. Soc. Econ. Sci. **107**, 98–106 (2020)
11. Zainalarifin, J., Effendi, H., Taryono, T.: Abundance and composition of solid waste in the Citarum River, West Java Province. IOP Conf. Ser. Earth Environ. Sci. **1266**, 012056 (2023)
12. Balai Besar Wilayah Sungai Pemali Jratun: Laporan Konservasi DAS Tuntang (2016)
13. Bi, N., Sun, Z., Wang, H., Wu, X., Fan, Y., Xu, C., Yang, Z.: Response of channel scouring and deposition to the regulation of large reservoirs: a case study of the lower reaches of the Yellow River (Huanghe). J. Hydrol. **568**, 972–984 (2018)
14. Lange, C.R., Lange, S.R.: Biomonitoring. Water Environ. Res. **69**(4), 900–915 (1997)
15. Morin, S., Gomez, N., Torn'es, E., Licursi, M., Rosebery, J.: Benthic diatom monitoring and assessment of freshwater environments: standard methods and future challenges. In: Aquatic Biofilms: Ecology, Water Quality and Wastewater Treatment, pp. 111–124 (2016)
16. Viso, R., Blanco, S.: River diatoms reflect better past than current environmental conditions. Water. **15**, 333 (2023)
17. Vidaković, D., Albert Serge Mayombo, N., Burfeid Castellanos, A., Kloster, M., Beszteri, B.: Diatom metabarcoding as a tool to assess the water quality of two large tributaries of the Danube River. Ecol. Indic. **168**, 112793 (2024)
18. Round, F.E., Crawford, R.M., Mann, D.G.: The Diatoms. Biology and Morphology of the Genera, p. 747. Cambridge University press, Cambridge (1990)
19. Rana, A.S., Verma, P., Mittal, P., Mahajan, M.: Investigation of diatom communities in highlands of western Himalayas: geographical and ecological patterns annals of Romanian. Soc. Cell Biol. **26**(1), 1347–1367 (2022)
20. Masouras, A., Karaouzas, I., Dimitriou, E., Tsirtsis, G., Smeti, E.: Benthic diatoms in river biomonitoring—present and future perspectives within the water framework directive. Water. **13**(4), 478 (2020)
21. Blanco, S.: What do diatom indices indicate? Modeling the specific pollution sensitivity index. Environ. Sci. Pollut. Res. Int. **31**(20), 29449 (2024)
22. Dalu, T., Cuthbert, R.N., Taylor, J.C., Magoro, M.L., Weyl, O.L., Froneman, P.W., Wasserman, R.J.: Benthic diatom-based indices and isotopic biomonitoring of nitrogen pollution in a warm temperate Austral river system. Sci. Total Environ. **748**, 142452 (2020)
23. Phuyal, P., Ranabhat, S., Khatri, S., Lamichhane, N., Pant, R.R., Thapa, L.B., Prasad Yadav, R.K.: Hydrochemical characteristics, water quality, and diatom assemblages in Dordi, a major tributary of Marshyangdi River, Central Nepal. Watershed Ecol. Environ. (2024)
24. Nisrina, N., Aprilianti, R., Junaedi, A.S.: Keanekaragaman Diatom Epilitik serta Potensinya Sebagai Bioindikator Kualitas Air di Sungai Gogor Kecamatan Wonosalam Kabupaten Jombang. Environ. Pollut. J. **3**(2), 674–682
25. Falah, M.H.A., Soeprobowati, T.R., Hadiyanto, H.: Epipelic diatom assemblages to determine water quality in the Tuntang River downstream, Demak, Central Java. IOP Conf. Ser. Earth Environ. Sci. **1137**, 012003 (2023)
26. Falah, M.H.A., Soeprobowati, T.R., Hadiyanto, H., Rahim, A., Noor, B.M., Permatasari, N.: Diatom stratigraphy as a flood record in the lower Tuntang River, Demak, Central Java. Evergreen. **10**(1), 272–282 (2023)
27. Soeprobowati, T.R., Tandjung, S.D., Sutikno, S., Hadisusanto, S., Suedy, S.W.A., Gell, P.: The minimum number of valves for diatom identification in Rawapening Lake Central Jawa. Biotropia. **23**(2), 96–104 (2016)
28. Battarbee, R.W.: Diatom analysis. In: Berglund, B.E. (ed.) Handbook of Holocene Paleoecology and Paleohydrology, pp. 527–570. Wiley & Sons, London (1986)
29. Soeprobowati, T.R., Hartuti, P., Jumari, J., Kenanga, S.: The relationship of water quality to epipelic diatom assemblages in Cebong Lake, Dieng Indonesia. Pol. J. Environ. Stud. **31**(1), 281–295 (2022)

30. Spaulding, S.A., Potapova, M.G., Bishop, I.W., Lee, S.S., Gasperak, T.S., Jovanoska, E., et al.: Diatoms.org: supporting taxonomists, connecting communities. Diatom Res. **36**(4), 291–304 (2021)
31. Guiry, M.D., Guiry, G.M.: Algae Base. World-Wide Electronic Publication, National University of Ireland, Galway. https://www.algaebase.org. Accessed 1 Jun 2023
32. Hammer, Ø., Harper, D.A.T., Ryan, P.D.: PAST: paleontological statistics software package for education and data analysis. Palaeontol. Electron. **4**(1), 1–9 (2001)
33. Liu, S., Yao, M., Chen, S., Yuan, X.: Surface sediment diatom assemblages response to water environment in Dongping Lake, North China. Water. **13**(3), 339 (2021)
34. Juggins, S.: C2 Data Analysis, Version 1.7.4., Software for Ecological and Palaeooecological Dana Analysis and Visualization. Newcastle University, Newcastle upon Tyne (2011)
35. Nolan, K.A., Callahan, J.E.: Beachcomber biology: the Shannon-Weiner species diversity index. In: O'Donnell, M.A. (ed.) Tested Studies for Laboratory Teaching. Proceedings of the 27th Workshop/Conference of the Association for Biology Laboratory Education (ABLE), vol. 27, pp. 334–338 (2006)
36. Odum, E.P.: Fundamental of Ecology. W.B. Sounders Company, Philladelphia, London, Toronto (1993)
37. Moore, J.C.: Diversity, taxonomic versus functional. In: Encyclopedia of Biodiversity, 2nd edn, pp. 648–656 (2012)
38. Krebs, C.J.: Ecology: The Experimental Analysis of Distribution and Abundance. Harper and Row Publisher Inc., New York (2009)
39. Brower, J.E., Zar, J.H.: Field and Laboratory Methods for General Ecology. Brown Co Publisher, Dubuque, IA (1998)
40. Hasrini, D.A., Soeprobowati, T.R., Jumari, J.: Epipelic diatoms to determine ecological status based on diatom index in mangrove ecosystem of Morosari Village, Indonesia. J. Ecol. Eng. **24**(11), 30–43 (2023)
41. Ma, C., Cui, R., Duan, Y., Zhang, N., Liu, Y., Sui, F., Fan, Y., Lu, X.: Differences in biodiversity and assembly mechanisms between planktonic and benthic diatom communities in riverine ecosystems: a case study in the Ashi river. Ecol. Indic. **157**, 111258 (2023)
42. Lemley, D.A., Adams, J.B., Bate, G.C.: A review of microalgae as indicators in South African estuaries. S. Afr. J. Bot. **107**, 12–20 (2016)
43. Balasubramaniam, J., Prasath, D., Jayaraj, K.A.: Microphytobenthic biomass, species composition and environmental gradients in the mangrove intertidal region of the Andaman Archipelago, India. Environ. Monit. Assess. **189**, 231 (2017)
44. Howarth, R.W., Chan, F., Swaney, D.P., et al.: Role of external inputs of nutrients to aquatic ecosystems in determining prevalence of nitrogen vs. phosphorus limitation of net primary productivity. Biogeochemistry. **154**, 293–306 (2021)
45. Poikane, S., Kelly, M., Cantonati, M.: Benthic algal assessment of ecological status in European lakes and rivers: challenges and opportunities. Sci. Total Environ. **568**, 603–613 (2016)
46. Lu, X., Zhang, Y., Liu, Y., Fan, Y.: Differences in planktonic and benthic diatoms reflect water quality during a rainstorm event in the Songhua River Basin of northeast China. Ecol. Indic. **144**, 109547 (2022)
47. Tekwani, N., Majdi, N., Mialet, B., Tornès, E., Urrea-Clos, G., Buffan-Dubau, E., Sabater, S., Tackx, M.: Contribution of epilithic diatoms to benthic–pelagic coupling in a temperate river. Aquat. Microb. Ecol. **69**, 47–57 (2013)
48. Hu, J., Song, Z., Zhou, J., et al.: Differences in diversity and community assembly processes between planktonic and benthic diatoms in the upper reach of the Jinsha River, China. Hydrobiologia. **849**, 1577–1591 (2022)
49. Hansson, A.: Quantifying the impact of periphytic algae on nutrient availability for phytoplankton. Freshw. Biol. **24**(2), 265–273 (1990)

50. Nunes, M., Lemley, D.A., Machite, A., Adams, J.B.: Benthic diatom diversity in microtidal mangrove estuaries. Mar. Pollut. Bull. **206**, 116706 (2024)
51. Benny, N., Thomas, L.C., Padmakumar, K.B.: Community structure of Microphytobenthos associated with mangrove ecosystems along the Southwest Coast of India. Estuar. Coasts. **44**, 1380–1391 (2021)
52. Hope, J.A., Paterson, D.M., Thrush, S.F.: The role of microphytobenthos in soft-sediment ecological networks and their contribution to the delivery of multiple ecosystem services. J. Ecol. **108**(3), 815–830 (2020)
53. Serôdio, J., Paterson, D.M.: Role of microphytobenthos in the functioning of estuarine and coastal ecosystems. In: Leal Filho, W., Azul, A.M., Brandli, L., Lange Salvia, A., Wall, T. (eds.) Life Below Water. Encyclopedia of the UN Sustainable Development Goals. Springer, Cham (2022)
54. Martínez de Fabricius, A., Maidana, N., Gómez, N., Sabater, S.: Distribution patterns of benthic diatoms in a Pampean river exposed to seasonal floods: the Cuarto River (Argentina). Biodivers. Conserv. **12**(12), 2443–2454 (2003)
55. Izagirre, O., Serra, A., Guasch, H., Elosegi, A.: Effects of sediment deposition on periphytic biomass, photosynthetic activity and algal community structure. Sci. Total Environ. **407**, 5694–5700 (2009)
56. Klamt, A.M., Hu, K., Huang, L., Chen, X., Liu, X., Chen, G.: An extreme drought event homogenises the diatom composition of two shallow lakes in southwest China. Ecol. Indic. **108**, 105662 (2020)
57. Kelly, M.G., Penny, C.J., Whitton, B.A.: Comparative performance of benthic diatom indices used to assess river water quality. Hydrobiologia. **302**, 179–188 (1995)
58. Lange-Bertalot, H., Hofmann, G., Werum, M., Kelly, M., Cantonati, M.: Freshwater Benthic Diatoms of Central Europe: Over 800 Common Species Used in Ecological Assessment, vol. 942, pp. 1–908. Koeltz Botanical Books, Schmitten-Oberreifenberg (2017)
59. Stubbington, R., Chadd, R., Cid, N., Csabai, Z., Miliša, M., Morais, M., Munné, A., Pařil, P., Pešić, V., Tziortzis, I., Verdonschot, R.C., Datry, T.: Biomonitoring of intermittent rivers and ephemeral streams in Europe: current practice and priorities to enhance ecological status assessments. Sci. Total Environ. **618**, 1096–1113 (2018)
60. Sarker, S., Riya, S.C., Rahman, M.J., Huda, A.S., Hossain, M.S., Das, N.: Spatial and temporal variability of phytoplankton dynamics in-relation to essential oceanographic variables in the south east coast of Bangladesh. J. Sea Res. **195**, 102438 (2023)
61. Sylvestre, F., Guiral, D., Debenay, J.: Modern diatom distribution in mangrove swamps from the Kaw Estuary (French Guiana). Mar. Geol. **208**(2–4), 281–293 (2004)
62. Kemp, A.E., Pike, J., Pearce, R.B., Lange, C.B.: The "Fall dump"—a new perspective on the role of a "shade flora" in the annual cycle of diatom production and export flux. Deep Sea Res. Part II Top. Stud. Oceanogr. **47**(9–11), 2129–2154 (1999)
63. Horner, R.A.: A Taxonomic Guide to Some Common Marine Phytoplankton. Biopress Ltd, Bristol (2002)
64. Furtado, J.A., Cavalcanti-Lima, L.F., Santos, V.H.M.D., Serejo, J.H.F., Pereira, L.C.C., Carvalho-Neta, R.N.F., Tchaicka, L.: Multimetric trophic status and harmful algal species in a harbor area of the Amazon Macrotidal Mangrove Coast. Ecohydrol. Hydrobiol. (2024)
65. Pramyoga, G., Soeprobowati, T.R.: Komunitas Fitoplankton di Telaga Menjer, Dieng, Jawa Tengah. In: Prosiding Seminar Nasional Hasil-Hasil Penelitian Pascasarjana UNDIP, pp. 118–122 (2016)
66. Hariyati, R., Soeprobowati, T.R., Chotidjah, A.: Persebaran Diatom Epipelik Secara Vertikal pada Ekosistem Mangrove Muara Sungai Banjir Kanal Timur Semarang. Buletin Anatomi dan Fisiologi. **17**(1), 72–89 (2009)

67. Castillejo, P., Chamorro, S., Paz, L., Heinrich, C., Carrillo, I., Salazar, J.G., Navarro, J.C., Lobo, E.A.: Response of epilithic diatom communities to environmental gradients along an Ecuadorian Andean River. C. R. Biol. **341**(4), 256–263 (2018)
68. Al-Hassany, J.S., Alrubai, G.H., Jasim, I.M.: The potential use of the diatom Nitzschia palea (Kützing) W. Smith For the removal of certain pollutants from Al-Rustumeyah wastewater treatment plant in Baghdad-Iraq. IOP Conf. Ser. Earth Environ. Sci. **779**, 012114 (2021)
69. Raji, K., Kumar, A., Padmavati, G.: Silicate as the Probable Causative Agent for the Periodic Blooms in the Coastal Waters of South Andaman Sea (2014)
70. Kelly, M.: Use of similarity measures for quality control of benthic diatom samples. Water Res. **35**(11), 2784–2788 (2001)
71. Bere, T., Tundisi, J.: Diatom-based water quality assessment in streams influence by urban pollution: effects of natural and two selected artificial substrates, Sao Carlos-SP, Brazil. Aquat. Sci. Technol. **15**, 54–63 (2011)
72. Han, L., Xu, B., Qi, F., Chen, Z.: Effect of nitrogen/phosphorus concentration on algal organic matter generation of the diatom Nitzschia palea: total indicators and spectroscopic characterization. J. Environ. Sci. **47**, 130–142 (2016)
73. Zgrundo, A., Wojtasik, B., Convey, P., Majewska, R.: Diatom communities in the high arctic aquatic habitats of northern Spitsbergen (Svalbard). Polar Biol. **40**(4), 873–890 (2017)
74. Holmes, M., Campbell, E., de Wit, M., Taylor, J.: The impact of drought in the Karoo - revisiting diatoms as water quality indicators in the upper reaches of the Great Fish River, Eastern Cape, South Africa. S. Afr. J. Bot. **149**, 502–510 (2022)
75. Gallo, L., Maurizio, B., Anna, C., Lucio, L.: Changes in epilithic diatom communities and periphytic biomass downstream of a reservoir on a Mediterranean river (Calabria region, S Italy). Turk. J. Bot. **39**(3), 555–569 (2015)
76. Szczepocka, E., Szulc, B.: The use of benthic diatoms in estimating water quality of variously polluted rivers. Oceanol. Hydrobiol. Stud. **38**(1), 17–26 (2009)
77. Wu, J.T.: A generic index of diatom assemblages as bioindicator of pollution in the Keelung River of Taiwan. Hydrobiologia. **397**, 79–87 (1999)
78. Salomoni, S.E., Rocha, O., Callegaro, V.L., Lobo, E.A.: Epilithic diatoms as indicators of water quality in the Gravataí river, Rio Grande do Sul, Brazil. Hydrobiologia. **559**(1), 233–246 (2006)
79. Korhonen, J.J., Köngäs, P., Soininen, J.: Temporal variation of diatom assemblages in oligotrophic and eutrophic streams. Eur. J. Phycol. **48**(2), 141–151 (2013)
80. Vilbaste, S., Truu, J.: Distribution of benthic diatoms in relation to environmental variables in lowland streams. Hydrobiologia. **493**(1), 81–93 (2003)
81. Mangadze, T., Bere, T., Mwedzi, T.: Epilithic diatom flora in contrasting land-use settings in tropical streams, Manyame Catchment, Zimbabwe. Hydrobiologia. **753**(1), 163–173 (2015)
82. Amutha, M., Muralidharan, M.: Diatom community structure along physicochemical gradients in upland river segments of Tamiraparani river system, South India. Int. J. Aquat. Biol. **5**(1), 12–21 (2017)
83. Lavoie, I., Vincent, W.F., Pienitz, R., Painchaud, J.: Benthic algae as bioindicators of agricultural pollution in the streams and rivers of southern Québec (Canada). Aquat. Ecosyst. Health Manag. **7**(1), 43–58 (2004)
84. Lavoie, I., Hamilton, P.B., Wang, Y.K., Dillon, P.J., Campeau, S.: A comparison of stream bioassessment in Québec (Canada) using six European and North American diatom-based indices. Nova Hedwig. Beih. **135**, 37–56 (2009)
85. Petersen, C.R., Jovanovic, N.Z., Grenfell, M.C., et al.: Responses of aquatic communities to physical and chemical parameters in agriculturally impacted coastal river systems. Hydrobiologia. **813**(1), 157–175 (2018)
86. Lavoie, I., Campeau, S., Darchambeau, F., et al.: Are diatoms good integrators of temporal variability in stream water quality? Freshw. Biol. **53**(4), 827–841 (2008)

87. Lu, X., Tian, L., Tao, S., et al.: The response of benthic diatom community to anthropogenic eutrophication of a river basin under agricultural influence in NE China. J. Freshw. Ecol. **3**(1), 495–511 (2022)
88. Zorzal-Almeida, S., Bartozek, E.C.R., Bicudo, D.C.: Homogenization of diatom assemblages is driven by eutrophication in tropical reservoirs. Environ. Pollut. **288**, 117778 (2021)
89. Mao, X., Liu, L., Zhao, H., Ge, Y., Jiang, G., Song, L., Ning, K., Zhao, H., Zhang, P.: Abrupt diatom assemblage shifts in Lake Baiyangdian driven by distinct hydrological changes and yet more so by gradual eutrophication. Limnologica. **105**, 126155 (2024)
90. Ge, Y., Mao, X., She, Z., Liu, L., Song, L., Li, Y., Liu, C.: Spatial heterogeneity of long-term environmental changes in a large agricultural wetland in North China: implications for wetland restoration. Catena. **219**, 106582 (2022)
91. Smol, J.P.: Pollution of Lakes and Rivers, a Paleoenvironmental Perspective, 2nd edn. Blackwell Publishing, Malden, MA (2008)
92. Karaouzas, I., Smeti, E., Vourka, A., Vardakas, L., Mentzafou, A., Tornés, E., Sabater, S., Muñoz, I., Skoulikidis, N.T., Kalogianni, E.: Assessing the ecological effects of water stress and pollution in a temporary river - implications for water management. Sci. Total Environ. **618**, 1591–1604 (2018)

Prediction of Flood Hydrograph of a Flood Affected Ungauged Station in Periyar River, India

Rugma Sunil[1], Rebecca Elsa Mathew[1], Jefrin Joji[1], Vishakh John Mathews[1], K. L. Priya[1(✉)], Athul M. Madhu[1], and Soufiane Haddout[2]

[1] Department of Civil Engineering, TKM College of Engineering, Kollam, Kerala, India
klpriyaram@gmail.com

[2] Department of Physics, Ibn Tofail University, Kenitra, Morocco

Abstract. Floods are among the most frequent and hazardous calamities worldwide. Climate change has intensified extreme precipitation and adversely impacted the health of rivers. One of the most significant flood events occurred in Kerala in 2018, causing devastating damage to many river basins. Among the most affected was the Periyar River. The limited number of gauging stations and the lack of an early warning system exacerbated the situation. To address this gap, we used the Hydrologic Engineering Center's Hydrologic Modeling System (HEC-HMS) to analyze the effects of rainfall on surface runoff and peak discharges of the Periyar River in Kerala. Additionally, we developed a flood inundation map for the river basin using HEC-RAS. For modeling, two routing methods, namely the Muskingum-Cunge method and the Kinematic Wave method were employed. The Kinematic Wave method was identified as the most suitable for predicting the flood hydrograph of the Periyar River. The hydrograph and discharge data were predicted for Aluva, a location along the Periyar River that is highly prone to severe flooding. The resulting hydrograph is suitable for flood forecasting and serves as a valuable tool for recommending appropriate mitigation measures to government agencies. Furthermore, the inundation map developed using HEC-RAS through steady and unsteady flow simulations provides a useful resource for zoning the watershed and formulating town planning policies.

Keywords: Flood routing · HEC-HMS · HEC-RAS · Periyar River

1 Introduction

With the alteration of the natural environment due to human interventions together with the effects of global climate change, recent years have witnessed floods occur more frequently and unpredictably across the globe. Urbanization and changes in demographic features within the river floodplain have led to increased exposure of communities to flood hazards. Developing countries in particular are facing the wrath of such disasters as the vulnerability level is very high across these regions due to poor socio-economic conditions and haphazard settlements. On the other hand, developed countries with

J. C. Casila et al. (Eds.): CCORE 2024, SPEES, pp. 234–249, 2025.
https://doi.org/10.1007/978-3-031-93887-0_22

significantly more built-up areas and infrastructures in place are subjected to huge economic losses if floods are not managed properly [1, 2]. Therefore, the necessity for proper design of control and mitigation measures for flood management is identified. Among the various components involved in flood management, flood forecasting is a crucial and inevitable process [3]. This requires a proper understanding of the underlying hydrological and hydro-dynamic processes of rivers and the associated catchment characteristics [4, 5]. Rainfall-runoff modeling (hydrological) anticipates evaluating the runoff from precipitation in a catchment [6] and hydraulic modeling aims to evaluate the magnitude of floods and the area inundated by them [7]. A combination of both will result in runoff simulation and estimating flood inundation levels [8].

Different flood simulation techniques can be used to determine the change in the shape of a hydrograph of a flood as it travels through natural rivers or artificial channels. Traditionally, the hydraulic (e.g. dynamic, kinematic, Muskingum-Cunge, diffusion wave models) and hydrologic (e.g. linear and nonlinear Muskingum models) routing procedures that are well known as distributed and lumped ways to hydraulic and hydrologic practitioners, respectively, can be utilized [9–11]. Hydrologic Engineering Center Hydrologic Modeling System (HEC-HMS) is one of the most sorted out models, which simulates the rainfall-runoff process for precise flood forecasting [12]. It is a physically-based model that uses rainfall data to predict streamflow [13, 14]. The sub-model of HEC-HMS consists of the Muskingum-Cunge routing model, wherein the downstream routing is performed. Zheng et al. [15] applied a Kinematic wave model for flood routing Tibet River and identified it to perform superior to dynamic wave approximation. Muskingum-Cunge model was identified to be suitable for river stretches having moderate values of bed slopes and roughness coefficient [16].

The likelihood of flooding and inundation of low-lying areas and associated flood risks during the occurrence of a storm can be interpreted from flood maps. This requires development of hydraulic models for performing hydraulic analysis of the selected stream or river. Hydrologic Engineering Center - River Analysis System (HEC-RAS) is a computer-based model for performing hydraulic analysis of river course [17]. The geometric data and flow data are used by the model as input parameters for hydraulic calculations [18]. HEC-RAS has the capability of performing one-dimensional and two-dimensional hydraulic calculations. Based on the purpose of the study, HEC-RAS provides different options for performing river analysis which are one-dimensional steady flow for water surface profile computation, one and two-dimensional unsteady flow simulation, quasi unsteady flow for sediment transport computation and water quality analysis. The location of flood plain and identification of potential risk zones during extreme storm events in Godavari River has been identified through flood modelling using HEC-RAS [19]. A flood modelling approach, combining GIS, HEC-HMS and HEC-RAS can be an effective strategy for flood modelling and has been applied for Hyderabad City, India [20] and Copper Slough Watershed in Champaign city [21].

Kerala had experienced an abnormally high rainfall for 2 months, starting from 1st June 2018 to 19th August 2018. This had resulted in severe flooding in thirteen out of fourteen districts in the State. The lack of stream gauge stations was a major limitation in estimating the risks associated with the extreme event. Thus, the main aim of the study was to predict the flood hydrograph of ungauged stations using HEC-HMS and to develop flood inundation maps of the Periyar River.

2 Study Area

The Periyar river basin is situated in the south-western part of India and it extends between 09°15′ N to 10° 20′ N and longitude 76° 10′ E to 77° 30′E. This river basin is the second largest basin of Kerala with a drainage area of 5398 sq. km of which 114 sq. km in the Anamalai fold lies in the State of Tamil Nadu and is 244 km long. It originates from the Sivagiri Malai (hills) in the Western Ghats at an elevation of 1830 m above MSL and finally drains into the Lakshadweep Sea. From its origin, Periyar traverses through an immense cliff of rocks in a northerly direction receiving several streamlets in its course. The dams on the Periyar River include the Idukki Dam, the Cheruthoni Dam and the Malankara Dam. The Periyar watershed is divided into 183 sub-watersheds and 448 micro-water sheds. One of the major river basins that were severely affected by the 2018 flood in Kerala state was the Periyar River Basin (PRB). The rainfall over Kerala during June, July, and 1st to 19th of August was 15%, 18%, and 164% respectively, which is above normal of 1650 mm [22]. Periyar River Basin (PRB) received a 145-year return period rainfall. This unexpectedly heavy rainfall had been devastating as the occurrence of a flood was unforeseen. The geographic location of Periyar basin is shown in Fig. 1.

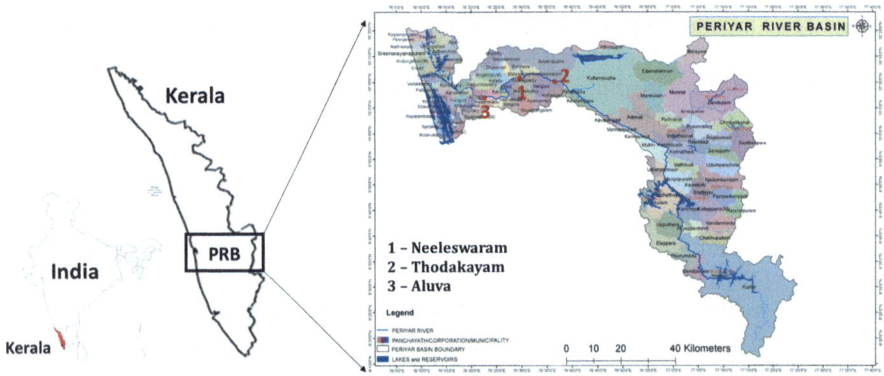

Fig. 1. Geographic location of Periyar River Basin

3 Theoretical Approach

The hydrologic models are used to estimate hydrological parameters by utilizing recorded data from both upstream and downstream sections of rivers and/or by employing robust optimization techniques to solve the one-dimensional conservation of mass and storage-continuity equations. Basic equation for hydrologic routing is:

$$\text{Continuity equation}: \frac{dS}{dt} = I(t) - Q(t) \tag{1}$$

where, *I (t)* is the Input, *Q (t)* the Output and *S (t)* the storage. *I (t)* is known parameter and *Q* and *S* are unknown parameters.

On the other hand, hydraulic models require several data related to river geometry and morphology and consume a lot of computer resources to solve the equations (St Venant's equation) numerically as given below:

$$\text{Equation of continuity}: \frac{\partial Q}{\partial x} + \frac{\partial A}{\partial t} - q = 0 \tag{2}$$

$$\text{Momentum equation}: \frac{\partial Q}{\partial t} + \frac{\partial}{\partial x}\left(\frac{Q^2}{A}\right) = gA\frac{\partial h}{\partial x} \tag{3}$$

However, semi-distributed models such as Muskingum–Cunge family procedures are available, which makes use of simple physical concepts and common river characteristics such as channel geometry, reach length, roughness coefficient, and slope which help to estimate the model parameters without complex and expensive numerical solutions. The kinematic wave method relates basin and flow characteristics directly to the two routing parameters, α and m.

$$Q = \alpha A^m \tag{4}$$

The parameters α and m are directly related to the shape of the channel, the boundary roughness, and the slope of the channel or overland flow surface. The kinematic wave approximation has been proven to be an accurate and efficient method of simulating storm water runoff from small basins for both overland flow and stream channel. The key to the understanding of the theoretical basis of the Muskingum-Cunge method is the recognition that the diffusion wave applies through a wider range than competing waves such as the kinematic and dynamic waves. Kinematic waves do not attenuate, while most flood waves attenuate at least a little bit; on the other hand, dynamic waves attenuate too much and, therefore, do not represent flood waves in typical cases. Diffusion waves, lying in the midrange of attenuation, are, by far, the most applicable wave from the standpoint of practice. The strength of the Muskingum-Cunge method is its theoretical basis as an analogy of the diffusion wave equation. The basic routing equation of the Muskingum-Cunge method is given below.

$${}^{n+1}Q_{j+1} = C_0{}^{n+1}Q_j + C_1{}^{n}Q_j + C_2{}^{n}Q_{j+1} \tag{5}$$

where j is a spatial index, n is a temporal index and C_0, C_1 and C_2 are calculated as follows:

$$\mathrm{Co} = \frac{\Delta t - 2KX}{2\,K(1 - X) + \Delta t} \tag{6}$$

$$\mathrm{C1} = \frac{\Delta t + 2KX}{2\,K(1 - X) + \Delta t} \tag{7}$$

$$\mathrm{C2} = \frac{2K(1 - X) - \Delta t}{2K(1 - X) + \Delta t} \tag{8}$$

The parameters K and X are calculated as follows:

$$\mathrm{K} = \frac{\Delta x}{c} \tag{9}$$

$$X = \frac{1}{2}\left(1 - \frac{q}{s_o c \Delta x}\right) \tag{10}$$

in which Δx = reach length (space interval); c = flood wave celerity; q = unit width discharge; and S_o = channel bed slope.

Flood inundation models are required to understand, assess and predict flood events and their impact in the nearby areas. The results from hydraulic models can be used in flood risk mapping, flood damage assessment, real time flood forecasting, flood related engineering, water resource planning, investigating flood plain erosion and sediment transport, floodplain ecology, river system hydrology. However accurate flood modelling at high spatio-temporal resolutions still remains a significant challenge in any hydrologic and hydraulic studies. This is mainly due to the complex and chaotic nature of flooding and uncertainty associated with the conceptualization of processes and the modelling parameters itself.

4 Methodology

The methodology adopted for the study is shown in Fig. 2. An understanding of the topographical, hydrological and climatic conditions of the study area and a proper set of data defining them are very important for analyzing the hydrologic and hydraulic situation. Data required for hydrologic modeling (HEC-HMS) include basin model maps, precipitation data, discharge data, length and bottom width, and Manning's coefficient. Data needed for hydraulic modeling (HEC-RAS) are Digital Elevation Model (DEM), land use and soil cover, basin model map, and discharge data. HEC-HMS has been used for hydrologic modeling, while QGIS was used as a platform for generating physical basin models for HEC-HMS. HEC-RAS has been used for mapping flood inundation of the basin.

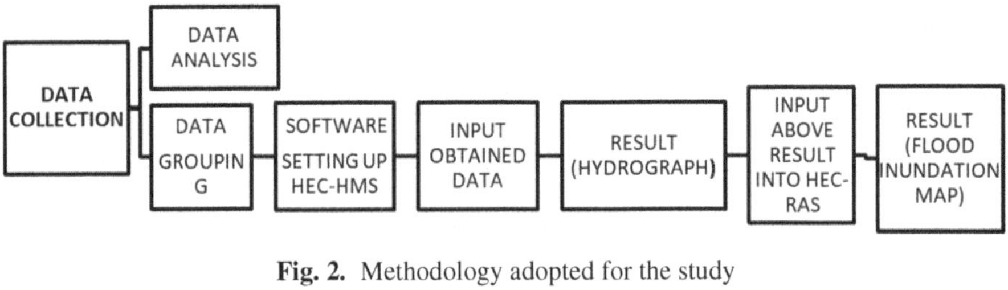

Fig. 2. Methodology adopted for the study

Hydrologic Modeling System (HEC-HMS) helps in simulating the hydrologic cycle (precipitation, evapotranspiration, infiltration, surface runoff and base flow) of a catchment by describing its physical and meteorological properties. It easily operates huge tasks in relation to hydrological studies, including losses, runoff transform, open channel routing, and analysis of meteorological data, rainfall-runoff simulation, and parameter estimation. The program uses the separate model to represent each component of the runoff process and computes runoff volume by subtracting losses (infiltration, storage, interception, evaporation, etc.) from precipitation.

The HEC-RAS system comprises a graphical user interface (GUI), separate analysis components, data storage and management capabilities, graphics, mapping and reporting facilities. The HEC-RAS system contains the river analysis components for: (1) One-dimensional steady flow water surface profile computations; (2) One-dimensional and/or two-dimensional unsteady flow simulation; (3) Quasi unsteady or fully unsteady flow movable boundary sediment transport computations; and (4) One dimensional water quality analysis. In addition to the four river analysis components, the system contains several hydraulic design features that can be invoked once the water surface profiles are computed. HEC-RAS has an extensive spatial data integration and mapping system (HEC-RAS Mapper). For the present study, steady and unsteady flow simulations were carried out. The steady flow component was modelled for supercritical flow regime water surface profiles. It was used to compare the variation in inundation during the years 2015–2019. The unsteady flow component is used to perform supercritical flow regime calculations in the unsteady flow computations module.

4.1 Data

Data used for hydraulic modeling were basin model map, precipitation data, discharge data, length and bottom width, and Manning's coefficient. The basin model map was prepared using QGIS software. The precipitation data was obtained from the NASA website (https://power.larc.nasa.gov/). Rainfall data of two stations, Neeleeswaram and Thodakayam were used for the study. The rainfall data of 2018 has been considered for flood routing and rainfall data of August 2016 was considered for validation and analysis of the selected method. Figure 3a shows the hyetograph for Neeleeswaram during August 2018. The daily discharge data for the station Neeleeswaram was collected from CWC Regional Office and is shown in Fig. 3b. Manning's coefficient was also obtained from CWC Regional Office and a value of 0.035 was adopted for the study. Data needed for hydrologic modeling were Digital Elevation Model (DEM), basin model map (Fig. 3c), Land Use Land Cover (LULC) map, and discharge data. The LULC map was obtained from Bhuvan website (http://bhuvan.nrsc.gov.in), from which HEC-RAS software assign the values of Manning's n as given in Table 1. Daily flow data of August 2018 was used for unsteady flow analysis and the maximum discharge data of 2015, 2016, 2017, 2018 and 2019 were used for steady flow analysis of Periyar River.

4.2 Hydrologic Model Development

Modeling. Rainfall-runoff modeling was carried out with the help of HEC-HMS. The modeling involved four steps: (1) Creating basin model; (2) Creating a meteorological model; (3) Creating a control specification; and (4) Simulation and validation.

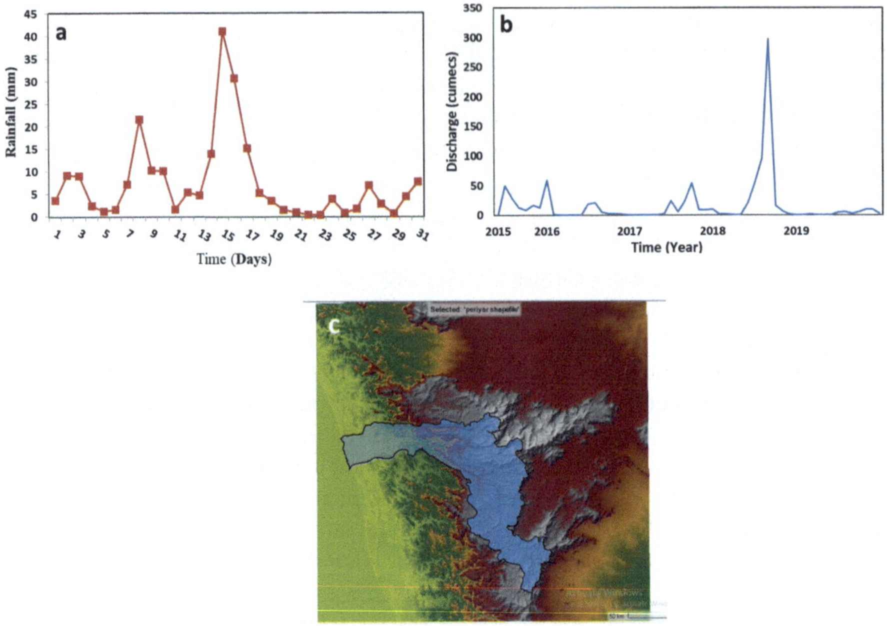

Fig. 3. (a) Rainfall hyetograph of station Neeleeswaram in August 2018; (b) Discharge data of station Neeleeswaram for 2015 to 2019; Basin and DEM maps of Periyar basin

Table 1. Manning's n for various land use types (Chow et al., 1988)

Land use type	Manning's *n*
Developed-high intensity	0.15
Developed-medium intensity	0.08
Developed-low intensity	0.1
Developed-open space	0.035
Cultivated crops	0.035
Pasture/hay	0.03
Grassland/herbaceous	0.035
Deciduous forest	0.16
Water bodies	0.035

Creating Basin Model. The methodology starts by creating a new basin model. In order to create, we have selected runoff-volume model, direct -runoff model, base flow model and routing model. For validation, we used SCS curve number (CN) for runoff volume model, SCS UH for direct-runoff model, exponential recession for base flow model, kinematic wave and Muskingum – Cunge for routing models. These models describe the fate of water that has not infiltrated or been stored on the watershed. The sub basin element and the reach elements are added to the basin map. The sub basin is used to represent the physical watershed. The reach is used to convey stream flow in the basin model.

Creating Meteorological Model. The meteorological model is used for estimating the precipitation for a sub-basin. It can utilize both point and gridded precipitation and has the capability to model frozen and liquid precipitation along with evapotranspiration. The meteorological model manager is used to add a new meteorological model. The precipitation method used is specified hyetograph.

Creation of Control Specification. A simulation time window sets the time spans and time intervals of a simulation run. A simulation time window is created by adding a control specification to the project. Here, a starting date and time, ending date and time and time interval have to be provided.

Simulation and Validation. A simulation run calculates the precipitation – runoff response in the basin model given input from the Meteorological model. The global summary table available contains peak flows and the time of peak flows for each hydrologic element in the basin model. The results were then optimized for obtaining parameters that gives more precise results. After completion of the building and parameterizing the model components, simulation and optimization of simulated results followed by validation and determination of flood hydrograph of the unknown point is performed.

Simulation and Optimization. The flood routing was performed using two routing methods Muskingum-Cunge method and Kinematic Wave routing method. Here, some of the parameters were entered by the trial-and-error method while some parameters were obtained from the Central Water Commission, Government of India as given in Table 2.

Table 2. HEC – HMS Parameter Summary.

Element Name	Channel	Length (Km)	Energy Slope	Manning's n	Bottom Width	Side Slope	Index Type	Area of Periyar Basin(km^2)
Neeleeswaram	Trapezoid	27	0.07	0.035	200	2	Index Flow	5398

Source: Central Water Commission GoI

The rainfall discharge modeling was carried out by using the precipitation data of Neeleeswaram, the downstream station (marked as 1 in Fig. 1) and discharge data of Thodakayam station, the upstream station (marked as 2 in Fig. 1) for the year 2018. Two modeling approaches were adopted using Muskingum-Cunge method and kinematic wave routing method. A global summary table and optimization trial results were obtained through modeling. Three Optimization runs were carried out to reduce the error in the results and to obtain optimized parameters which are used as input parameters during the flood routing of ungauged station Aluva (marked as 3 in Fig. 1). From optimization trials, the percentage differences in peak flow (m^3/s) and discharge volumes (m3) were compared with observed discharge data and discharge volume during that time period and trials with least errors were selected.

Validation and Results. Validation was carried out to identify the best suited method for flood routing. This was executed by analyzing the methods with precipitation and discharge data of August 2016, thereby ensuring precision and accuracy. The routing method and the corresponding optimized parameters were selected for the generation of flood hydrograph of Aluva in Ernakulam district.

4.3 Hydraulic Model Development

Modeling. The methodology used for carrying out Hydraulic modeling involved 4 major steps: (1) Creating RAS geometry; (2) Adding river attributes; (3) Adding flow data and boundary conditions; and (4) Simulation and Analysis.

Creating RAS Geometry. The geometry in HEC-RAS contains information on cross-sections, hydraulic structures, river banks, roughness coefficient, etc. The geometric data for HEC-RAS geometry file are created in the RAS Mapper component of HEC-RAS. Before geometry generation, it is necessary to create the basic conditions to develop geospatial features. These include specifying a coordinate system. The coordinate system defined by the projection file needs to be a projected coordinate system.

Add River Attributes. The centerline of Periyar River was drawn from Thodakayam (upstream) to Aluva (downstream).

Add Flow Data and Boundary Conditions. HEC-RAS allows for two main types of simulation: steady and unsteady state analysis. In steady state analysis, constant boundary conditions are entered, whilst for unsteady it is included a hydrograph.

Simulation andAnalysis. After completion of all the above steps the steady and unsteady simulation models were run one after the other. The visualization of inundated areas and cross-section details can be performed in the RAS Mapper. The results are then stored in the system by exporting them using GIS tools.

The operations performed in the HEC-RAS software include unsteady flow analysis and steady flow analysis.

Unsteady Flow Analysis. The unsteady flow component is used to perform supercritical flow regime calculations in the unsteady flow computations module. The hydraulic calculations for cross-sections that were developed for the steady flow component were incorporated into the unsteady flow module. Unsteady flow was used to find the inundation boundary and water surface profile during August 2018. Unsteady flow analysis was also used to observe inundation maps for different computation time such as 2 h, 12 hand 24 h. The software pre-computes hydraulic properties for all possible water surface elevations at each cross section and it uses flow and stage data simultaneously at all sections and at each time step. The unsteady flow analysis was carried out using the discharge data of August 2018 with the computation time to 2 h, 12 h and 24 h so as to compare the results obtained by each computation time. The mapping output interval and hydrograph output interval of 1 day was adopted. The flow regime for steady flow was set as super-critical.

Steady Flow Analysis. Steady flow analysis is carried out using HEC-RAS to detect the change in flooding pattern during 5 consecutive years, from 2015 to 2019. The software computes exact hydraulic properties at a section for each trial water surface elevation from the elevation/station points and n-values using iterative convergence section-by-section for each flow. The maximum discharge of the selected years is given in Table 3.

Table 3. Observed maximum discharge (2015–2019)

Year	Maximum flow (m^3/s)
2015	1726
2016	653
2017	1433
2018	6210
2019	4112

5 Results and Discussion

5.1 Hydrologic Simulation

Hydrologic simulation was performed using Muskingum-Cunge method and Kinematic Wave method and the results are given in Table 4a. An error of −3.79% in peak discharge and an error of 76.7% in volume were observed. Optimization of these results was performed to minimize the errors and three trials were carried out (Fig. 4). A summary of the performance of the two methods for optimization trials is given in Table 4b. The results obtained during trial 1 for Muskingum-Cunge method and trials 1 and 3 for Kinematic Wave method were validated in order to adopt the appropriate flood routing method and parameters and it was observed that trial 3 of Kinematic-Wave routing method showed the least deviation from the observed flow both in terms of discharge volume and peak discharge.

Table 4a. Computed results of station Neeleeswaram

Method	Peak discharge (m^3/s)	Time of peak	Volume (m^3)
Muskingum-Cunge	6057.57	16.8.2018, 00:00	4775.48
Kinematic Wave	6058.28	16.8.2018, 00:00	4776.48

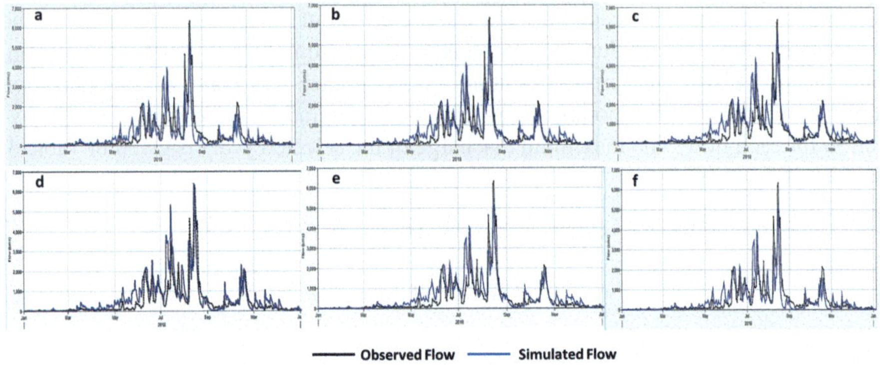

Fig. 4. Simulated discharge using (a to c) Muskingum-Cunge method for three optimization trials; (d to f) Kinematic wave method for three optimization trials

Table. 4b. Summary of optimization trials of discharge and volume

Trial	Parameter	Simulated	Observed	% error
Muskingum-Cunge method				
1	Volume (m^3)	2703.31	2701.09	0.08
	Peak Flow (m^3/s)	5244.2	6296.7	−16.7
2	Volume (m^3)	3420.33	2701.09	26.63
	Peak flow (m^3/s)	5525.4	6296.7	−12.2
3	Volume (m^3)	4015.94	2701.09	48.68
	Peak flow (m^3/s)	5809.2	6296.7	−7.7
Kinematic Wave method				
1	Volume (m^3)	3588.53	2701.09	32.85
	Peak Flow (m^3/s)	6391.1	6296.7	1.5
2	Volume (m^3)	3411.64	2701.09	26.31
	Peak flow (m^3/s)	5260.2	6296.7	−16.5
3	Volume (m^3)	2699.63	2701.09	−0.05
	Peak flow (m^3/s)	5169.7	6296.7	−17.9

5.2 Prediction of Ungauged Station

Using the Kinematic Wave routing method, the flood hydrograph of ungauged station namely, Aluva, which is more prone to flooding was estimated. For the analysis, the precipitation data of Neeleeswaram and the optimized parameters obtained from trial 3 of kinematic-wave routing method were adopted. The hydrologic modelling was executed using kinetic wave routing method. Table 5 shows the global summary of routing at Aluva during August 2018. The peak discharge for Aluva is obtained as 5960.77 m^3/s and the volume is 2915.58 m^3 and also the corresponding time for peak discharge is 16th August 2018. Figure 5 shows the predicted flood hydrograph of Aluva obtained.

Table 5. Global summary of routing at Aluva during August 2018

Hydrologic Element	Drainage Area (km^2)	Peak Discharge (m^3 /s)	Time of Peak	Volume (m^3)
Neeleeswaram	5398	5964.23	16.8.2018, 00:00	2912.03
Aluva	5398	5960.77	16.8.2018, 00:00	2915.58

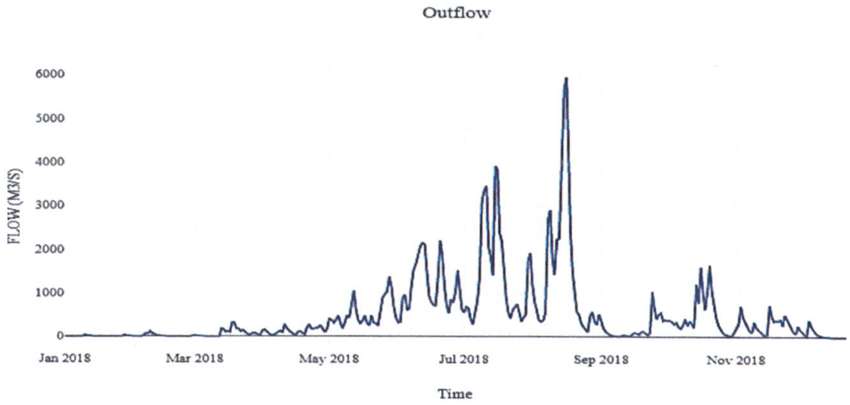

Fig. 5. Predicted flood hydrograph of ungauged station, Aluva

5.3 Hydraulic Simulation

Unsteady Flow Analysis. Figure 6a shows the inundation map of a portion of Periyar basin starting from Thodakayam. The map clearly depicts the consequences of flood events at Aluva, North Paravur and Nedumbassery while places at the other end such as Thodakayam and Malayattoor were least affected, which was attributed to the elevated topography. The cross-sectional plots of the most affected areas such as Nedumbassery Airport, and Malayattoor are also presented.

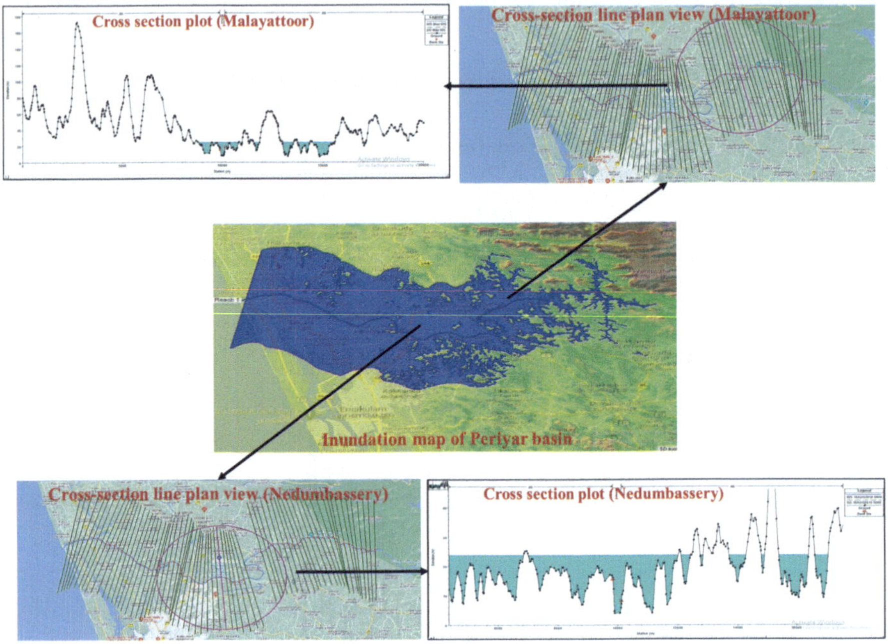

Fig. 6. Inundation map and cross-sectional plots of Periyar River Basin

Steady Flow Simulation. Steady flow describes conditions in which depth and velocity at a given channel location do not change with time. The computed results from the year 2015–2019 are shown in Fig. 7a–e respectively. The results clearly show that maximum flows in the years 2015, 2016, and 2017 have not resulted in flooding. However, the floods in 2018 had a significant impact on nearby areas of the Periyar River, while those in 2019 were less severe but has still affected adjacent low-lying areas. A discernible trend in flooding over these 5 years is not evident, necessitating a detailed study incorporating historical data and meteorological parameters to identify vulnerable zones in Kerala. Such insights can inform the implementation of appropriate flood management techniques to prepare for future events. Proposed mitigation methods include flood zoning, retention ponds, and detention ponds, which should be deployed following a thorough analysis of low-lying areas and the feasibility of each technique.

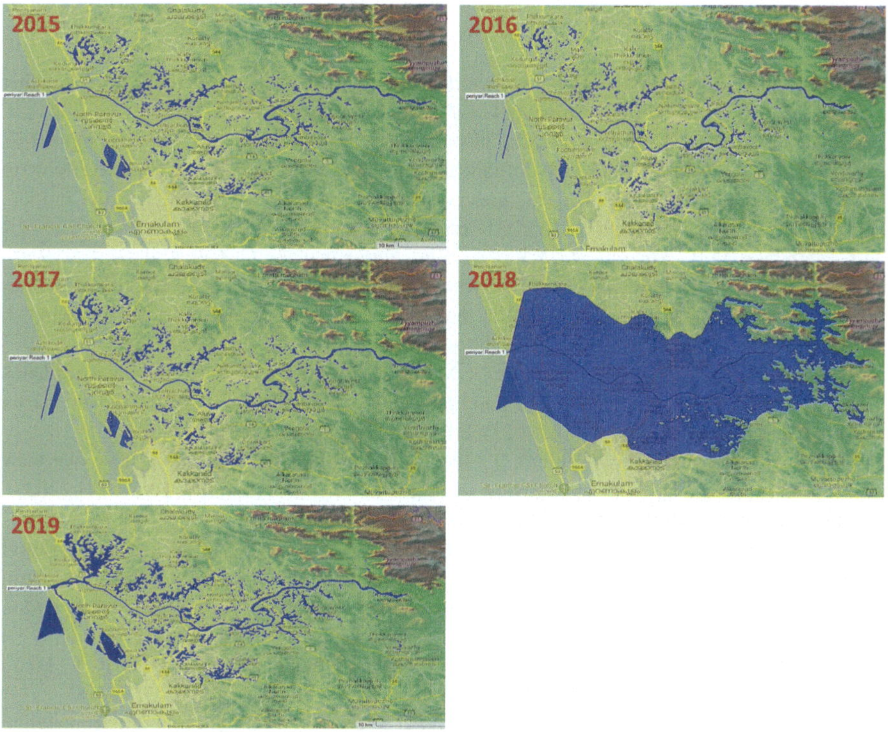

Fig. 7. Inundation maps of Periyar River Basin from steady flow simulations during 2015 to 2019

6 Conclusions

This study aimed at developing hydrological models of Periyar river basin to evaluate the effect of rainfall on surface runoff and peak discharges. Two methods, namely Muskingum-Cunge method and Kinematic Wave routing method were adopted to identify the best suited method for hydrologic modeling of Periyar basin. The study identified that Kinematic Wave method could simulate peak discharge and volume during such extreme events, with a percentage error of 17.9 and 0.05 respectively. The hydrologic modeling was extended to an ungauged station, Aluva for simulating the peak discharge and was estimated to be 5960.77 m3/s. The resulting flood hydrograph generated could be useful for future projections of flow during similar extreme events. The hydraulic modeling performed for mapping the inundation of the basin revealed that the downstream locations were more prone to flooding and the vulnerable regions to flooding were identified and mapped for various years, from 2015 to 2019. The study was carried out using 1D model, where the variations in the geometrical cross sections were neglected, and the simulations of flood considering 2D models can be attempted in future studies. The results of the present study are useful for flood forecasting and thereby flood rescue operations or mitigation measures can be adopted. Similarly, hydrograph can be plotted for other ungauged stations.

References

1. Dewan, T.H.: Societal impacts and vulnerability to floods in Bangladesh and Nepal. Weather Clim. Extr. **7**, 36–42 (2015). https://doi.org/10.1016/j.wace.2014.11.001
2. Gunathilake, M.B., Karunanayake, C., Gunathilake, A.S., Marasingha, N., Samarasinghe, J.T., Bandara, I.M., Rathnayake, U.: Hydrological models and artificial neural networks (ANNs) to simulate streamflow in a tropical catchment of Sri Lanka. Appl. Comput. Intell. Soft Comput., 1–9 (2021). https://doi.org/10.1155/2021/6683389
3. Gunathilake, M.B., Amaratunga, Y.V., Perera, A., Karunanayake, C., Gunathilake, A.S., Rathnayake, U.: Statistical evaluation and hydrologic simulation capacity of different satellite-based precipitation products (SbPPs) in the Upper Nan River Basin, Northern Thailand. J. Hydrol. Reg. Stud. **32**, 100743 (2020). https://doi.org/10.1016/j.ejrh.2020.100743
4. Gangani, P., Mangukiya, N.K., Mehta, D.J., Muttil, N., Rathnayake, U.: Evaluating the efficacy of different DEMs for application in flood frequency and risk mapping of the Indian coastal River Basin. Climate. **11**(5), 114 (2023). https://doi.org/10.3390/cli11050114
5. Kumar, V., Sharma, K.V., Caloiero, T., Mehta, D.J., Singh, K.: Comprehensive overview of flood modeling approaches: a review of recent advances. Hydrology. **10**(7), 141 (2023). https://doi.org/10.3390/hydrology10070141
6. Abdessamed, D., Abderrazak, B.: Coupling HEC-RAS and HEC-HMS in rainfall–runoff modeling and evaluating floodplain inundation maps in arid environments: case study of Ain Sefra city, Ksour Mountain. SW of Algeria. Environ. Earth Sci. **78**, 586 (2019). https://doi.org/10.1007/s12665-019-8604-6
7. Zin, W.W., Kawasaki, A., Takeuchi, W., San, Z.M.L.T., Htun, K.Z., Aye, T.H., Win, S.: Flood hazard assessment of Bago River Basin, Myanmar. J. Disaster Res. **13**, 14–21 (2018)
8. Natarajan, S., Radhakrishnan, N.: An integrated hydrologic and hydraulic flood modeling study for a medium-sized ungauged urban catchment area: a case study of Tiruchirappalli city using HEC-HMS and HEC-RAS. J. Inst. Eng. Ser. A. **101**, 381–398 (2020). https://doi.org/10.1007/s40030-019-00427-2
9. Cahyono, C., Adidarma, W.K.: Influence analysis of peak rate factor in the flood events' calibration process using HEC–HMS. Model. Earth Syst. Environ. **5**, 1705–1722 (2019). https://doi.org/10.1007/s40808-019-00625-8
10. Kulkarni, A., Kale, G.: Identifying best combination of methodologies for event based hydrological modeling using HEC-HMS software: a case study on the Panchganga River basin, India. Sustain. Water Resour. Manag. **8**, 123 (2022). https://doi.org/10.1007/s40899-022-00691-4
11. Sengül, S., Ispirli, M.N.: Predicting snowmelt runoff at the source of the mountainous Euphrates River Basin in Turkey for water supply and flood control issues using HEC-HMS modeling. Water. **14**, 284 (2022). https://doi.org/10.3390/w14030284
12. Filianoti, P., Gurnari, L., Zema, D.A., et al.: An evaluation matrix to compare computer hydrological models for flood predictions. Hydrology. **7**, 42 (2020). https://doi.org/10.3390/hydrology7030042
13. Chiang, S., Chang, C.-H., Chen, W.-B.: Comparison of rainfall-runoff simulation between support vector regression and HEC-HMS for a rural watershed in Taiwan. Water. **14**, 191 (2022). https://doi.org/10.3390/w14020191
14. Shekar, N.C.S., Vinay, D.C.: Performance of HEC-HMS and SWAT to simulate streamflow in the sub-humid tropical Hemavathi catchment. J. Water Clim. Change. **12**, 3005–3017 (2021). https://doi.org/10.2166/wcc.2021.072
15. Zheng, H., Huang, E., Luo, M.: Applicability of kinematic wave model for flood routing under unsteady inflow. Water. **12**(9), 2528 (2020). https://doi.org/10.3390/w12092528

16. Barati, R., Akbari, G.H., Rahimi, S.: Flood routing of an unmanaged river basin using Muskingum–Cunge model; field application and numerical experiments. Casp. J. Appl. Sci. Res. **2**(6), 8–20 (2013)
17. Vashist, K., Singh, K.K.: HEC-RAS 2D modeling for flood inundation mapping: a case study of the Krishna River Basin. Water Pract. Technol. **18**(4), 831–844 (2023). https://doi.org/10.2166/wpt.2023.048
18. Brunner, G.W.: HEC-RAS River Analysis System Hydraulic Reference Manual. Hydrologic Engineering Center, Davis (2016)
19. Kute, S., Kakad, S., Bhoy, V., Walunj, A.: Flood modeling of river Godavari using HEC-RAS. Int. J. Res. Eng. Technol. **3**(09), 81–87 (2014)
20. Rangari, V.A., Sridhar, V., Umamahesh, N.V., et al.: Floodplain mapping and management of urban catchment using HEC-RAS: a case study of Hyderabad City. J. Inst. Eng. (India) Ser. A. **100**, 49–63 (2019)
21. Thakur, B., Parajuli, R., Kalra, A., Ahmad, S., Gupta, R.: Coupling HEC-RAS and HEC-HMS in Precipitation Runoff Modeling and Evaluating Flood Plain Inundation Map, pp. 240–251. World Environmental and Water Resources Congress, Sacramento (2017)
22. Sudheer, K.P., Bhallamudi, M., Narasimham, B., Thomas, J., Bindhu, V.M., Vema, V., Kurian, C.: Role of dams on the floods of August 2018 in Periyar River Basin, Kerala. Curr. Sci. **116**(5), 780–794 (2019)

Analyzing Historical and Potential Climate Change Impacts on Monthly Streamflow Using SWAT+ in the Marikina River Basin, Philippines

Jazmine Joyce A. Zapata[1(✉)], Joan Cecilia C. Casila[1], Tonni Agustiono Kurniawan[2], and Soufiane Haddout[3]

[1] Land and Water Resources Engineering Division, Institute of Agricultural and Biosystems Engineering, College of Engineering and Agro-Industrial Technology, University of the Philippines Los Baños, Laguna 4031, Philippines
jazapata@up.edu.ph

[2] College of the Environment and Ecology, Xiamen University, Xiamen 361102, Fujian, China

[3] Department of Physics, Ibn Tofail University, Kenitra, Morocco

Abstract. Climate change poses significant risks to flood-prone regions, such as the Marikina River Basin in the Philippines, by altering key hydrological variables like rainfall and temperature. Understanding these impacts is crucial for developing effective flood risk management and drought mitigation strategies. This study assesses the potential effects of climate change on streamflow by comparing historical observations with future climate projections using the SWAT+ model. Daily weather data (1995–2012) from the Philippine Atmospheric, Geophysical and Astronomical Services Administration (PAGASA) and the National Aeronautics and Space Administration – Prediction of Worldwide Energy Resource (NASA-POWER) were used for streamflow model calibration and validation. Calibration with observed monthly discharge data from the Sto. Niño Station (2005–2008) yielded an NSE of 0.742 and a PBIAS of −0.198%, while validation with data from 2010–2012 resulted in an NSE of 0.555 and a PBIAS of 18.296%. The calibrated model parameters were then applied to future climate projections. Climate scenarios RCP4.5 and RCP8.5 were assessed under normal, dry, and wet year conditions by modifying rainfall and temperature inputs. Results indicate significant variability, with wet years showing increased discharge and dry years showing reduced discharge. Under normal conditions, RCP4.5 projected lower discharge, whereas RCP8.5 indicated higher discharge and more extreme fluctuations. The findings provide valuable insights for flood risk management and drought mitigation in the Marikina River Basin, supporting future climate adaptation efforts in tropical urbanizing basins.

Keywords: SWAT+ · climate change · Marikina River basin

1 Introduction

Climate change is one of the most urgent environmental challenges today, significantly impacting hydrological cycles and water resources globally. Water resources in any basin are influenced by variables such as precipitation, soil, land use, natural disasters, and

J. C. Casila et al. (Eds.): CCORE 2024, SPEES, pp. 250–263, 2025.
https://doi.org/10.1007/978-3-031-93887-0_23

human-induced events [1]. Climate change has caused multiple extreme weather and environmental phenomena, including sea level rise, droughts, flooding, unpredictable fluctuations in rainfall, and increased temperatures [2]. These climate extremes can lead to damage to land property and loss of income in agriculture where there is a scarcity of irrigation during droughts and overflow of rivers during floods [3, 4].

The Intergovernmental Panel on Climate Change (IPCC) emphasizes that changes in the hydrological cycle are among the most significant consequences of climate change that potentially lead to more frequent and severe floods, droughts, and shifts in water availability [5]. To address these uncertainties, the IPCC [5] developed four Representative Concentration Pathways (RCPs), which represent varying degrees of changes in Earth's energy balance due to CO_2 levels. The Department of Science and Technology-Philippine Atmospheric, Geophysical and Astronomical Services Administration (DOST-PAGASA), an agency responsible for climate projections in the Philippines, suggested two of these greenhouse gas (GHG) emission trajectory options, for the future: the RCP4.5 and RCP8.5. Their reports projected the percentage increases in temperature and changes in rainfall for the twenty-first century or years 2036 to 2065 [6, 7].

Streamflow, or discharge, is directly influenced by precipitation, temperature, and land use [8]. Climate change can alter these variables, leading to significant changes in streamflow patterns. For example, higher temperatures can increase evaporation rates, reducing water availability [9]. Changes in precipitation can either increase the frequency of floods or prolonged drought periods, depending on whether rainfall becomes more intense or irregular [10]. Understanding how climate change affects streamflow is crucial for water resource management, ecosystem conservation, and disaster risk reduction [11].

The SWAT+, a completely revised version of the Soil and Water Assessment Tool (SWAT), was developed to solve the present and future challenges in water resources modeling and management [12]. It is a comprehensive hydrological model designed to simulate the effects of land management practices on water, sediment, and agricultural chemical yields in complex watersheds [13]. Applying SWAT+ to assess the potential impacts of climate change on the streamflow of a watershed can provide valuable insights into future water resource scenarios [14].

The Marikina River Basin, located in Metropolitan Manila, Philippines, is a vital part of the larger Pasig-Marikina-Laguna de Bay watershed. The watershed has been extensively studied for various parameters, including streamflow and sediment dynamics [15], in response to different phenomena, as it is a critical and highly flood-prone area in the Philippines. This river basin is crucial for Metro Manila, supplying water for domestic use, agriculture, and industry. The topography, land use, and climatic conditions of the basin make it highly susceptible to the impacts of climate change, particularly variations in streamflow that can exacerbate flooding or lead to water shortages.

While SWAT and SWAT+ have been widely used globally to assess climate change impacts on streamflow, there is a lack of focused studies on the Marikina River Basin. Research in other regions has shown the effectiveness of SWAT and its different versions in predicting changes in streamflow under various climate scenarios, contributing to better water management practices [16–18]. However, localized studies are essential due to the unique climatic, geographical, and socio-economic conditions of each watershed. This study aims to bridge this gap by applying SWAT+ to the Marikina River Basin, providing specific insights relevant to this critical area.

This study is crucial for understanding climate change impacts on a key Philippine water resource. It can inform water management policies, aid in flood control planning, and support sustainable development. Through SWAT+, the study aims to analyze the impact of climate change on the streamflow of Marikina River Basin. The analysis will be based from the RCP4.5 and RCP8.5 scenarios from the DOST-PAGASA. Overall, this study contributes significantly to sustainable water management, infrastructure planning, and environmental conservation of the Marikina River Basin.

2 Methodology

2.1 Study Area

The Marikina River Basin, located in Eastern Metropolitan Manila, spans approximately 53,000 hectares. This study focuses on the Upper Marikina Basin, with coordinates ranging from 121.0727° to 121.3264° E longitude and 14.5618° to 14.8355° N latitude. The topography of the basin features a combination of valleys and mountains, especially in the upstream region. Flat, low-lying areas are situated in the western section, while gentle rolling hills and some rugged ridges are found in the eastern section. The discharge of the Marikina River significantly contributes to the overall flow of the Pasig River, often causing floods in nearby areas during severe storms.

The basin has a Type I climate, with a maximum rain period from June to September. The annual rainfall of the basin averages 2,750 mm, influenced by various weather systems throughout the year. One of the highest recorded rainfall events in Metro Manila occurred during Typhoon Ketsana, locally known as Tropical Storm Ondoy last September 2009. According to DPWH, the peak streamflow during this period reached around 1,838 m^3/s. Figure 1 shows the map of the Marikina River Basin.

2.2 Input Data

In analyzing streamflow, SWAT+ requires specific datasets to be imported for data analysis. Similar to the study of Serrano et al. [19], the Digital Elevation Model (DEM) used in the study is the Interferometric Synthetic Aperture Radar (IFSAR) DEM that was resampled to a resolution of 10m. It is a Digital Terrain Model (DTM), which provides an accurate representation of terrains, helping to improve the accuracy and efficiency of the simulation.

The land cover map used in this study is derived from the 2010 Land Cover Map by NAMRIA. For streamflow calibration and validation, and analysis on climate change impact, only a single data for land cover is used. The soil map was obtained from the Food and Agriculture Organization of the United Nations (FAO), providing global soil data for SWAT+. These datasets are used by SWAT+ in analyzing hydrological processes such as runoff, evapotranspiration, infiltration, and water retention.

The weather data, which includes daily rainfall, maximum and minimum temperatures, solar radiation, humidity, and wind speed, are crucial for modeling streamflow in

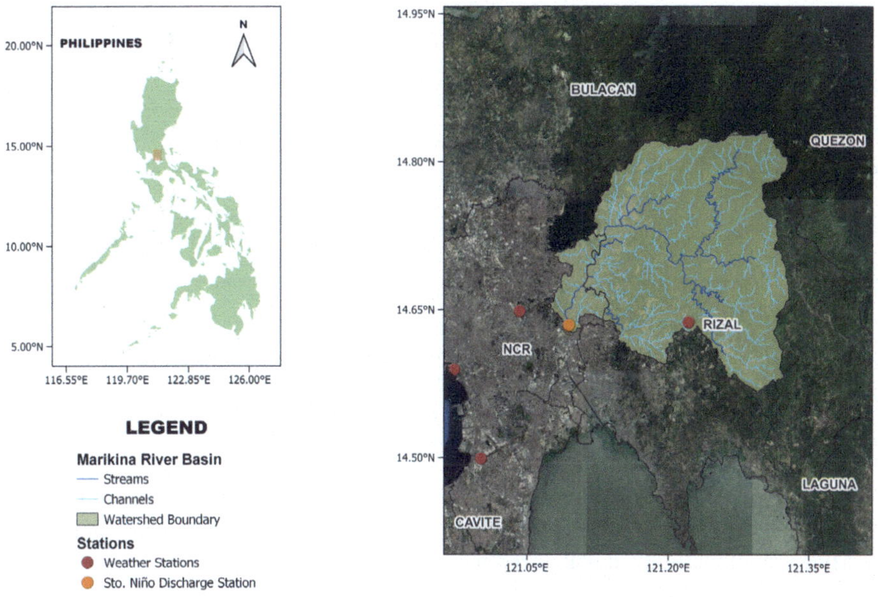

Fig. 1. Study area

the watershed. These data were obtained from the Boso-Boso, NAIA, Port Area, and Science Garden stations of the Philippine Atmospheric, Geophysical, and Astronomical Services Administration (PAGASA), except for solar radiation. Following the approach of Dapin and Ella (2020), daily solar radiation data were sourced from the National Aeronautics and Space Administration – Prediction of Worldwide Energy Resource (NASA-POWER) using the same coordinates as the PAGASA weather stations. For streamflow calibration and validation, discharge data from the Sto. Niño Station were used. The locations of these stations are shown in Fig. 1.

The discharge data was essential for calibrating the streamflow generated by SWAT+. Monthly discharge data from 2005 to 2008 and 2010 to 2012 was obtained from the Department of Public Works and Highways (DPWH). The Sto Niño Station, near the Marikina Bridge, was selected for this study.

2.3 Streamflow Model Development

The study followed the general methodology in SWAT+ for the development of the streamflow model of Marikina River, Philippines. Figure 2 shows the process of applying the RCP4.5 and RCP8.5 scenarios to the calibrated and validated streamflow.

SWAT+ Model Setup. In this study, the SWAT+ model was employed to simulate streamflow in the Marikina River Basin. Specifically, QSWAT+, a plugin in Quantum Geographic Information System (QGIS), was utilized. The SWAT+ model predicted key components of the hydrological cycle in this study. The model utilized the Natural

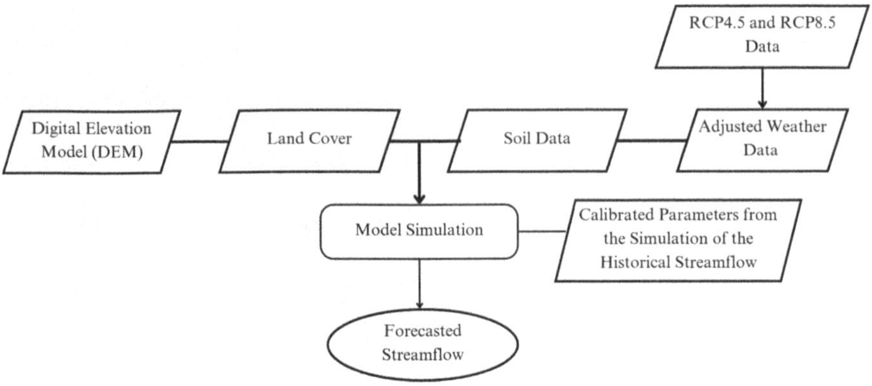

Fig. 2. Schematic diagram for the climate change impact projection.

Resources Conservation Service-Curve Number (NRCS-CN) method and the Penman-Monteith method to assess surface streamflow and potential evapotranspiration, while the Muskingum method simulated channel routing.

Calibration and Validation of the Simulated Streamflow. Since each watershed has its unique characteristics, calibrating and validating historical streamflow data is essential for accurately analyzing future streamflow. This process helps identify the key parameters that need adjustment in the model. These parameters were then used in the analysis of climate change impact.

The calibration of simulated streamflow utilized actual discharge data from the Sto. Niño Station, through the DPWH. The calibration period was from 2005 to 2008, and the validation period was from 2010 to 2012. Manual calibration was performed using the SWAT+ Toolbox, while the model performance was evaluated using the Nash-Sutcliffe Efficiency (NSE) and Percent of Bias (PBIAS). NSE measures the relative magnitude of residual variance compared to measured variance [20], ranging from $-\infty$ to 1, with higher values indicating greater accuracy. PBIAS measures the average tendency of simulated data to be greater or smaller than observed data [21], with an optimal value of 0.0. Positive PBIAS values indicate model underestimation, while negative values indicate overestimation. Model simulation is considered satisfactory if NSE $>$ 0.50 and PBIAS is within $\pm 25\%$ for streamflow [22].

Calibration and validation were conducted to establish a reliable set of parameters for simulating future climate change scenarios. These processes also enabled the assessment of streamflow variations between historical and projected conditions.

Climate Change Scenarios Parameterization. QSWAT+ allows the simulation of various scenarios to simulate different watershed conditions. In this study, climate change scenarios were generated by modifying only the rainfall and temperature inputs. Climate projections for the mid-twenty-first century (2036–2065) under RCP4.5 and RCP8.5 were used, representing moderate and high carbon emissions, respectively. Each RCP scenario included three setups: driest year, normal year, and wettest year. Minimum and maximum temperatures were adjusted based on the *Philippine Climate Extreme Report*

(2020) [7], with values shown in Table 1. Rainfall adjustments followed the DOST-PAGASA [6] report, *Observed Climate Trends and Projected Climate Change in the Philippines*, with projected monthly changes presented in Table 2. These modifications were applied on the historical rainfall and temperature data, and therefore represented in the model as the future climate change scenarios.

Table 1. Adjustments for temperatures (°C) in different scenarios.

Description	Moderate emission – RCP4.5			High emission – RCP8.5		
	Driest Year	Normal Year	Wettest Year	Driest Year	Normal Year	Wettest Year
Minimum Temperature	+0.7	+0.7	+1	+1.7	+1.7	+1.8
Maximum Temperature	+1.3	+1.2	+1.1	+1.8	+1.6	+1.6

Table 2. Percent change in rainfall data for different seasons and scenarios.

Description	Moderate emission – RCP4.5			High emission – RCP8.5		
	Driest Year	Normal Year	Wettest Year	Driest Year	Normal Year	Wettest Year
DJF (December–January–February)	−0.1	+17.7	+55.5	+2.7	+27.8	+53.4
MAM (March–April–May)	+0.7	+6.9	+25.7	−7.2	+4.8	+19.8
JJA (June–July–August)	−21.3	−10.1	−0.4	−17.0	−6.1	+7.7
SON (September–October–November)	−10.8	−6.0	+7.7	−8.0	+3.9	+19.9

Analysis of Climate Change Impact. To evaluate the climate change impacts on streamflow, a baseline scenario was first established. In line with the studies of Chawanda et al. [23], Dapin and Ella [4], and Singson et al. [24], this baseline is defined by a year that most accurately represents the actual discharges observed during the simulation period. The simulated streamflow of the chosen baseline was then compared on the simulated streamflow of the future climate change scenario, considering the adjustments on the rainfall and temperature.

In this study, the monthly average of actual discharge was calculated for the periods 2005 to 2008 and 2010 to 2011. The years 2009 and 2012 were excluded to avoid possible alteration due to the unusually high discharge caused by Typhoon Ketsana and the Southwest Monsoon, respectively. The year with the highest NSE relative to the overall monthly average was selected as the baseline.

3 Results and Discussion

3.1 Calibration and Validation of Streamflow

In SWAT+ modeling, it is necessary to calibrate different parameters to achieve satisfactory results. In this study, the calibration and validation are also crucial since it defines the needed parameters to be adjusted in the climate change scenarios. These parameters were deemed significant through the NSE and PBIAS of the manual calibration, and in consideration of other studies [4, 16] that have similar watershed characteristics. The calibration parameters and their corresponding calibrated values are shown in Table 3.

Through the SWAT+ Toolbox, the NSE and PBIAS of the model were determined by simulating the model and comparing it to the actual data. Figure 3 shows the comparison between the actual discharge at the Sto. Niño Station and the simulated discharge of the channel with the same coordinate.

As shown in Fig. 3, the calibrated discharges closely matched the peaks of the actual discharges, with the calibrated streamflow slightly exceeding the observed peak values. This is particularly during September 2005, where the calibrated discharge of 114 m^3/s was approximately 35.15 m^3/s higher than the actual discharge. However, the model tended to underestimate some baseflow values. The discrepancies in peak flow and baseflow events can be attributed to errors in the meteorological data, which are unevenly distributed across the study area. Additionally, the use of monthly data instead of daily data could influence the intensity of streamflow in certain months. The reliability of the calibration results is further strengthened by the obtained NSE of 0.742 and PBIAS of −0.198%. Both results are considered acceptable, indicating that the calibrated model is capable to predict future scenarios efficiently.

Validation assesses the model's ability to realistically simulate future output data based on given weather inputs. In this study, monthly streamflow data from 2010 to 2012 were used for validation. Figure 4 presents a comparison between the actual and validated streamflow at the Sto. Niño Station and its corresponding channel. This process further confirms the model's reliability and applicability for future climatic predictions and scenario analyses.

As shown in Fig. 4, most of the actual peak discharges align closely with the simulated discharges. A noticeable peak discharge occurs in August 2012, which corresponds to the 2012 Southwest Monsoon. This event caused widespread destruction in Marikina due to intense rainfall and strong winds, resulting in high streamflow and significant flooding.

The validated model achieved an NSE of 0.555 and a PBIAS of 18.296%. While there is a decrease compared to the calibration results, the NSE still falls within acceptable limits, confirming the model's validity. The positive PBIAS value suggests that the simulated discharges are generally underestimated relative to the actual values. This discrepancy, particularly observed in the peak discharge of August 2012, is typical in hydrological modeling, as highlighted in studies by Dapin and Ella [4], Serrano et al. [19], and Singson et al. [24].

One limitation of SWAT+ is its ability to accurately simulate peak discharges. During calibration, the model tended to overestimate peak discharges relative to actual measurements. However, when the model was tested with a different dataset using the same

Table 3. Calibrated SWAT+ parameters for the monthly streamflow.

SWAT+ Parameter	Description	Change Type	Range of Values	Calibrated Values
CN2	SCS runoff curve number for moisture condition 2*	Percent	35–95	30
ALPHA	Baseflow alpha factor (days)	Replace	0–1	0.9
CANMX	Maximum canopy storage (mm)	Replace	0–100	5
EPCO	Plant uptake compensation factor	Replace	0–1	0.05
ESCO	Soil evaporation compensation factor	Replace	0–1	0.05
CHK	Effective hydraulic conductivity in main channel alluvium (mm/hr)	Replace	−0.01–500	1
CHN	Manning's "n" value for the main channel*	Replace	−0.01–0.3	0.001
K	Saturated hydraulic conductivity (mm/hr)	Relative Value	0.0001–20000	200
AWC	Available Water Capacity of soil layer	Percent	0.01–1	−40
SURLAG	Surface runoff lag coefficient	Replace	0.05–24	10
REVAP_CO	Groundwater "revap" coefficient	Replace	0.02–0.2	0.02
CRK	Potential or maximum crack volume of the soil profile expressed as a fraction of the total soil volume	Replace	0–1	0.8

* dimensionless

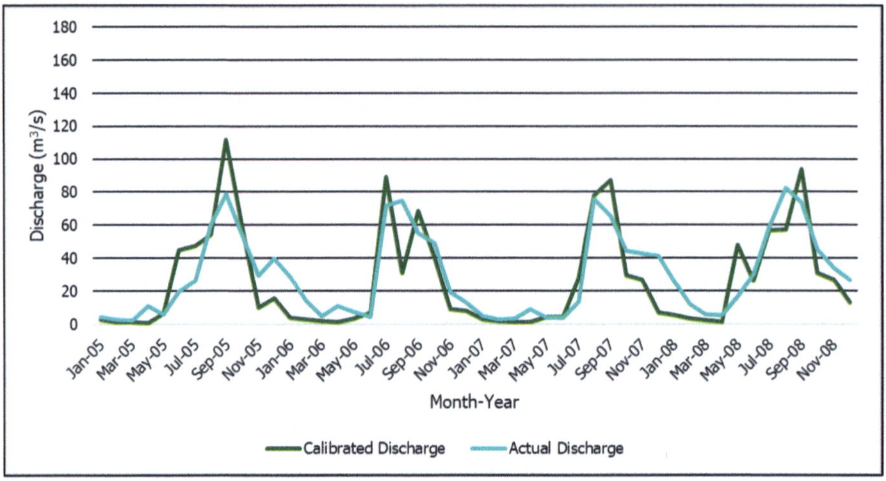

Fig. 3. Comparison of the actual and calibrated discharges.

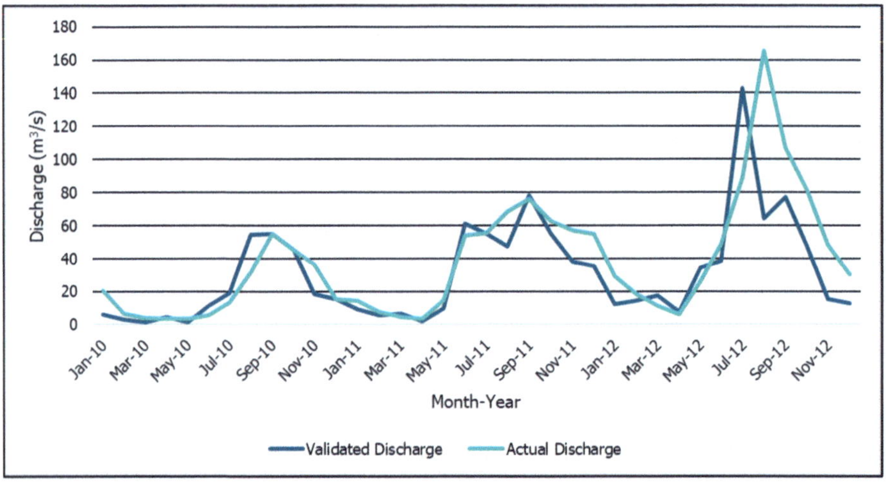

Fig. 4. Comparison of the actual and validated discharges.

adjusted calibration parameters, it occasionally underestimated the discharge. Despite these inconsistencies, the model generally provided a good match for peak discharges.

Given that both the calibrated and validated models produce satisfactory results, they can be confidently used for forecasting when provided with the appropriate weather input data.

3.2 Assessment of Climate Change Impacts

Baseflow Scenario. To determine the base flow, the monthly discharges from the years 2005 to 2008 and 2010 to 2011, along with their monthly average were determined.

These discharges are shown in Fig. 5. Also, the NSE of the monthly discharges, with respect to the overall average monthly discharge, is shown in Table 4.

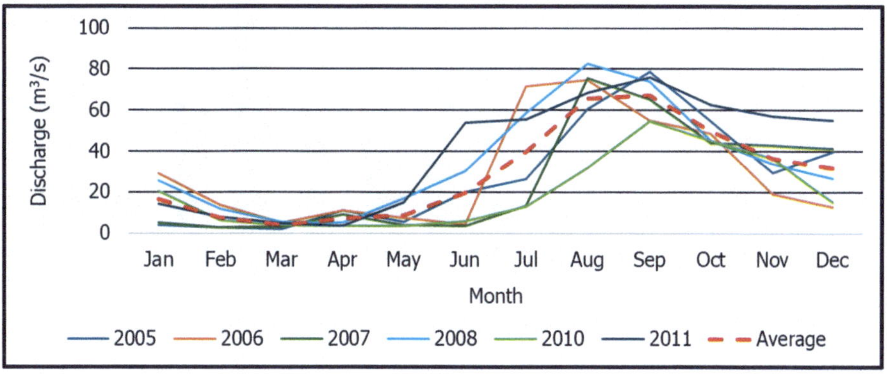

Fig. 5. Comparison of monthly discharges for the baseflow scenario.

Table 4. NSE of the monthly discharges based on the monthly average data.

Year	NSE
2005	0.88
2006	0.58
2007	0.75
2008	0.82
2010	0.55
2011	0.53

As seen in Table 4, the year 2005 has the highest NSE. Therefore, the baseline scenario that was chosen for the analysis of climate change impact is 2005.

Streamflow Analysis. The impacts of climate change were evaluated under dry, normal, and wet scenarios for RCP4.5 and RCP8.5. The data used in the analysis was the simulated historical streamflow obtained from the calibration in 2005. The climate change scenarios adjustments in temperatures and percent changes in rainfall were applied on the historical streamflow. Using the same calibration parameters, along with adjusted rainfall and temperature data, streamflow was simulated for each scenario. The resulting discharge variations under different conditions are presented in Fig. 6a, b.

For the RCP4.5 scenario, streamflow from the normal setup closely matches the simulated discharge. However, from June to September, the normal setup significantly decreased compared to the simulated streamflow. The driest setup consistently shows lower streamflow than the baseline, while the wet setup exceeds the baseline, especially during peak discharges, with September having a 9.91% increase.

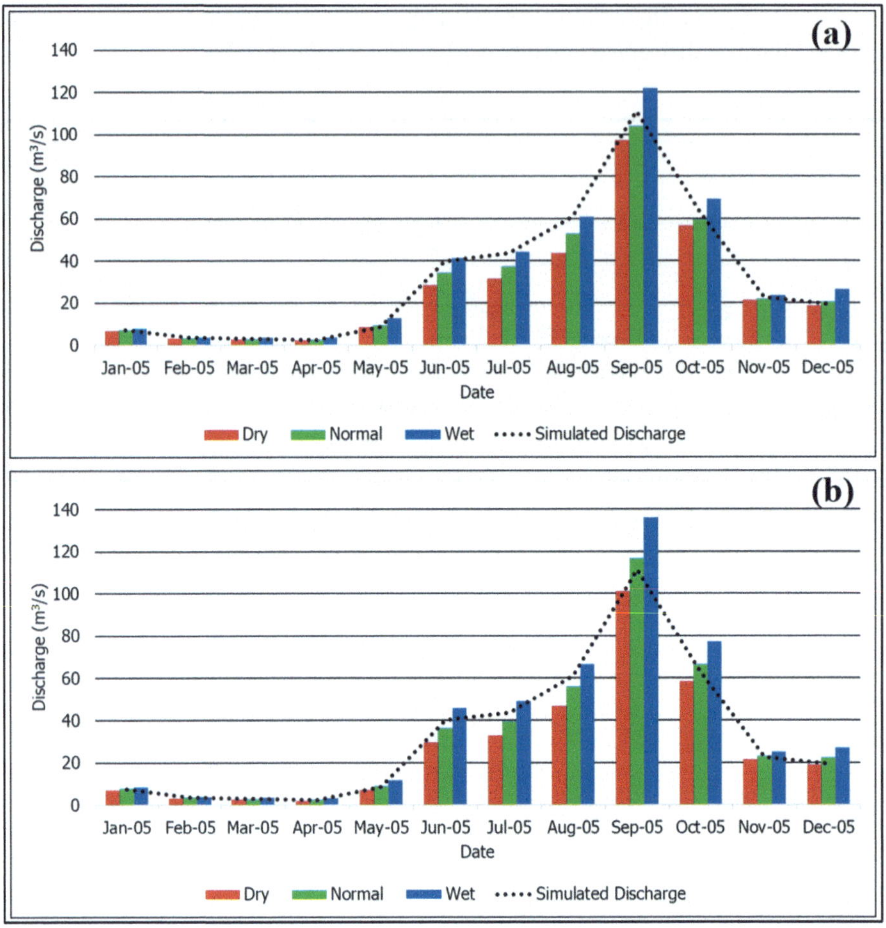

Fig. 6. Year 2005 monthly streamflow under a) RCP4.5 and b) RCP8.5 climate change scenarios.

For RCP8.5, the wet scenario shows a more significant increase in streamflow, especially in September, which has a 22.52% increase—the largest increase among all setups. The dry setup has lower streamflow, while the normal setup shows an overall increase compared to the simulated discharge. Table 5 summarizes the average percent change over the period.

Table 5 indicates that under both RCP scenarios, streamflow decreases more drastically in dry years, suggesting reduced water availability due to higher temperatures and erratic rainfall. This effect is more pronounced under RCP8.5, which can be attributed to the higher emissions leading to a greater reduction in water availability. In wet years, streamflow is projected to increase, predicting more intense rainfall events. Although higher streamflow may improve water availability, it could also lead to more frequent flooding, complicating water resource management.

Table 5. Percent changes for the different scenarios.

Conditions	Percent change (%)	
	RCP4.5	RCP8.5
Driest Setup	−10.95	−11.66
Normal Setup	−1.70	3.26
Wettest Setup	17.65	21.89

In normal years, streamflow remains stable, with slight decreases in RCP4.5 and slight increases in RCP8.5. These variations are likely due to minor shifts in rainfall patterns and temperature-driven changes in water demand.

Overall, the analysis reveals that the Marikina River Basin's streamflow patterns are highly sensitive to the shifts in rainfall and temperature patterns. These results enhance the need for comprehensive and adaptive management strategies to mitigate the harmful impacts of climate change on the Marikina River Basin.

A similar trend in discharge was also observed by Singson et al. [24]. For the RCP4.5, it was predicted that there would be a decrease of discharge for dry and normal years, but for wet years, the discharge increased. For the RCP8.5 projection, the model estimated a decline in discharge during the dry years. However, there will be a 0.53% and 21.88% increase in inflow during normal and wet years, respectively.

4 Conclusion

This study offers valuable insights into the potential impacts of future climate change on streamflow in the Marikina River Basin, Philippines, using SWAT+ modeling. The findings emphasize key changes in streamflow patterns under RCP4.5 and RCP8.5 climate scenarios, providing useful information for water resource management and disaster prevention strategies. Using dry, normal, and wet scenarios for the two climate scenarios, the study projected the impact of climate change on streamflow. The results indicated significant variability, with wet years showing higher discharge and dry years showing reduced discharge. RCP8.5 scenarios generally predicted more extreme changes compared to RCP4.5.

This variability underscores the need for adaptive water resource management strategies. The findings indicate an increased risk of flooding during extreme weather events and potential water shortages during dry periods, emphasizing the importance of robust flood management and water conservation measures.

For further investigations, it is recommended to project the late twenty-first-century climate change scenarios to better understand and investigate the trends under different climate extremes. In modeling, it is advisable to use more weather stations, ideally located within the watershed boundary, to improve the accuracy of the weather data. Additionally, extending the calibration and validation periods over a longer range of years would enhance the robustness of the model. If available, daily discharge data is recommended be used to examine the streamflow on a day-to-day basis more precisely.

Given the extensive capabilities of SWAT+, it would be beneficial to investigate the behavior of sediments, nutrients, and plant uptake within the basin, provided appropriate calibration data is available. This comprehensive approach would contribute significantly to the understanding of the hydrological and ecological dynamics of the watershed.

Overall, this research provides valuable insights into the hydrological behavior of the Marikina River Basin under current and future climate scenarios.

Acknowledgment. The authors would like to thank the professors and students who assisted during the modeling and completion of this study.

References

1. Abbasa, N., Wasimi, S., Al-Ansari, N.: Assessment of climate change impacts on water resources of Al-Adhaim, Iraq using SWAT model. Engineering. **8**(10), 716–732 (2016). https://doi.org/10.4236/eng.2016.810065
2. Bolan, S., Padhye, L.P., Jasemizad, T., Govarthanan, M., Karmegam, N., Wijesekara, H., Amarasiri, D., Hou, D., Zhou, P., Biswal, B.K., Balasubramanian, R., Wang, H., Siddique, K.H., Rinklebe, J., Kirkham, M., Bolan, N.: Impacts of climate change on the fate of contaminants through extreme weather events. Sci. Total Environ. **909**, 168388 (2024). https://doi.org/10.1016/j.scitotenv.2023.168388
3. Ataç, B.A., Hartmann, T.: Climate change and property rights: challenges for land policy. In: Advances in Chemical Pollution, Environmental Management and Protection, pp. 1–22. Springer, Heidelberg (2022). https://doi.org/10.1016/bs.apmp.2022.10.013
4. Dapin, I.G., Ella, V.B.: Simulating climate variability impacts on streamflow of the Layawan River watershed using the SWAT model. Int. Agric. Eng. J. **29**(2), 9–23 (2020)
5. IPCC: Climate Change 2014: Impacts, Adaptation, and Vulnerability. Contribution of Working Group II to the Fifth Assessment Report of the Intergovernmental Panel on Climate Change. Cambridge University Press, Cambridge (2014)
6. DOST-PAGASA: Philippine Climate Extremes Report 2020: Observed and Projected Climate Extremes in the PAGASA. DOST-PAGASA, Quezon City, Philippines (2018)
7. DOST-PAGASA: Philippine Climate Extremes Report 2020: Observed and Projected Climate Extremes in the Philippines to Support Informed Decisions on Climate Change Adaptation and Risk Management. DOST-PAGASA, Quezon City, Philippines (2021)
8. Bai, X., Zhao, W., Liu, H., Zhang, Y., Yang, Q., Liu, J., Chang, X.: Effects of precipitation changes and land-use alteration on streamflow: a comparative analysis from two adjacent catchments in the Qilian Mountains, arid northwestern China. Front. Environ. Sci. **11** (2023). https://doi.org/10.3389/fenvs.2023.1097049
9. Greve, P., Seneviratne, S.I.: Assessment of future changes in water availability and aridity. Geophys. Res. Lett. **42**(13), 5493–5499 (2015). https://doi.org/10.1002/2015gl064127
10. Tabari, H.: Climate change impact on flood and extreme precipitation increases with water availability. Sci. Rep. **10**(1) (2020). https://doi.org/10.1038/s41598-020-70816-2
11. Arnell, N.: Climate change and global water resources. Glob. Environ. Chang. **9**, S31–S49 (1999). https://doi.org/10.1016/s0959-3780(99)00017-5
12. Bieger, K., Arnold, J.G., Rathjens, H., White, M.J., Bosch, D.D., Allen, P.M., Volk, M., Srinivasan, R.: Introduction to SWAT+, a completely restructured version of the soil and water assessment tool. J. Am. Water Resour. Assoc. **53**(1), 115–130 (2016). https://doi.org/10.1111/1752-1688.12482

13. Koltsida, E., Mamassis, N., Kallioras, A.: Hydrological modeling using the soil and water assessment tool in urban and peri-urban environments: the case of Kifisos Experimental Subbasin (Athens, Greece). Hydrol. Earth Syst. Sci. **27**(4), 917–931 (2023). https://doi.org/10.5194/hess-27-917-2023
14. Zhang, L., Nan, Z., Xu, Y., Li, S.: Hydrological impacts of land use change and climate variability in the headwater region of the Heihe River Basin, northwest China. PLoS One. **11**(6), e0158394 (2016). https://doi.org/10.1371/journal.pone.0158394
15. Casila, J.C.C., Andres, H.L., Haddout, S., Yokoyama, K.: Sediment transport modeling in the Pasig River, Philippines post Taal volcano eruption. Geosciences. **14**(2), 45 (2024). https://doi.org/10.3390/geosciences14020045
16. Abbaspour, K., Rouholahnejad, E., Vaghefi, S., Srinivasan, R., Yang, H., Kløve, B.: A continental-scale hydrology and water quality model for Europe: calibration and uncertainty of a high-resolution large-scale SWAT model. J. Hydrol. **524**, 733–752 (2015). https://doi.org/10.1016/j.jhydrol.2015.03.027
17. Gassman, P.W., Reyes, M.R., Green, C.H., Arnold, J.G.: The soil and water assessment tool: historical development, applications, and future research directions. Trans. ASABE. **50**(4), 1211–1250 (2007)
18. Nkwasa, A., Chawanda, C.J., Van Griensven, A.: Regionalization of the SWAT+ model for projecting climate change impacts on sediment yield: an application in the Nile basin. J. Hydrol. Reg. Stud. **42**, 101152 (2022). https://doi.org/10.1016/j.ejrh.2022.101152
19. Serrano, J., Jamilla, J.M., Hernandez, B.C., Herrera, E.: Evaluating hydrologic model performance of global and local weather data inputs. Int. Arch. Photogramm. Remote Sens. Spatial Inf. Sci. **XLVI-4/W6-2021**, 265–271 (2021). https://doi.org/10.5194/isprs-archives-xlvi-4-w6-2021-265-2021
20. Nash, J.E., Sutcliffe, J.V.: River flow forecasting through conceptual models part I—a discussion of principles. J. Hydrol. **10**, 282–290 (1970)
21. Gupta, H.V., Sorooshian, S., Yapo, P.O.: Status of automatic calibration for hydrologic models: comparison with multilevel expert calibration. J. Hydrol. Eng. **4**, 135–143 (1999)
22. Moriasi, D.N., Arnold, J.G., Van Liew, M.W., Bingner, R.L., Harmel, R.D., Veith, T.L.: Model evaluation guidelines for systematic quantification of accuracy in watershed simulations. Trans. ASABE. **50**(3), 885–900 (2007). https://doi.org/10.13031/2013.23153
23. Chawanda, C.J., Nkwasa, A., Thiery, W., Van Griensven, A.: Combined impacts of climate and land-use change on future water resources in Africa. Hydrol. Earth Syst. Sci. **28**(1), 117–138 (2024). https://doi.org/10.5194/hess-28-117-2024
24. Singson, C.L., Alejo, L.A., Balderama, O.F., Bareng, J.L.R., Kantoush, S.A.: Modeling climate change impact on the inflow of the Magat reservoir using the soil and water assessment tool (SWAT) model for dam management. J. Water Clim. Change. **14**(3), 633–650 (2023). https://doi.org/10.2166/wcc.2023.240

Agro-Climate Resilience

Transforming Packaging Sustainability: Biodegradable Coatings Enriched with Essential Oils from Food Waste

Chetan Chauhan[1](✉), Shanta Kumari[2], and Sudershan Singh[1]

[1] Department of Chemistry, Sardar Patel University, Mandi, H.P, India
chetan@spumandi.ac.in

[2] Department of Economics, Eternal University, Baru Sahib, H.P, India

Abstract. Food packaging plays a vital role in mitigating nutrients losses by extending shelf life and maintaining product quality. However, the non-biodegradable nature of the conventional plastic packaging poses significant environmental challenges. Consequently, the global shift towards sustainable packaging solutions has led to innovative approaches in the utilization of biodegradable coatings derived from natural sources that offer promising properties for food packaging applications. Moreover, extraction and incorporating essential oils (EOs) from global waste including food waste present an opportunity for resource utilization into various coating materials that enhances their functional properties, improving food preservation. This review article explores the plausible potential of EOs to be used as biodegradable coatings from waste to extend shelf life and further enhance food storage quality by mitigating the microbial spoilage, prevent moisture loss, and act as gas barriers. Additionally, they stabilize various food parameters such as pH, color, acidity, and firmness, reducing weight loss. Various packaging systems has been designed to utilize waste-derived EOs for preserving both plant and animal-based foods. Biodegradable packaging from natural polymers and incorporating EOs extracted from food waste represents a revolutionized sustainable approach for waste management and food preservation. We have examine recent literature on EOs extracted from waste food items along with upcoming challenges like volatility and degradation, affecting shelf life and compatibility. Future research should focus on cost-effective extraction, advanced encapsulation, safety standardization, and eco-friendly materials to enhance the sustainability of EO-based packaging.

Keywords: Biodegradable Coatings · Essential Oils (EOs) · Food Packaging

1 Introduction

Packaging plays a crucial and indispensable role in the transportation, preserving food quality, ensuring safety, and extending shelf life and also providing essential information about food and processing (Fig. 1) [1, 2]. However, the predominant use of non-biodegradable plastics in packaging over several decades has engendered significant

J. C. Casila et al. (Eds.): CCORE 2024, SPEES, pp. 267–286, 2025.
https://doi.org/10.1007/978-3-031-93887-0_24

environmental concerns. Petrochemical polymers, owing to their availability, affordability, visual appeal, barrier qualities, lightness, and transparency, have been favored materials for packaging [3]. Nevertheless, the surge in plastic packaging has led to an alarming accumulation of non-biodegradable waste, posing severe threats to both ecosystems and human health. Table 1 lists the types of contemporary food packaging based on materials, functionality, and technology.

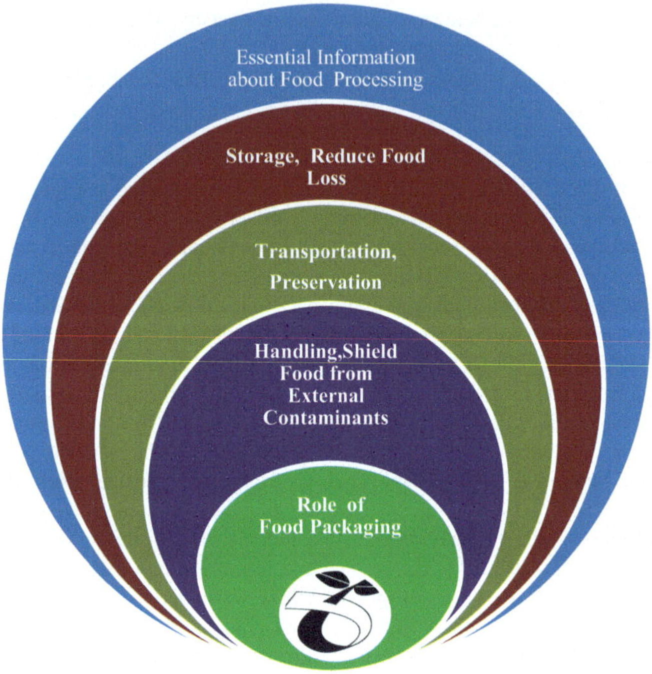

Fig. 1. Essential role of packaging in food industry

EPA (U.S. Environmental Protection Agency) reported that packaging-related materials, including glass, metal, plastic, paper, and paperboard, accounted for only 31% of Municipal solid waste (MSW). In contrast, non-packaging materials such as newspapers, phone books, and office paper generate more than double the MSW volume [16]. Advances in food technology including food processing and packaging are essential to maintaining one of the safest food supplies in the world. Packaging extends the benefits of food processing, allowing food to be transported safely over long distances while remaining fresh and wholesome at the point of consumption. However, modern packaging technology must strike a balance between ensuring food protection and addressing challenges such as energy and material costs, growing environmental and social awareness, and stringent regulations on pollution and waste management [17]. Therefore innovations in food packaging systems are poised to address the evolving market demands of the consumes including consumer preferences for safe, healthier, high-quality food products to minimize the environmental impact (Fig. 2). Emerging technologies in active and

Table 1. Classification of Food packaging based on materials and functionality

Type of Packaging		Examples	Advantages	Limitations	References
Based on Material	Plastic Packaging	Bottles (PET), pouches, cling wraps	Lightweight, durable, moisture-resistant, cost-effective	Non-biodegradable, contributes to plastic pollution	[1, 3]
	Glass Packaging	Jars for pickles, bottles for beverages	Recyclable, non-reactive, excellent barrier properties	Fragile, heavy, higher transport costs	[4]
	Metal Packaging	Aluminum cans, tin-plated steel cans	Heat-resistant, light-resistant, excellent protection against contaminants	Prone to corrosion, high production energy costs	[5]
	Paper and Cardboard	Cartons for cereals, boxes for frozen foods	Eco-friendly, biodegradable, cost-effective	Not moisture-resistant unless treated, less durable	[6]
	Biodegradable Packaging	Plant-based polymers, bagasse plates	Sustainable, reduces environmental impact	Limited durability, higher production costs	[7]
Based on Technology	Active Packaging	Oxygen scavengers, ethylene absorbers	Extends shelf life by interacting with food environment	Higher cost, complex technology	[8]
	Modified Atmosphere (MAP)	Fresh produce, meat, dairy products	Prolongs freshness, prevents oxidation	Requires special equipment, not suitable for all food types	[9, 10]
	Vacuum Packaging	Processed meat, cheese, coffee	Prevents oxidation and spoilage, reduces volume	Not effective for fragile or liquid products	[10]
	Edible Packaging	Candy wrappers, fruit coatings	Reduces waste, can be consumed with food	Limited applications, regulatory challenges	[11]
	Aseptic Packaging	UHT milk, juices, soups	Long shelf life, maintains quality without refrigeration	High initial equipment cost, requires complex production processes	[12]
	Smart Packaging	Sensors for seafood, QR codes	Monitors food quality, ensures safety, enhances traceability	High cost, requires consumer awareness	[13]
	Portion-Control Packaging	Single-serve snacks, coffee pods	Convenient, reduces food waste	Higher packaging material usage per unit	[14, 15]

intelligent packaging offer sustainable solutions to extend shelf life cycle and improve the quality and safety of food products. Simultaneously, advancements are being made to enhance the passive properties of food packaging materials, such as, barrier effectiveness, thermal and mechanical strength as well the stability [18] The rise of sustainable packaging system is a significant move toward mitigating the environmental impacts by incorporating edible or bio materials including the plant-based extracts, and nanomaterials. By integrating active and smart packaging technologies, it is possible to develop a versatile multifunctional food packaging system that ensures compatibility with food used. Achieving this synergy represents the ultimate goal for the future of food packaging technology, combining innovation with sustainability to meet both consumer and environmental needs.

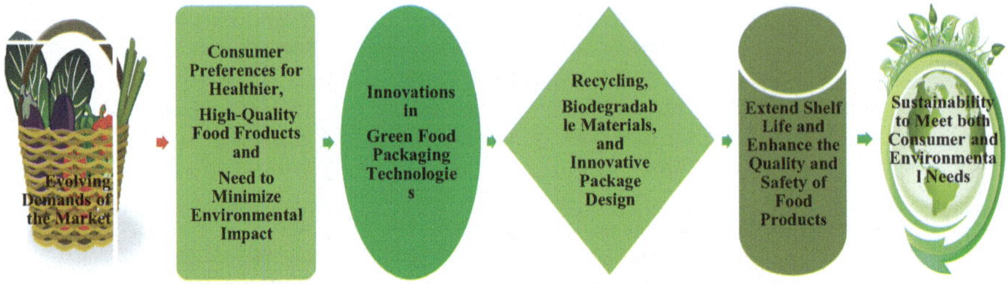

Fig. 2. Evolving demands of market and need of sustainability in food packaging.

Active packaging technologies present innovative opportunities for the food industry to enhance food preservation. Key active packaging systems currently in use include oxygen scavengers, carbon dioxide emitters and absorbers, moisture absorbers, ethylene absorbers, ethanol emitters, flavor-releasing and absorbing systems, time-temperature indicators, and antimicrobial films [2]. The ultimate objective of smart food packaging technologies, in addition to maintaining the quality and safety of food for the consumer, is to consider reduction of environmental impacts. In contemporary era, one of the solutions is to explore eco-friendly biomaterials instead of conventional food packaging systems like petroleum-based materials that impose various limitations (Fig. 3). However, the upcoming challenges of biomaterials in the food packaging industry should be carefully monitored by evaluated the techniques to eliminate or minimize limitations. Numbers of studies have been carried to improve the properties of biopolymer-based packaging materials to produce a eco friendly product for the food industry [19]. The threats during the application of plastic packaging extend beyond pollution to encompass potential health hazards posed by the spreading of contaminant like additives and monomers into food, feed and various other strata of environment. Consequently, there's a growing momentum and contemporary research drive to transition towards sustainable solution for packaging [1]. In the current scenario as evidenced from number of studies that consumer attitudes, underscore a willingness to pay premiums for sustainable solution for packaging [17–20]. In the recent years the food packaging industry is continuously evolving through innovative approaches that ingredient to provide effective, safer, sustainable solutions that match the consumer demands. Among the group of packaging options, biodegradable based coatings represent cutting-edge technology for preserving food commodities with integrity of nutrient profile for consumers. Such coatings, comprised of thin layers of diverse range of materials, are applied directly to food surface or packaging materials to improve preservation, protection, and functional properties [1]. Notably, they degrade into simple substances upon exposure to the environment, rendering them sustainable and environmentally friendly alternatives for food preservation. Biodegradable based coatings typically characteristic components such as polysaccharides, proteins, essential oils and lipids, selected for sustainable attributes such as biodegradability, non-toxicity, and recyclability. Application as thin layers on food products, coatings offer numerous merits, including aroma retention, prevention of chemical reactions, gas barrier properties, mitigation of spoilage, contamination, and ultimately extension of shelf life of packaged food. Moreover, such coatings address wide range of concerns including dehydration, browning, and oxidative rancidity, without adverse effects, while promoting controlled release of bioactive compounds integrated into the coatings [1]. In recent years, there has been a transition towards exploring natural compounds, particularly essential oils (EOs), as bioactive agents in packaging, driven by environment concerns surrounding during the application of synthetic alternatives [1]. EOs, extracted from various plant parts through processes such as extraction and steam distillation, offer antibacterial, antifungal, and antioxidant properties, making them valuable for food preservation [21]. Traditionally used as flavoring agents, EOs has expanded their role to include food preservation, owing to their safety for human consumption compared to synthetic chemical preservatives [1]. Active packaging incorporates functional components into the packaging system to improve safety extend shelf life of packed food. The inclusion of bioactive substances, like plant extracts, peptides, essential oils, and bacteriocins, rich in antioxidant and antimicrobial properties, can help preserve food quality. A key challenge for the packaging industry today is finding ways to combine environmentally friendly materials with active components, while maintaining the functionality of the packaging with smaller environmental footprint [22]. Integration of biodegradable materials with EOs derived from natural agro-industrial by-products contributes to a effective food packaging sustainable solutions [23].

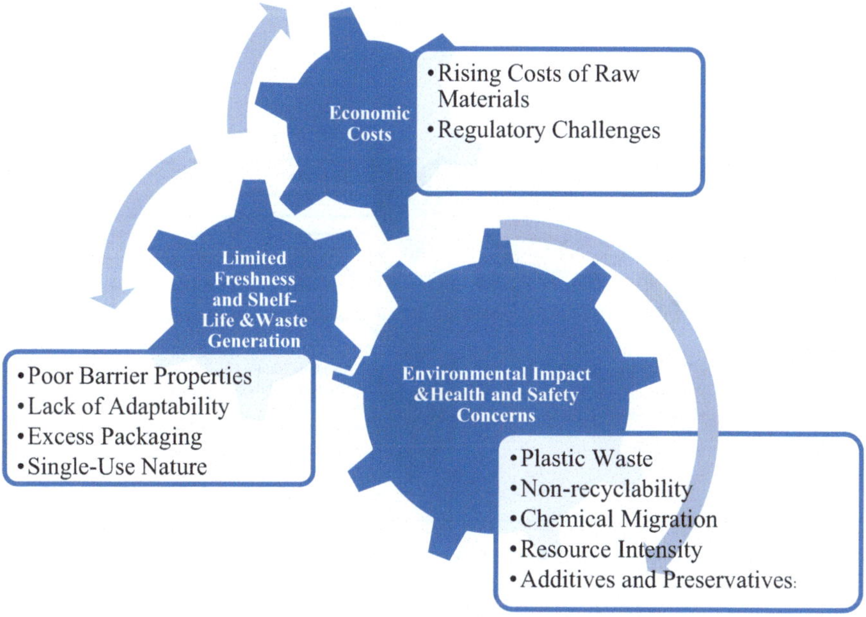

Fig. 3. Limitations of conventional food packaging.

The objective of current review paper is to explore the potential of biodegradable coatings incorporating EOs extracted from food waste as a sustainable solution for food packaging. By addressing the challenges posed by conventional plastic packaging, this article aims to highlight the advantages of EOs in improving food preservation, extending shelf life, and reducing environmental impact. The expected outcome is to provide insights into the feasibility of utilizing waste-derived EOs in packaging systems, identify current limitations, and suggest future research directions for enhancing the effectiveness and scalability of eco-friendly food packaging technologies.

2 Essential Oils

The global food processing industry faces various upcoming environmental challenges associated with ineffective waste management [1]. A huge amount of biomass generated by such industries remains underutilized despite its potential as a source of beneficial compounds, including EOs [1]. Exploring food waste for EO extraction presents a promising strategy for waste valorization and incorporation into biodegradable coatings, enhancing their preservation capabilities. Overall, the integration of EOs derived from food waste into biodegradable coatings offers a sustainable approach to preserving perishable food commodities, imparting desired properties for enhanced shelf life and quality maintenance. EOs extracted from fruits and vegetables are categorized and generally recognized as safe (GRAS) by the FDA (Food and Drug Administration). EOs consist of number of bioactive compounds, including monoterpenes, sesquiterpenes, phenylpropanoids, esters, acids, aldehydes, ketones, and alcohols, which confer upon them medicinal, antioxidant, antimicrobial and other characteristics (Fig. 4, Table 2) [1, 24].

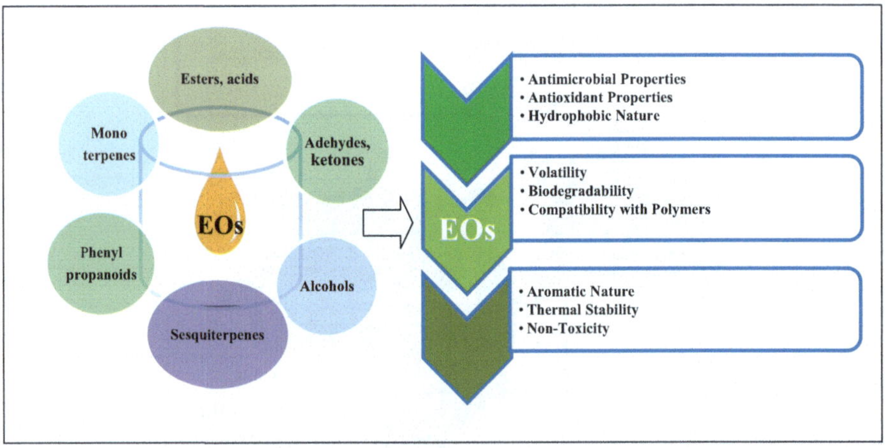

Fig. 4. Bioactive component of EOs to be well suited for food packaging

Table 2. Characteristic of Essential oils (EOs)

Characteristic	Description	References
Volatility	Essential oils are highly volatile and evaporate quickly at room temperature, allowing them to disperse easily in the air.	[25]
Hydrophobic Nature	Essential oils are hydrophobic, meaning they do not mix with water but dissolve readily in organic solvents.	[26]
Complex Composition	Mixture of chemical compounds such as terpenes, aldehydes, esters, and alcohols.	[27]
Antimicrobial Properties	Most essential oils exhibit antibacterial, antifungal, antiviral activity against microbes, making them potential candidate in food preservation and medicine.	[28]
Aromatic Nature	Known for their strong, pleasant aroma, essential oils are widely used in perfumes, aromatherapy, and flavoring agents.	[29]
Light Sensitivity	Essential oils are sensitive to light, and prolonged exposure can degrade their quality and effectiveness.	[30]
Thermal Sensitivity	High temperatures can alter the chemical structure of essential oils, affecting their potency and efficacy.	[31]
Antioxidant Properties	Rich in compounds that prevent oxidation, essential oils are used in skincare and food preservation.	[32]
Low Density	Most essential oils are less dense than water, causing them to float when mixed.	[33]
Color and Appearance	Ranges from colorless to pale yellow or dark brown, depending on the source and composition.	[34]
Therapeutic Properties	Many essential oils, such as lavender and peppermint, exhibit therapeutic benefits, including relaxation and pain relief.	[35]

2.1 Essential Oils Derived from Different Food Waste

Exploring food waste as a raw material for (EOs represents a sustainable approach to addressing the dual challenges of food waste management and the demand for eco-friendly packaging solutions [36]. Food waste, such as fruit peels, seeds, and other by-products, often contains a rich composition of bioactive compounds, including terpenes, flavonoids, and phenolics, which are valuable for their antimicrobial, antioxidant, and preservative properties [36]. Such bioactive natural compounds make food waste an excellent source for EOs, enabling their use in biodegradable\food packaging systems while reducing environmental pollution and resource depletion. Repurposing food waste not only prevents the unnecessary disposal of organic material but also aligns with circular economy principles, transforming waste into value-added products [37]. Moreover, sourcing EOs from food waste supports sustainable manufacturing practices by minimizing reliance on primary agricultural resources, thereby reducing the carbon footprint associated with production. This approach offers a cost-effective solution for industries, as waste-derived raw materials are abundant and otherwise underutilized. These innovative approaches trough exploration of food waste to produce EOs and integrating them into biodegradable food packaging align with Sustainable Development Goals (SDGs) led down by United Nations and cited below [38]:

SDG 12: Responsible Consumption and Production**:** By converting food waste into functional EOs for packaging, this approach promotes sustainable production patterns, reduces food waste, and encourages resource efficiency. It aligns with the goal of halving global food waste and achieving sustainable management of natural resources [38].

SDG 13: Climate Action: Utilizing food waste reduces methane emissions from landfills, a significant contributor to climate change. Additionally, the reduced reliance on petroleum-based polymers for packaging mitigates greenhouse gas emissions [38].

SDG 3: Good Health and Well-being: By developing antimicrobial and antioxidant food packaging solutions, this approach enhances food safety, reduces contamination, and minimizes health risks associated with conventional plastic packaging [38].

SDG 9: Industry, Innovation, and Infrastructure: Advancing the integration of waste-derived EOs into innovative packaging systems fosters research and innovation in sustainable industries, contributing to the development of resilient, eco-friendly infrastructure [38].

SDG 15: Life on Land: Preventing food waste and reducing plastic pollution through biodegradable packaging helps protect terrestrial ecosystems, minimize soil contamination, and reduce harm to biodiversity [38].

By addressing above cited SDGs, the exploration of food waste for EO production with diverse bioactive compositions has been found to well suited holistic approach to sustainability, offering environmental, economic, and societal benefits while evolving into eco-friendly food packaging technologies (Table 3).

EO is liquid obtained as extracts from various parts of plants including medicinal or herbal plants. EO is preferred explored to be applied as a food packaging material due to various inherited characteristics of bioactive composition, which offer effective antioxidant and antimicrobial effects against pathogenic microorganisms that are responsible for food spoilage (Table 4) [48–49]. Integration of biopolymers with EOs, a complex network is created within the polymeric matrix that reduces the volatility, controls the release, and enhances both the thermal and mechanical stability of EOs, making them more suitable for use in food packaging and processing applications. This study provides a comprehensive review of the techniques used for encapsulating EOs, the natural polymers used to load EOs, and the functional properties of EO-loaded biopolymeric particles, highlighting their potential antioxidant and antimicrobial benefits. Furthermore, the widespread use of EO-loaded biopolymers in the food industry is discussed, though research on their application in specific areas such as confectionery processing (e.g., biscuits, chocolates) is still limited. Further studies are needed to explore and expand the potential applications of EO-loaded biopolymeric particles in food processing industries. However, the effectiveness of EOs is often limited by the high volatility of their bioactive compounds, causing them to be released rapidly [50].

Table 3. Broad spectrum application of various EOs extracted from waste food items

Waste food source	Examples	Key EOs	Applications	References
Citrus Peels	Orange, lemon, lime, grapefruit peels	Limonene, citronellal, linalool	Natural preservatives, antimicrobial agents, and flavor enhancers	[39]
Herb and Spice Residues	Basil stems, oregano leaves, rosemary	Carvacrol, thymol, eugenol	Antioxidants, antimicrobials, and fragrances	[40]
Fruit Seeds and Pits	Mango seeds, avocado pits, grape seeds	Oleic acid, linoleic acid, tocopherols	Cosmetic formulations, biodegradable coatings, and nutritional additives	[41]
Vegetable Waste	Onion skins, carrot peels, garlic peels	Allicin, diallyl sulfide, beta-carotene	Food packaging, pest control, and natural dyes	[39]
Nut Shells and Husks	Walnut shells, almond husks, coconut shells	Phenolic compounds, tannins	Protective coatings, adhesives, and antimicrobial solutions	[42]
Coffee Grounds	Spent coffee grounds	Caffeine, diterpenes, volatile aromatics	Odor neutralizers, natural colorants, and sustainable coatings	[43]
Floral Waste	Marigold petals, rose petals, jasmine blooms	Geraniol, nerol, citral	Aromatherapy, flavorings, and antimicrobial packaging	[44]
Banana Peels	Discarded banana peels	Isoamyl acetate, octyl acetate	Fragrance, biodegradable films, and pest repellents	[45]

(*continued*)

Table 3. (*continued*)

Waste food source	Examples	Key EOs	Applications	References
Pomegranate Peels	Discarded peels from juice extraction	Punicic acid, ellagic acid	Antimicrobial agents, biodegradable packaging, and cosmetics	[46]
Other Food Processing Waste	Pineapple cores, papaya seeds, watermelon rinds	Bromelain, papain, various terpenes	Enzymatic applications, packaging films, and bio-based coatings	[47]

2.2 Extraction Methods for EOs

Essential oils are liquid extracts derived from aromatic plants, offering a wide range of applications across various industries [48–49]. EOs are extracted using number of methods, each with its own advantages, which affects the biological and physicochemical potential of the EOs (Table 5) [60–69]. Essential oils contain over 200 constituents, including both volatile and non-volatile components. Their antimicrobial, anticancer, anti-inflammatory, and antiviral properties make them highly effective and efficient in various therapeutic applications. Number of advanced and classical methods has been explored for extraction of EOs. Advanced techniques such as supercritical fluid extraction, subcritical extraction, and solvent-free microwave extraction are favored for their reduced the extraction times, low level energy consumption, minimal solvent use, and minimize CO_2 emissions. Classical methods such as hydrodistillation, steam distillation, hydrodiffusion, and solvent extraction are also widely used, though advanced methods are increasingly recognized for their efficiency and environmental benefits [69].

2.3 Essential Oils (EOs)-Based Food Packaging Coating

Essential oils (EOs) are concentrated oil extracts with a variety of applications in cosmetics, aromatherapy, food industry, and medicine [70–77]. The application of EOs as coating or packaging plays a vital role in preserving their potency, stability, and efficacy. There are number of coating types of EOs used food packaging as shown in the Table 6.

3 Challenges of Essential Oils (EOs)-Based Food Packaging

The use of essential oils (EOs) in food packaging holds promise for advancing food preservation and sustainability.

However, critical challenges which can limit their shelf life and functional properties must be addressed to ensure its feasibility, scalability, and wide adoption [78–83]. Regulatory compliance for EO-based materials and ensuring environmental sustainability,

Table 4. Characteristics of EOs to be well suited for food packaging

Characteristics	Relevance to Food Packaging	References
Antimicrobial Potential	Inhibits the growth of bacteria, fungi, and molds on food surfaces, enhancing food safety and extending shelf life.	[28, 51]
Antioxidant Potential	Reduces oxidative spoilage in food by neutralizing free radicals, maintaining freshness and nutritional quality.	[31, 52]
Volatility	Controlled release of essential oils through packaging materials ensures prolonged protection and effectiveness.	[25, 53]
Biodegradability	Essential oils are eco-friendly and degrade naturally, contributing to sustainable packaging solutions.	[54]
Hydrophobic Nature	Provides a barrier against moisture and humidity, preventing food spoilage due to water exposure.	[55]
Compatibility with Polymers	Easily incorporated into biodegradable and synthetic polymers for active food packaging solutions.	[56]
Aromatic Nature	Adds pleasant scents to packaging, masking undesirable odors and enhancing consumer appeal.	[57]
Thermal Stability	Essential oils with high thermal stability retain their properties during heat-based packaging processes like extrusion or coating.	[58]
Non-Toxicity	Safe for direct and indirect contact with food, ensuring compliance with food safety regulations.	[59]

given the limitations of certain biodegradable matrices, further complicates their adoption. Followings are the critical issues in the wide adoption and application of EO-based food packaging (Fig. 5)

3.1 Prone to Degradation and High Volatility

EOs are inherently volatile and prone to degradation due to oxidative process, light exposure, and at elevated temperatures. Such inherited characteristics can limit their shelf life and reduce their efficacy as antimicrobial agents in food packaging system. Additionally, the rapid release of bioactive compounds from EOs can compromise the controlled, sustained functionality needed for food preservation. Advanced techniques, such as encapsulation and film-coating technologies, are necessary to stabilize EOs,

Table 5. Extraction methods for EOs

Extraction Method	Description	Advantages	Limitations	References
Steam Distillation	Steam is passed through food waste material to vaporize volatile components, which are condensed into essential oils.	Cost-effective, widely used, suitable for high-temperature materials.	Not ideal for heat-sensitive compounds; requires significant water and energy.	[60, 61]
Hydro distillation	Food waste is submerged in water, and heat is applied to release the essential oils.	Simple method, suitable for small-scale extractions.	Time-consuming and inefficient for some raw materials.	[62]
Cold Pressing	Physical pressing of food waste (e.g., citrus peels) to extract oil without heat.	Retains natural aroma and composition of the oil.	Limited to high-oil-content materials like citrus peels.	[63]
Solvent Extraction	Solvents like ethanol or hexane dissolve the oil, which is then separated.	Efficient for extracting oils from delicate or dry materials.	Requires solvent recovery and purification; potential toxicity.	[34]
Supercritical CO2 Extraction	Apply CO_2 as a solvent to extract oils at low temperatures.	Environmentally friendly, no solvent residues, preserves sensitive compounds.	High initial cost and requires specialized equipment.	[21]
Microwave-Assisted Extraction (MAE)	Uses microwaves to heat food waste material, releasing the essential oil.	Faster extraction and energy-efficient.	May degrade heat-sensitive compounds; requires optimization.	[64]
Ultrasound-Assisted Extraction (UAE)	Uses ultrasonic waves to create cavitation bubbles, breaking cell walls and releasing oils.	Faster and more efficient than traditional methods.	Requires specialized equipment and optimization.	[65]

(continued)

Table 5. (*continued*)

Extraction Method	Description	Advantages	Limitations	References
Enzymatic Extraction	Enzymes break down cell walls, releasing essential oils.	Mild process, suitable for heat-sensitive materials.	Time-consuming and expensive enzymes.	[66]
Expression or Mechanical Pressing	Physically presses food waste without the use of heat or solvents.	Retains oil's natural properties, solvent-free.	Limited to materials with high oil content.	[67]
Fermentation	Microorganisms ferment food waste to release essential oils.	Enhances yield and quality of oils.	Time-intensive and requires controlled conditions.	[68]

Table 6. Coating types of essential oils (EOs) used food packaging

Coating Type	Description	Common Bio-polymers/Materials	Key Essential Oils Used	Benefits	References
Edible Coatings	Coatings applied directly to food products or in food packaging to enhance preservation and safety.	Starch, Gelatin, Casein, Lipids	Oregano, Thyme, Clove, Cinnamon, Rosemary	• Prevents microbial growth • Acts as a barrier against oxygen and moisture • Natural and biodegradable	[70, 71]
Microencapsulation	Encapsulates essential oils in microscopic capsules for controlled release and protection from degradation.	Gelatin, Chitosan, Alginate	Rosemary, Oregano, Thyme	• Controlled release of bioactive compounds • Protection from oxidation and evaporation	[72, 73]
Chitosan-based Coatings	Biopolymer coatings made from chitosan, often used in combination with essential oils for antimicrobial effects.	Chitosan	Rosemary, Oregano, Garlic, Cinnamon	• Antimicrobial and antioxidant properties • Biodegradable and non-toxic	[74]

(*continued*)

Table 6. (*continued*)

Coating Type	Description	Common Bio-polymers/Materials	Key Essential Oils Used	Benefits	References
Film Coatings with Essential Oils	Polymers infused with essential oils, used as films for food packaging.	Polyethylene, PLA, PVA	Oregano, Clove, Lemon, Cinnamon	• Active protection against spoilage • Gradual release of essential oils • Barrier to moisture and oxygen	[75, 76]
Nanocomposite Coatings	Coatings incorporating nanoparticles for enhanced mechanical properties, alongside essential oils for active food preservation.	Nano cellulose, Silica, Clay	Rosemary, Oregano, Clove	• Enhanced barrier properties • Sustained release of essential oils • High mechanical strength	[77]

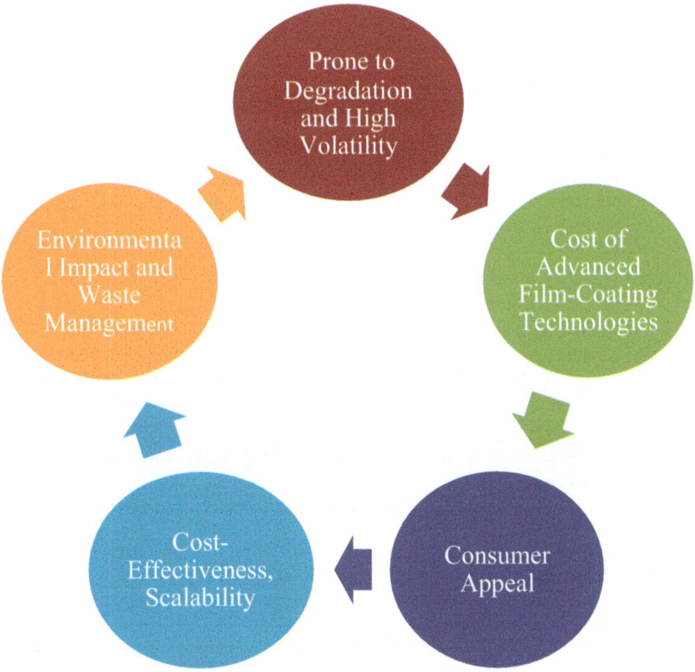

Fig. 5. Critical issues in the application of EO-based food packaging

control their control release, and maintain their properties. However, these technologies increase the complexity and cost of packaging production.

3.2 Cost of Advanced Film-Coating Technologies

The integration of EOs into food packaging requires advanced film-coating technologies to ensure compatibility with packaging materials and to enable sustained release of active compounds. These technologies often involve encapsulation techniques like microencapsulation, nanoemulsion, or layer-by-layer assembly, which are resource-intensive and costly. The need for specialized equipment and skilled labor further increases production costs, posing a barrier to widespread adoption, particularly for small-scale manufacturers or developing markets.

3.3 Consumer Appeal

EO-based packaging may face resistance from consumers due to its sensory attributes. The strong aromas and flavors of certain EOs can interact with food products, altering their taste or smell, which may lead to reduced consumer acceptance. Furthermore, consumer preferences for packaging aesthetics, transparency, and perceived value may conflict with the functional aspects of EO-based materials. Designing packaging systems that strike a balance between sensory appeal and functionality is essential to drive market acceptance.

3.4 Cost-Effectiveness, Scalability, and Sustainability

The cost-effectiveness of EO-based food packaging depends on the availability and sustainability of raw materials, as well as the efficiency of extraction and manufacturing processes. While food waste offers a sustainable source of EOs, large-scale extraction processes remain expensive and resource-intensive. Additionally, ensuring the scalability of EO-based packaging systems requires the development of affordable and efficient technologies. The availability of raw materials, especially in regions with limited agricultural diversity, may further challenge sustainability efforts. Addressing these issues through innovations in raw material sourcing, energy-efficient extraction, and low-cost production methods will be critical for long-term viability.

3.5 Environmental Impact and Waste Management

EO-based packaging systems are designed to reduce environmental impact by incorporating biodegradable and renewable materials. However, the sustainability of these systems depends on the entire lifecycle of the packaging, including waste management. Biodegradable materials used in EO-based packaging may require specific conditions for degradation, such as industrial composting facilities, which are not universally available. Additionally, the production processes for EO-based materials, including encapsulation and polymer modification, may generate secondary waste or consume significant resources, potentially offsetting their environmental benefits. A thorough lifecycle assessment is necessary to ensure that EO-based packaging systems genuinely contribute to sustainability goals.

Addressing these challenges will require interdisciplinary approaches involving material science, food technology, consumer behavior studies, and policy development.

Innovations in EO stabilization, cost reduction, and sensory optimization can pave the way for more feasible applications. Simultaneously, policymakers should focus on supporting research, incentivizing sustainable practices, and establishing guidelines for regulatory compliance. By overcoming these barriers, EO-based food packaging can evolve into a viable and sustainable alternative to conventional systems, aligning with global goals for food security and environmental preservation.

4 Conclusion

The integration of essential oils (EOs) extracted from food waste into biodegradable coatings offers a sustainable solution to address effectively the food preservation and packaging challenges in food industry. Such EOs based coatings has inherited potential to utilize the antimicrobial and antioxidant properties of EOs to extend food shelf life while reducing the reliance on conventional packaging system to address the environmental, food safety and waste concerns. Leveraging food waste for essential oil extraction aligns with SDGs and circular economy principles, transforming waste into value-added materials for global sustainable solution. While contemporary challenges remain in optimizing EO stability, release rates, and commercial scalability, advancements in extraction methods and biopolymer matrices can revolutionize food packaging technologies for a greener future.

References

1. Khalid, S., Naeem, M., Talha, M., Hassan, S.A., Ali, A., Maan, A.A., Bhat, Z.F., Aadil, R.M.: Development of biodegradable coatings by the incorporation of essential oils derived from food waste: a new sustainable packaging approach. Packag. Technol. Sci. **37**(3), 167–185 (2024)
2. Donkor, L., Kontoh, G., Yaya, A., Bediako, J.K., Apalangya, V.: Bio-based and sustainable food packaging systems: relevance, challenges, and prospects. Appl. Food Res. **3**(2), 100356 (2023)
3. Smith, E., Bilec, M.M., Khanna, V.: Evaluating the global plastic waste management system with Markov chain material flow analysis. ACS Sustain. Chem. Eng. **11**(6), 2055–2065 (2023)
4. Mesidis, J., Lockshin, L., Corsi, A.M., Page, B., Cohen, J.: Measuring the effect of product and environmental messaging attributes on alternative wine packaging choices. J. Clean. Prod. **431**, 139502 (2023)
5. Bandara, R., Indunil, G.M.: Food packaging from recycled papers: chemical, physical, optical properties and heavy metal migration. Heliyon. **8**(10), e10767 (2022)
6. Kannan, S., Parashar, B., Chaurasia, A.: E-commerce and green packaging: sustainable business trends. In: Green Innovations for Industrial Development and Business Sustainability, pp. 22–39. CRC Press (2024)
7. Jahangiri, F., Mohanty, A.K., Misra, M.: Sustainable biodegradable coatings for food packaging: challenges and opportunities. Green Chem. **26**, 4934 (2024)
8. Caner, C., Yüceer, M., Harte, B.: Trends in sustainability and innovative food packaging materials: an overview. Akademik Gıda Özel Sayı: Yeşil Dönüşüm / Special Issue: Green Transformation, 65–77 (2024)
9. Al Shoffe, Y., Johnson, L.K.: Opportunities for prediction models to reduce food loss and waste in the postharvest chain of horticultural crops. Sustainability. **16**(17), 13045 (2024)

10. Gorris, L.G.M., Peppelenbos, H.W.: Modified atmosphere and vacuum packaging to extend the shelf life of respiring food products. HortTechnology. **2**(3), 303–309 (1992)
11. Yemenicioğlu, A.: Recent developments shaping the future of antimicrobial edible food packaging: a review. Int. J. Food Sci. Technol. **59**(12), 9646–9665 (2024)
12. David, J.R., Coronel, P.M., Simunovic, J. (eds.): Handbook of Aseptic Processing and Packaging. CRC Press (2022)
13. Patel, A., Nagera, K., Anjana, R.: Development of a framework to improve sustainable manufacturing and service economics using ISM for food industries. Int. J. Biotechnol. Biomater. **10**(1), 1–10 (2024)
14. Thapliyal, D., Karale, M., Diwan, V., Kumra, S., Arya, R.K., Verros, G.D.: Current status of sustainable food packaging regulations: global perspective. Sustainability. **16**(13), 5554 (2024)
15. Swetha, B.S., Prabhakar, I., Ainapur, S.D.: Food waste reduction and sustainable food systems: strategies, challenges, and future directions. J. Sci. Res. Rep. **30**(5), 328–336 (2024)
16. U.S. Environmental Protection Agency (EPA): Municipal Solid Waste in the United States: 2005 Facts and Figures, pp. 1–50. U.S. Environmental Protection Agency, Washington, DC (2005)
17. Marsh, K., Bugusu, B.: Food packaging—roles, materials, and environmental issues. J. Food Sci. **72**(3), R39–R55 (2007)
18. Han, J.W., Ruiz-Garcia, L., Qian, J.P., Yang, X.T.: Food packaging: a comprehensive review and future trends. Compr. Rev. Food Sci. Food Saf. **17**(4), 860–877 (2018)
19. Ozdemir, M., Floros, J.D.: Active food packaging technologies. Crit. Rev. Food Sci. Nutr. **44**(3), 185–193 (2004)
20. Taherimehr, M., Yousefnia Pasha, H., Tabatabaeekoloor, R., Pesaranhajiabbas, E.: Trends and challenges of biopolymer-based nanocomposites in food packaging. Compr. Rev. Food Sci. Food Saf. **20**(6), 5321–5344 (2021)
21. Chakravarty, I., Parmar, V.M., Mandavgane, S.A.: Current trends in essential oil (EO) production. Biomass Convers. Biorefinery, 1–24 (2021)
22. de Oliveira-Filho, J.G., Lemes, A.C., Braga, A.R.C., Egea, M.B.: Biodegradable eco-friendly packaging and coatings incorporated with natural active compounds. In: Food Packaging, pp. 171–206. CRC Press (2020)
23. Carpena, M., Nuñez-Estevez, B., Soria-Lopez, A., Garcia-Oliveira, P., Prieto, M.A.: Essential oils and their application on active packaging systems: a review. Resources. **10**(1), 7 (2021)
24. Angane, M., Swift, S., Huang, K., Butts, C.A., Quek, S.Y.: Essential oils and their major components: an updated review on antimicrobial activities, mechanism of action, and their potential application in the food industry. Foods. **11**(3), 464 (2022)
25. Jugreet, B., Sharmeen, S.S., Kannan Rengasamy, R.R., Mahomoodally, F.M.: Chemistry, bioactivities, mode of action and industrial applications of essential oils. Trends Food Sci. Technol. **101**, 89–105 (2020)
26. Adorjan, B., Buchbauer, G.: Biological properties of essential oils: an updated review. Flavour Fragr. J. **25**(6), 407–426 (2010)
27. Ribeiro-Santos, R., Andrade, M., de Melo, N.R., Sanches-Silva, A.: Use of essential oils in active food packaging: recent advances and future trends. Trends Food Sci. Technol. **61**, 132–140 (2017)
28. Bhavaniramya, S., Vishnupriya, S., Al-Aboody, M.S., Vijayakumar, R., Baskaran, D.: Role of essential oils in food safety: antimicrobial and antioxidant applications. Grain Oil Sci. Technol. **2**(2), 49–55 (2019)
29. Jackson-Davis, A., White, S., Kassama, L.S., Coleman, S., Shaw, A., Mendonca, A., et al.: A review of regulatory standards and advances in essential oils as antimicrobials in foods. J. Food Prot. **86**(2), 100025 (2023)

30. Fernández-López, J., Viuda-Martos, M.: Introduction to the special issue: application of essential oils in food systems. Food Secur. **7**(4), 56 (2018)
31. Al-Maqtari, Q.A., Rehman, A., Mahdi, A.A., Al-Ansi, W., Wei, M., Yanyu, Z., et al.: Application of essential oils as preservatives in food systems: challenges and future prospectives—a review. Phytochem. Rev., 1–38 (2021)
32. Laina, K.T., Drosou, C., Krokida, M.: Evaluation of functional extrudates enriched with essential oils for enhanced stability. Food Bioprod. Process. **147**, 264–276 (2024)
33. Bibow, A., Oleszek, W.: Essential oils as potential natural antioxidants, antimicrobial, and antifungal agents in active food packaging. Antibiotics. **13**(12), 1168 (2024)
34. Patel, M., Patel, K., Bera, K., Prajapati, B.: Herbal formulations for the treatment of fungal infection. In: Herbal Formulations, Phytochemistry and Pharmacognosy, pp. 1–20. Elsevier (2024)
35. Turek, C., Stintzing, F.C.: Stability of essential oils: a review. Compr. Rev. Food Sci. Food Saf. **12**(1), 40–53 (2013)
36. Korte, I., Kreyenschmidt, J., Wensing, J., Bröring, S., Frase, J.N., Pude, R., Konow, C.: Can sustainable packaging help to reduce food waste? A status quo focusing plant-derived polymers and additives. Appl. Sci. **11**(11), 5307 (2021)
37. Narisetty, V., Adlakha, N., Singh, N.K., Dalei, S.K., Prabhu, A.A., Nagarajan, S., Kumar, A.N., Nagoth, J.A., Kumar, G., Singh, V., Kumar, V.: Integrated biorefineries for repurposing of food wastes into value-added products. Bioresour. Technol. **363**, 127856 (2022)
38. Bexell, M., Jönsson, K.: Responsibility and the United Nations' sustainable development goals. Forum Dev. Stud. **44**(1), 13–29 (2017)
39. Osaili, T.M., Dhanasekaran, D.K., Zeb, F., Faris, M.E., Naja, F., Radwan, H., et al.: A status review on health-promoting properties and global regulation of essential oils. Molecules. **28**(4), 1809 (2023)
40. Grover, S., Aggarwal, P., Kumar, A., Kaur, S., Yadav, R., Babbar, N.: Utilizing citrus peel waste: a review of essential oil extraction, characterization, and food-industry potential. Biomass Convers. Biorefinery, 1–22 (2024)
41. Kosalec, I., Cvek, J., Tomić, S.: Contaminants of medicinal herbs and herbal products. Arch. Ind. Hyg. Toxicol. **60**(4), 485–501 (2009)
42. Kumar, H., Dhalaria, R., Guleria, S., Sharma, R., Kumar, D., Verma, R., et al.: Non-edible fruit seeds: nutritional profile, clinical aspects, and enrichment in functional foods and feeds. Crit. Rev. Food Sci. Nutr. **64**(33), 13298–13317 (2024)
43. Nolasco, A., Squillante, J., Velotto, S., D'Auria, G., Ferranti, P., Mamone, G., et al.: Exploring the untapped potential of pine nut skin by-products: a holistic characterization and recycling approach. Food Secur. **13**(7), 1044 (2024)
44. Ahmed, H., Abolore, R.S., Jaiswal, S., Jaiswal, A.K.: Toward circular economy: potentials of spent coffee grounds in bioproducts and chemical production. Biomass. **4**(2), 286–312 (2024)
45. Pires Jr., E.D.O., Di Gioia, F., Rouphael, Y., García-Caparrós, P., Tzortzakis, N., Ferreira, I.C., et al.: Edible flowers as an emerging horticultural product: a review on sensorial properties, mineral and aroma profile. Trends Food Sci. Technol. **137**, 31–54 (2023)
46. Rather, J.A., Akhter, N., Ayaz, Q., Mir, S.A., Singh, A., Goksen, G., et al.: Fruit peel valorization, phytochemical profile, biological activity, and applications in food and packaging industries: comprehensive review. Curr. Food Sci. Technol. Rep. **1**(2), 63–79 (2023)
47. Monika, P., Chandraprabha, M.N., Hari Krishna, R., Vittal, M., Likhitha, C., Pooja, N., Chaudhary, V.: Recent advances in pomegranate peel extract mediated nanoparticles for clinical and biomedical applications. Biotechnol. Genet. Eng. Rev. **40**(4), 3379–3407 (2024)
48. Almeida, H.H., Fernandes, I.P., Amaral, J.S., Rodrigues, A.E., Barreiro, M.F.: Unlocking the potential of hydrosols: transforming essential oil byproducts into valuable resources. Molecules. **29**(19), 4660 (2024)

49. Misharina, T.A., Terenina, M.B., Krikunova, N.I.: Antioxidant properties of essential oils. Appl. Biochem. Microbiol. **45**, 642–647 (2009)
50. Hossen, M.A., Shimul, I.M., Sameen, D.E., Rasheed, Z., Dai, J., Li, S., Qin, W., Tang, W., Chen, M., Liu, Y.: Essential oil–loaded biopolymeric particles on food industry and packaging: a review. Int. J. Biol. Macromol. **130765** (2024)
51. Burt, S.: Essential oils: their antibacterial properties and potential applications in foods—a review. Int. J. Food Microbiol. **94**(3), 223–253 (2004)
52. Kalemba, D.A.A.K., Kunicka, A.: Antibacterial and antifungal properties of essential oils. Curr. Med. Chem. **10**(10), 813–829 (2003)
53. Rout, S., Tambe, S., Deshmukh, R.K., Mali, S., Cruz, J., Srivastav, P.P., et al.: Recent trends in the application of essential oils: the next generation of food preservation and food packaging. Trends Food Sci. Technol. **129**, 421–439 (2022)
54. Gómez-Estaca, J., López-de-Dicastillo, C., Hernández-Muñoz, P., Catalá, R., Gavara, R.: Advances in antioxidant active food packaging. Trends Food Sci. Technol. **35**(1), 42–51 (2014)
55. Mohd Hatta, F.A.: Advancements in bioactive intelligent packaging: enhancing food safety and sustainability. Food Sci. Nutr. Cases, fsncases20240019 (2024)
56. Khalid, S., Naeem, M., Talha, M., Hassan, S.A., Ali, A., Maan, A.A., et al.: Development of biodegradable coatings by the incorporation of essential oils derived from food waste: a new sustainable packaging approach. Packag. Technol. Sci. **37**(3), 167–185 (2024)
57. Sharma, S., Barkauskaite, S., Jaiswal, A.K., Jaiswal, S.: Essential oils as additives in active food packaging. Food Chem. **343**, 128403 (2021)
58. Ju, J., Chen, X., Xie, Y., Yu, H., Guo, Y., Cheng, Y., et al.: Application of essential oil as a sustained release preparation in food packaging. Trends Food Sci. Technol. **92**, 22–32 (2019)
59. Vaishnavi, B.C., Ananya, S.J., Jitendra, N.Y., Rajendra, B.S.: Strategies to improve stability of essential oils. J. Pharm. Sci. Res. **13**(7), 416–425 (2021)
60. Barbieri, C., Borsotto, P.: Essential oils: market and legislation. Potential Essent. Oils, 107–127 (2018)
61. Aziz, Z.A., Ahmad, A., Setapar, S.H.M., Karakucuk, A., Azim, M.M., Lokhat, D., et al.: Essential oils: extraction techniques, pharmaceutical and therapeutic potential—a review. Curr. Drug Metab. **19**(13), 1100–1110 (2018)
62. Perović, A.B., Karabegović, I.T., Krstić, M.S., Veličković, A.V., Avramović, J.M., Danilović, B.R., Veljković, V.B.: Novel hydrodistillation and steam distillation methods of essential oil recovery from lavender: a comprehensive review. Ind. Crop. Prod. **211**, 118244 (2024)
63. Drinić, Z., Pljevljakušić, D., Živković, J., Bigović, D., Šavikin, K.: Microwave-assisted extraction of O. vulgare L. spp. hirtum essential oil: comparison with conventional hydro-distillation. Food Bioprod. Process. **120**, 158–165 (2020)
64. Ferreira, G., Sobral, L., Barreto, D.W., Calado, V.: Thermal and compositional analysis of Orange essential oil obtained from citrus industry waste. J. Adv. Therm. Sci. Res. **7**, 48–55 (2020)
65. Ray, A., Dubey, K.K., Marathe, S.J., Singhal, R.: Supercritical fluid extraction of bioactives from fruit waste and its therapeutic potential. Food Biosci. **52**, 102418 (2023)
66. Cardoso-Ugarte, G.A., Juárez-Becerra, G.P., Sosa Morales, M.E., López-Malo, A.: Microwave-assisted extraction of essential oils from herbs. J. Microw. Power Electromagn. Energy. **47**(1), 63–72 (2013)
67. Singh, T., Pandey, V.K., Singh, R., Dash, K.K., Kovács, B., Mukarram, S.A.: Ultrasound assisted extraction of phytochemicals from Piper betel L. Ultrason. Sonochem. **106**, 106894 (2024)
68. Berenguer, C.V., Andrade, C., Pereira, J.A., Perestrelo, R., Câmara, J.S.: Current challenges in the sustainable valorisation of agri-food wastes: a review. Processes. **11**(1), 20 (2022)

69. Srivastav, P. P., & Karunanithi, S. (Eds.) Emerging Methods for Oil Extraction from Food Processing Waste. (2024)
70. Oliveira, A.C.D., Gobato, C., Pereira, K.N., Carvalho, M.V., Santos, J.V., Pinho, G.D., Kamimura, E.S.: Application of essential oils as natural antimicrobials in lactic acid bacteria contaminating fermentation for the production of organic cachaça. Int. J. Food Microbiol. **424**, 110742 (2024)
71. Dhall, R.K.: Advances in edible coatings for fresh fruits and vegetables: a review. Crit. Rev. Food Sci. Nutr. **53**(5), 435–450 (2013)
72. Díaz-Montes, E., Castro-Muñoz, R.: Edible films and coatings as food-quality preservers: an overview. Foods. **10**(2), 249 (2021)
73. Thirumalaisamy, R., Jayanthi, B., Rathika, R., Selvankumar, T.: Spontaneous emulsification techniques of green/food grade nanoemulsions. In: Bio-Based Nanoemulsions for Agri-Food Applications, pp. 137–152. Elsevier (2022)
74. Sousa, V.I., Parente, J.F., Marques, J.F., Forte, M.A., Tavares, C.J.: Microencapsulation of essential oils: a review. Polymers. **14**(9), 1730 (2022)
75. Hashemi, H., Ebrahimi, B., Awlqadr, F.H., Ov, N.S., Ebdali, H., Sarlak, Z., Mohammadi, R.: Gelatin-chitosan based composite films enriched with *Satureja kermanshahensis* Jamzad essential oil nanoemulsion and carbon dots: characterization and functional properties. J. Polym. Environ., 1–16 (2024)
76. Vermeiren, L., Devlieghere, F., Debevere, J.: Effectiveness of some recent antimicrobial packaging concepts. Food Addit. Contam. **19**(S1), 163–171 (2002)
77. Liu, Z., Wang, S., Liang, H., Zhou, J., Zong, M., Cao, Y., Lou, W.: A review of advancements in chitosan-essential oil composite films: better and sustainable food preservation with biodegradable packaging. Int. J. Biol. Macromol. **133242** (2024)
78. Malucelli, L.C., Magalhães, W.L.E.: General overview and future trends of nanotechnology in the food industry. In: Food Applications of Nanotechnology, pp. 1–17 (2019)
79. Gutierrez, J., Barry-Ryan, C., Bourke, P.: Antimicrobial activity of plant essential oils using food model media: efficacy, synergistic potential and interactions with food components. Food Microbiol. **26**(2), 142–150 (2009)
80. Wang, J., Zhao, F., Huang, J., Li, Q., Yang, Q., Ju, J.: Application of essential oils as slow-release antimicrobial agents in food preservation: preparation strategies, release mechanisms and application cases. Crit. Rev. Food Sci. Nutr. **64**(18), 6272–6297 (2024)
81. Bumbudsanpharoke, N., Ko, S.: Nanoclays in food and beverage packaging. J. Food Sci. Technol. **56**, 1438–1449 (2019)
82. Kilcast, D.: Shelf-life evaluation of foods. Int. J. Food Sci. Technol. **36**(8), 856–856 (2001)
83. Sardar, N.R., Akbari, S.H., Modi, R.B., Tiwari, M., Tagalpallewar, G.P.: Application of essential oils in food packaging: a concise review. Eur. J. Nutr. Food Saf. **16**(3), 60–67 (2024)

Author Index

J. C. Casila et al. (Eds.): CCORE 2024, SPEES, pp. 287–288, 2025.
https://doi.org/10.1007/978-3-031-93887-0

The manufacturer's authorised representative in the EU is Springer Nature Customer Service Centre GmbH, Europaplatz 3, 69115 Heidelberg, Germany. If you have any concerns regarding our products, please contact ProductSafety@springernature.com

Printed and bound by CPI Group (UK) Ltd, Croydon, CR0 4YY

30/09/2026

02225588-0007